INTRODUCTION TO SYMBOLIC LOGIC AND ITS APPLICATIONS

BY

RUDOLF CARNAP

University of California

Translated by

William H. Meyer, University of Chicago

and

John Wilkinson, Wesleyan University

Dover Publications, Inc.

New York

Published in Canada by General Publishing Company, Ltd., 30 Lesmill Road, Don Mills, Toronto, Ontario.
Published in the United Kingdom by Constable and Company, Ltd.

This Dover edition, first published in 1958, is a new English translation of *Einführung in die symbolische Logik* as published by Julius Springer in 1954.

International Standard Book Number: 0-486-60453-5

Library of Congress Catalog Card Number: 58-12611

Manufactured in the United States of America
Dover Publications, Inc.
180 Varick Street
New York, N. Y. 10014

For

INA

in deep gratitude

PREFACE TO THE ENGLISH EDITION

I wish to express my gratitude to my two translators, Professor William H. Meyer of the University of Chicago and Professor John Wilkinson of Wesleyan University, who between them provided the basic translation, revised it, made many improvements in wording and arrangement, and supplied additional explanations. The translation owes its existence to their generous devotion of time and interest. Translating a technical book requires a good knowledge of the subject matter in addition to linguistic abilities and sensitivities. In my opinion, the translators happily combined these abilities and performed an excellent job.

Except for numerous minor corrections and changes made either by me or by the translators, the translation follows in general the German original. In the following places, however, I made major changes or additions. In **20** ff., the explanations of the terms 'language', 'syntactical system', and 'semantical system' have been changed and made more exact. A new section, **26b**, has been added on the formalization of syntax and semantics. To the first explication of linear order in **31,** represented by Russell's concept of a series (D5), I have now added a second explication, represented by the concept of a simple order (D8, based on D6 and D7). This second concept has certain advantages and has recently seen increased use. The concept of a simple order is employed in some of the definitions of **38**. In **42a**, the distinction between the basic language L and the axiomatic language L′ is new. In **42b**, the distinction between interpretations and models has been made sharper. There are several changes in the axiom system of set theory (**43**). In **43a**, the axiom of regularity (A9) has been added. The original **43b** is omitted (it gave a second version of the system, with eight primitives, among them seven functors). The new **43b** is an expansion of a part of the original **43a,** with an altered form of the axiom of restriction (A10). Also, **43c** is newly added; here another version of the axiom system is described, which uses only individual variables. In the axiom system of neighborhoods (**46**), **46b** contains a new second version; and the definitions in **46c** are now based on this simpler version.

The bibliography (**56**) has been brought up to date. In chapters A, B, and C, many new exercises have been added; I wish to thank my student, David B. Kaplan, for his efficient help in this connection.

For the most part, the terminology in this English edition is based on terms used by me in classes and in recent publications. Suggestions for some other terms I owe to the translators and other colleagues. I went over the whole translation carefully and bear the sole responsibility for the accuracy of the content.

RUDOLF CARNAP

University of California
at Los Angeles

May 1957

PREFACE TO THE GERMAN EDITION

During the past century logic has assumed an entirely new form, that of symbolic logic (or mathematical logic, or logistic). The use of symbols is, of course, the most striking feature of the new logic. Nevertheless, its essential characteristics lie in other directions: precision of formulation, greatly extended scope (especially in the theory of relations and of high-level concepts), manifold applications of its new methods. In consequence the last decades have seen an ever-increasing interest in symbolic logic, notably among mathematicians and philosophers, but also among those working in quite specialized fields who give attention to the analysis of the concepts of their disciplines.

Today, and particularly in the United States, symbolic logic is a recognized subject for teaching and research. The majority of American scholars who write on epistemology, analysis of language, scientific method, foundations of mathematics, axiomatic method, and the like, regard symbolic logic as an indispensable tool.

It is my hope that this book will reinforce, among German-speaking peoples, the general interest in symbolic logic.

What chiefly differentiates the present book from other logic texts (mostly in English) may be summarized under the following heads. In addition to the elementary portions of the theory, whose treatment is customary in most books, there is also a detailed presentation of the more advanced topics (especially the logic of relations) required for the application of logic. Further, the entire second part of the present book is given over to the application of symbolic logic. In this second part we first explain the construction of various language forms that must be considered in the application of logic; thereafter, we give in symbolic form axiom systems from different fields. Finally, in accordance with modern views, the present book outlines the theories of formal language systems (logical syntax) and interpreted language systems (semantics).

It may be thought that these last theories transcend the natural limits of an introductory text. However, I consider it important for anyone who would make the new symbolic methods his own that he learn from the very beginning to think from the point of view of the construction of deductive systems: in so doing, he gains for himself the insight that symbolism is a language conforming to exact rules whose use can sharpen the forms of his own thinking. It is this deliberate consideration of logical syntax and semantics which—apart from essentially greater length—

mainly distinguishes the present book from my former *Abriss der Logistik* (Wien 1929, 114 p.), now out of print and in many respects out of date because of rapid developments in the field.

The present book can be used as the text of a two-semester course in symbolic logic. The first semester, the introductory part of the course, could e.g. be based on Chapter A together with several illustrative applications drawn from Part II (see my explanations in **42e**). The second semester of the course could center chiefly on Chapter C supplemented by other applications from Part II; and to these matters can be added (to a degree desired by the instructor) considerations of syntactical and semantical theory, based either on the sketch provided in Chapter B or on the fuller presentations found in other books. Of course, the whole field of modern logic—including the theory of formal and interpreted language systems—is so extensive that two one-year courses are far more appropriate to it.

RUDOLF CARNAP

Institute for Advanced Study,
Princeton, N.J.

January 1954

CONTENTS

PART ONE

System of symbolic logic

PART TWO

Application of symbolic logic

PART ONE

SYSTEM OF SYMBOLIC LOGIC

Chapter A

The simple language A

1. THE PROBLEM OF SYMBOLIC LOGIC

1a. The purpose of symbolic language. Symbolic logic (also called mathematical logic or logistic) is the modern form of logic developed in the last hundred years. This book presents a system of symbolic logic, together with illustrations of its use. Such a system is not a theory (i.e. a system of assertions about objects), but a *language* (i.e. a system of signs and of rules for their use). We will so construct this symbolic language that into it can be translated the sentences of any given theory about any objects whatever, provided only that some signs of the language have received determinate interpretations such that the signs serve to designate the basic concepts of the theory in question. So long as we remain in the domain of pure logic (i.e. so long as we are concerned with building this language, and not with its application and interpretation respecting a given theory), the signs of our language remain uninterpreted. Strictly speaking, what we construct is not a language but a schema or skeleton of a language: out of this schema we can produce at need a proper language (conceived as an instrument of communication) by interpretation of certain signs.

Part Two of this book sees a variety of such interpretations, and the symbolic formulation (axiomatically, for the most part) of theories from various domains of science. All this is *applied logic*. Part One of the book attends to *pure logic*: here we describe the structure of the symbolic language by specifying its rules. In the present Chapter A, the first of the three chapters comprising Part One, we describe a simple symbolic language A containing the following sorts of signs (to be explained later): sentential constants and variables, individual constants and variables, predicate constants and variables of various levels and types, functor constants and variables, sentential connectives, and quantifiers. The third chapter, Chapter C, presents a more comprehensive language C. In Chapter B a symbolic language B is represented both as a syntactical system and as a semantical system.

If certain scientific elements—concepts, theories, assertions, derivations, and the like—are to be analyzed logically, often the best procedure is to translate them into the symbolic language. In this language, in contrast to ordinary word-language, we have signs that are unambiguous and formulations that are exact: in this language, therefore, the purity and correctness of a derivation can be tested with greater ease and accuracy. A derivation is counted as pure when it utilizes no other presuppositions than those specifically enumerated. A derivation in a word-language often involves presuppositions which were not made explicitly, but which entered unnoticed. Numerous examples of this are afforded by the history of geometry, especially in connection with attempts to derive Euclid's axiom of parallels from his other axioms.

A further advantage of using artificial symbols in place of words lies in the brevity and perspicuity of the symbolic formulas. Frequently a sentence that requires many lines in a word-language (and whose perspicuity is consequently slight) can be represented symbolically in a line or less. Brevity and perspicuity facilitate manipulation and comparison and inference to an extraordinary degree. The twin advantages of exactness and brevity appear also in the usual mathematical notations. Had the mathematician been confined to words and denied the use of numerals and other special symbols, the development of mathematics to its present high level would have been not merely more difficult, but psychologically impossible. To appreciate this point, one need only attempt to translate into the word-language e.g. so elementary a formula as "$(x+y)^3=x^3+3x^2y+3xy^2+y^3$" ("The third power of the sum of two arbitrary numbers equals the sum of the following summands: ..."). The symbolic method gives mathematics an advantage in its investigation of numbers, numerical functions, etc.; symbolic logic seeks this same advantage in full generality for its treatment of concepts of any kind.

In the course of constructing our symbolic language systems, it frequently happens that a new precisely-defined concept is introduced in place of one which is familiar but insufficiently precise. Such a new concept is called an *explicatum* of the old one, and its introduction an *explication*. (The concept to be explicated is sometimes called the *explicandum*.) E.g. the concept of L-truth (to be defined technically later (**5b**) on the basis of exact rules) is an explicatum of the concept of logical or necessary truth, which is defined with insufficient exactness despite its frequent occurrence in philosophy and traditional logic. Again, the concept of the inductive cardinal numbers (**37c**) is an explicatum for the concept of finite number that has been widely used in mathematics, logic and philosophy, but never exactly defined prior to Frege. [For a more complete exposition of the methods of explication and the requirements an adequate explicatum must meet, see Carnap [Probability], Chapter I.]

1b. The development of symbolic logic. Symbolic logic was founded around the middle of the last century and carried on into the present more by mathematicians than philosophers (cf. references to the literature, **57**). The reason for this lies in the historical fact that during the past century mathematicians became increasingly more conscious of the need to reexamine and reconstruct the foundations of the whole edifice of mathematics. Finding the traditional (i.e. aristotelian-scholastic) logic a totally inadequate instrument for this purpose, the mathematicians set about to develop a system of logic that was at once more appropriate, more accurate and more comprehensive.

The resulting new symbolic logic (especially in the systems of Frege, Whitehead-Russell, and Hilbert) clearly evinced a suitability to the first task set it, viz. to provide a basis for *the reconstruction of mathematics* (arithmetic, analysis, function theory, and the infinitesimal calculus). Further, in its *logic of relations* the new symbolic logic developed an abstract theory of arbitrary order-forms, and thereby created the possibility of representing and logically analyzing theories in which relations play an essential role, e.g. the various geometries, physical theories (especially in reference to space and time), epistemology and, latterly, even certain branches of biology. This development was a particularly significant advance beyond traditional logic. For traditional logic had neglected relations almost completely and hence proved entirely useless in connection with the axiomatic method (e.g. in geometry) that has become so important in recent decades. Still another merit of symbolic logic—minor, but nonetheless valuable—is that it achieved the complete solution of certain contradictions, viz. the so-called logical *antinomies* (cf. **21c**), whose analysis and elimination were beyond the reach of the old logic.

For *literature* on matters treated here, see the references, **57**. In the text of this book, citations of the literature are phrased with the help of abbreviated titles in square brackets; cf. the bibliography, **56**. ('[P.M.]' is used without author names for: Whitehead and Russell, *Principia mathematica*; and similarly for several of my own works.)

Regarding *terminology*. In the domain of symbolic logic the expressions "algebraic logic", "algebra of logic", etc., were employed at an earlier date but are no longer customary today. In addition to "symbolic logic" and "mathematical logic", the designation "logistics" is often used, especially on the European continent; it is short and permits the formulation of the adjective "logistic". The word "logistics" originally signified the art of reckoning, and was proposed by Couturat, Itelson and Lalande independently in 1904 as a name for symbolic logic (according to the assertion of Ziehen, *Lehrbuch der Logik*, p. 173, note 1, and Meinong, *Die Stellung der Gegenstandstheorie*, p. 115).

Concerning results of the new symbolic logic in comparison with traditional logic, cf. Russell [World], Chap. II; Carnap [Neue Logik]; Menger [Logic]. On the special importance of the logic of relations, cf. Russell, *ibid.*

Concerning *the reconstruction of mathematics* on the basis of the new logic, cf. the basic older works: Frege [Grundlagen] and [Grundgesetze]; Peano [Formulaire]; as chief work, [P.M.]; and also Russell [Introduction]; a more recent work: Hilbert and Bernays [Grundlagen]; for an easy presentation of the basic ideas: Carnap, "Die Mathematik als Zweig der Logik", *Blätter f. dt. Philos.* 4, 1930; Carnap, "Die logizistische Grundlegung der Mathematik", *Erkenntnis* 2, 1931.

2. INDIVIDUAL CONSTANTS AND PREDICATES

2a. Individual constants and predicates. The theoretical treatment of any domain of objects consists in setting up sentences concerning the objects of the domain (sentences ascribing certain properties and relations to the objects in question), and in establishing rules according to which other sentences can be derived from those given. The basic objects treated of in a given language system are called the *individuals* of the system; and their totality, the *domain of individuals* (briefly, the *domain*) of the system. This domain is sometimes called the *universe of discourse*. To form sentences concerning the individuals of a given domain there must first of all be available in the language two kinds of signs: 1. names for the individuals of the domain—we call these *individual constants*; 2. designations for the properties and relations predicated of the individuals—we call these *predicates*.

For individual constants we use the letters '*a*', '*b*', '*c*', '*d*', '*e*'. E.g. if our language were to be applied to the domain comprising the heavenly bodies, '*a*' might perhaps designate the sun, '*b*' the moon, etc. Again, if the domain were a certain group of people, '*a*' might be taken as an abbreviation for 'Charles Smith', '*b*' for 'John Miller', etc. So long as our considerations are purely logical, we shall not trouble ourselves as to what special domain of individuals our language might be applied, and what particular individuals of that domain might be designated by '*a*', '*b*', etc. It is only when we move away from pure logic (i.e. from consideration of the skeleton language to be constructed in what follows) that we speak of the interpretation of the separate individual constants and predicates. We do this last e.g. in the second part of this book, where several systems are presented as applications; we do it also in the first part, in connection with illustrative examples.

For *predicates* we use the letters '*P*', '*Q*', '*R*', '*S*', '*T*'. In connection with illustrative applications, we also use for predicates various letter groups with first letter capitalized (cf. the examples in **2c** below); these letter groups are based on words of the word-language.

E.g. in a certain application '*P*' might designate the property Spherical. [I prefer this mode of expression to the more elaborate turn of phrase "the property of being spherical". Similarly, I write "the property Prime Number", "the property Odd", etc. Again, I use "the class Spherical" in place of "the class of spherical individuals"; and analogously, "the class Blue", etc. And again, I say "the relation Greater" rather than "the relation that obtains between x and y when x is greater than y"; and similarly "the relation Similar", "the relation Father", etc.] Now suppose that, in addition to designating the property Spherical by '*P*', we take '*a*' to designate the sun and '*b*' to designate the moon. Then in our symbolic language we write the sentence '$P(a)$' for "the sun is spherical". Similarly, '$P(b)$' is the translation into our symbolic language of the English sentence "the moon is spherical". To give a symbolic translation of the sentence "the sun is greater than the moon", we need a sign for the relation Greater. Taking '*R*' for this relation, we write '*R*(*a*,*b*)' as our symbolic translation of "the sun is greater than the moon". Again, if *a* and *b* are persons (i.e. '*a*' and '*b*' are interpreted as personal names), and '*S*' is taken to designate the relation Similar, then '*S*(*a*,*b*)' means "*a* is similar to *b*". Likewise, we can translate the sentence "*a* is jealous of *b* with respect to *c*" into '*T*(*a*,*b*,*c*)' if we use '*T*' to designate the triadic or three-place relation Jealous.

In the sentences '*P*(*a*)' and '*R*(*b*,*c*)', the '*a*' and '*b*' and '*c*' are called *argument-expressions*. Further, '*b*' is said to stand in the first *argument-position*, '*c*' in the second. We say '*P*' is a one-place (or monadic) predicate, and '*R*' a two-place (or dyadic) predicate. Generally, a predicate is said to be *n*-adic (or *n*-place, or of degree *n*) in case it has *n* argument-positions. Predicates of degree higher than two can be introduced whenever they are needed in connection with a given domain of objects. We say that '*P*(*a*)' is a *sentence-completion* or *full-sentence* of the predicate '*P*'; similarly, '*R*(*b*,*c*)' is a sentence-completion of '*R*'. The examples given here illustrate the use of single letters as predicates and argument-expressions, but not such a use of letter groups (this occurs in **2c**) and compound expressions. When single letters are so used we usually omit parentheses and commas, and write simply '*Pa*', '*Rab*', '*Tabc*', etc.

Regarding terminology. 1. In ordinary word language there is no word which comprehends both properties and relations. Since such a word would serve a useful purpose, let us agree in what follows that the word "attribute" shall have this sense. Thus a one-place attribute is a property, and a two-place (or a many-place) attribute is a relation. 2. Similarly, it is useful to have a comprehensive term for the *designations* of one- and many-place attributes. For this, let us follow Hilbert and use the word "*predicate*". (Heretofore, this word has been confined mostly to properties or to designations of them, and has not included many-place attributes or predicates.) Thus a one-place predicate is a sign for a one-place attribute (i.e. for a property); and in general an *n*-place predicate is a sign for an *n*-place attribute. 3. Let us always distinguish clearly between *signs* and *what is designated*. Failure to observe this distinction has in the past occasioned much confusion in logic and in philosophy generally (cf. [Syntax] **42**). In speaking *about* an expression, let us always put the expression in quotation marks or use some special designation for it, e.g. a German letter as in **21a**. We make but one exception to this practice: we omit quotation marks in case the expression stands on a line either alone or with a designating number or letter; see e.g. our enumeration of the formulas in T8-2. Suppose e.g. '*Pa*' is taken as a symbolic translation of "*a* is old"; then we say: "*P* (but not: '*P*') is a one-place attribute, viz. the property Old; this attribute is designated by a one-place *predicate* '*P*' ". Similarly, we say: "the two-place *relation R* exists between such and such persons", "the two-place *predicate* '*R*' occurs in such and such a sentence". And similarly: "the individual *a*...", "the name '*a*'...".

2b. Sentential constants. It is often burdensome to work with sentences that are entirely written out like '*Pa*' or '*Rbc*', especially if they are even longer or are repeated frequently in the same connection. We therefore use on occasion the letters '*A*', '*B*', '*C*' as abbreviations for any sentences whatever of the symbolic language. These letters are called *sentential constants* (or: propositional constants). E.g. in a certain case '*A*' might be taken as an abbreviation for '*Pa*'; as soon as '*P*' and '*a*' are interpreted, '*A*' is also interpreted. In our use of a sentential constant we will for the most part leave open what particular sentence it stands for as an abbreviation.

2c. Illustrative predicates. To facilitate framing examples in connection with the further construction of our symbolic language system, we list here various predicates, functors (cf. **18**) and individual constants for particular domains of individuals.

1. The domain: physical *things*

moon	the moon
Book(*a*)	*a* is a book
Blue(*a*)	*a* is blue
Sph(*a*)	*a* is spherical

2. The domain: *human beings* (presently alive)

Ml(*a*)	*a* is male
Fl(*a*)	*a* is female
Stud(*a*)	*a* is a student
Fa(*a*,*b*)	*a* is father of *b*
Mo(*a*,*b*)	*a* is mother of *b*
Par(*a*,*b*)	*a* is a parent of *b*
Bro(*a*,*b*)	*a* is brother of *b*
Hus(*a*,*b*)	*a* is husband of *b*
Friend(*a*,*b*)	*a* is friend of *b*

3. The domain: *natural numbers* (0, 1, 2, etc.)

0, 1, 2,...	(in their usual signification)
Even(*a*)	*a* is an even number
Prime(*a*)	*a* is a prime number
Gr(*a*,*b*)	*a* is greater than *b* $(a>b)$
Sm(*a*,*b*)	*a* is smaller than *b* $(a<b)$
Pred(*a*,*b*)	*a* is the (immediate) predecessor of *b* $(a+1=b)$
Sq(*a*,*b*)	*a* is the square of *b* $(a=b^2)$
sq(*a*)	the square of *a* (a^2)
prod(*a*,*b*)	the product of *a* and *b* $(a \cdot b)$

'*sq*' and '*prod*' are functors, cf. **18**.

3. SENTENTIAL CONNECTIVES

3a. Descriptive and logical signs. The individual constants and predicates we have become acquainted with up to now are mostly (viz. in the first two of the three domains considered in **2c**) non-logical signs or, better, *descriptive signs*. Such signs designate things or processes in the world, or properties or relations of things, or the like. Determinate meaning is attached to descriptive signs only when we apply them, i.e. only when we go outside pure logic. Thus we must distinguish between descriptive signs and *logical signs* which do not themselves refer to anything in the world of objects, but do serve (along with descriptive signs) in sentences about empirical objects. The use of logical signs is determined by the logical rules of the language; on the other hand, meaning is arbitrarily attached to descriptive signs when they are applied to a given domain of individuals.

Among the logical signs are the parentheses '(' and ')' and the comma ',' as in e.g. '*Fa*(*a*,*b*)'. However, these signs have only a subordinate role, analogous to that of punctuation marks. More important as logical signs are the *connectives*, which are used to form compound sentences from simpler sentences (e.g. from sentence-completions of predicates). In what follows we introduce the connective signs and specify how they shall be used, thereby determining their meaning. This determination is accomplished in a two-step fashion: 1. by specifying truth-conditions for compound sentences; and 2. by specifying English translations of the connectives. Specifications of this latter sort, while easier to grasp, are of course less exact because the English words to be employed correspond in some cases only approximately to the connective meanings and moreover the usage of these words is itself often ambiguous. Specification of truth-conditions for a connective consists in an agreement which fixes the conditions under which a compound sentence (formed by means of the connective and the sentences that enter as components) is to be considered true in terms of the truth and falsity of its components.

3b. Connective signs. Suppose we have two sentences, '*A*' and '*B*'. Then the sentence '$(A)\vee(B)$' is called the *disjunction* (or alternation, or logical sum) of the sentences '*A*' and '*B*'. We agree that the disjunctive sentence '$(A)\vee(B)$' is true if and only if at least one of the two sentences '*A*' and '*B*' is true, i.e. if either '*A*' is true, or '*B*' is true, or both of them are true. The sign '$\vee$' of disjunction corresponds with fair exactness to the English word "or" in those cases where "or" stands between two sentences and is used (as it most frequently is) in the non-exclusive sense; when "or" is used in the exclusive sense, the sentence "*A* or *B*" has the meaning: "either *A* or *B*, but not *A* and *B*". Accordingly, '$(Pa)\vee(Qb)$' means: "*a* is *P* or *b* is *Q*, or both". Again, '[*Stud*(*a*)]$\vee$[*Fl*(*a*)]' means "*a* is either a student or a female, or both (i.e. a woman student)". We remark in this connection that the parentheses which enclose the sentential parts of a compound will be written indifferently as round brackets and as square brackets.

Next, let us agree that the sentence '$(A).(B)$'—the *conjunction* (or logical product) of '*A*' and '*B*'—is true just in case both '*A*' and '*B*' are true. The sign '.' of conjunction thus corresponds to the English word "and", where "and" stands between sentences. Hence '$(Pa).(Qb)$' means "*a* is *P* and *b* is *Q*", and '[*Stud*(*a*)].[*Fl*(*a*)]' means "*a* is a woman student".

Whereas the signs of disjunction and conjunction join together two sentences, the sign '$\sim$' of negation is used in connection with but *one* sentence. We say that the sentence '$\sim(A)$' is true just in case '*A*' is not true, i.e. '*A*' is false. Thus the negation sign corresponds to the English word "not". Regarding this translation, however, we must observe that while the connective refers to the entire sentence, the word "not" generally refers to but a portion of the entire sentence.

Accordingly, '$\sim[P(a)]$' means "*a* is not *P*"; and '$\sim[Even(3)]$' means "3 is not even".

The sentence '$(A) \supset (B)$' is an abbreviation for '$[\sim(A)] \vee (B)$'. Hence '$(A) \supset (B)$' is true just in case either '*A*' is false, or else '*B*' is true, or both. In many cases, '$(A) \supset (B)$' corresponds to the English sentence "if *A*, then *B*". There is an important difference between the two sentences, however. In English, the if-sentence is used only when there is a connection (perhaps of a logical or causal sort) between the two sentential parts of the compound. In the symbolic language the $\supset$-sentence is used without any such limitation. Thus, if '*A*' is "my desk is black", then '$[Blue(moon)] \supset (A)$' is true whether '*A*' is true or false, because '*Blue*(*moon*)' is false. (In English, however, the sentence "If the moon is blue, then my desk is black" would scarcely be considered an appropriate correct sentence. It falls rather among the many sentences of that word-language which are not customarily included either with the true sentences or with the false sentences—and this, even though sufficient knowledge is at hand to decide the truth or falsity of the sentential parts. Sentences of this sort simply do not occur in a well-constructed language). Similarly, the sentence '$(A) \supset [Sph(moon)]$' is true whether '*A*' is true or false, because '*Sph*(*moon*)' is true. We shall become acquainted later (in **9c**) with a class of sentences whose '$\supset$' can always be translated appropriately by "if—then". Note however that the often-inappropriate if-translation for '$(A) \supset (B)$' can be avoided by using instead "not *A*, or *B*"; this last translation is always appropriate.

The sign '$\supset$' is frequently called the *implication sign*, and '$(A) \supset (B)$' read "*A* implies *B*". It is to be emphasized, however, that '$\supset$' is not to be given the usual signification of "implication" and "implicate", viz. (logical) entailment; nor is '$(A) \supset (B)$' to be read " '*B*' is a consequence of '*A*' " or " '*B*' is deducible from '*A*' ". So much should be clear from our previous examples. [One should therefore be on his guard against translating '$(A) \supset (B)$' as "from *A* follows *B*".] The name "implication sign" for '$\supset$' goes back to the erroneous interpretation just given; in the past, this designation has occasioned much obscurity (cf. [Syntax] **69**, at the end). Since it is in general use, we retain "implication sign" as a technical expression, taking care to separate it clearly from the original meaning of the words. [The technical meaning here in mind for '$\supset$' is sometimes called "material implication" in contrast to "logical implication", which is the relation holding between '*A*' and '*B*' when '*B*' is a logical consequence of '*A*'. To avoid confusing these two possibilities, we have decided to call '$(A) \supset (B)$' a "conditional sentence" or a *conditional* rather than an "implication", and to read it "If *A*, then *B*".] Also, in connection with the conditional '$(A) \supset (B)$' we find it convenient to retain the name "*antecedent*" for the first component '(A)' and the name "*consequent*" for the second component '(B)'.

The sentence '$(A) \equiv (B)$' is called the *biconditional* (or: equivalence) of

'A' and 'B', and is counted as true just in case 'A' and 'B' are both true or else both false. This sentence is often called simply a "biconditional", an "equivalence" or a "material equivalence". It refers, of course, strictly to the equality of truth-values (cf. **4a**), and not to the identity of meaning of its two members (this last relation is called "logical equivalence", cf. **6a**). We read '$(A) \equiv (B)$' as "A is equivalent to B" or "A if and only if B".

3c. Omission of parentheses. Up to this point we have taken only sentences of the simplest form to serve as components in our sentential compounds. However, sentences which are themselves compounds can occur as components in a sentential composition, e.g. the compound '$\sim(A)$' in the sentence '$[\sim(A)] \vee (B)$', and the compound '$(A) \vee (B)$' in the sentence '$[(A) \vee (B)] . (C)$'. Since compositions of this sort can lead to a great accumulation of parentheses, it is out of practical expediency that we establish the following *rules for omitting parentheses.* The rules are stated so as to apply not only to sentences but also to sentential formulas, i.e. to sentences and other similar expressions (cf. **7a**).

It is considered permissible to omit the parentheses that enclose a component formula provided one of the following conditions is satisfied:

1. The component formula so enclosed is of simplest form, i.e. it contains no other sentential formula as a proper part. [Examples: '$A \vee B$', '$\sim Pa$'.]

2. The component formula so enclosed is a compound formed with a connective more cohesive than the connective that has the component as a member. For this purpose we count '$\sim$' more cohesive than '$\vee$' and '$.$', and the latter two more cohesive than '$\supset$' and '$\equiv$'. [Examples: '$(\sim A) \vee B$' can now be written '$\sim A \vee B$', similarly '$(\sim A) . B$' can be written '$\sim A . B$' because '$\sim$' is more cohesive than the other connectives. Again, '$A \vee B \supset C . D$' may be written in place of '$(A \vee B) \supset (C . D)$' because '$\vee$' and '$.$' are more cohesive than '$\supset$'. Likewise, we may write '$A . B \equiv C \vee D$' for '$(A . B) \equiv (C \vee D)$'.]

3. The component formula so enclosed is a disjunction and is itself the first member of a disjunction; or it is a conjunction and also the first member of a conjunction. [Examples: Instead of '$(A \vee B) \vee C$' we write '$A \vee B \vee C$'. We shall see later (T8-6m) that '$A \vee (B \vee C)$' can be transformed into '$(A \vee B) \vee C$'; thus '$A \vee (B \vee C)$' may also be written '$A \vee B \vee C$'. Analogously, instead of '$(A . B) . C$' we write '$A . B . C$', and we do the same for '$A . (B . C)$'.]

3d. Exercises. Many different phrases in English translate into the same logical connectives. E.g. let 'A' and 'B' be sentences; then "If A, then B" and "B provided that A" may both be symbolized by '$A \supset B$' (although strictly speaking, the latter is somewhat weaker than the former). Using the symbols 'A', 'B', '(',')', '$\vee$' '$.$', '$\supset$', '$\equiv$', '$\sim$', symbolize the following: **1.** "B if A." — **2.** "A on the condition that B." — **3.** "B unless A." — **4.** "Assuming A, B." — **5.** "The condition that A is both necessary and sufficient for B." — **6.** "Neither A nor B." — **7.** "B only if A." — **8.** "Not B provided that if A, then B." — **9.** "Neither B nor A only if B and A." — **10.** "On the condition that A, not B only if B then A." — **11.** "If A, then if B then A." — **12.** "A, or B and A." — **13.** "Not B, but (i.e., and) if A then B."

4. TRUTH-TABLES

4a. Truth-tables. We term Truth and Falsity the two possible *truth-values* of a sentence. Since every sentence is either true or else false, two independent sentences 'A' and 'B' can show four possible combinations of truth-values: either both sentences are true, or only the first, or only the second, or neither. Designating Truth by 'T' and Falsity by 'F', these four cases may be indicated thusly: TT, TF, FT, FF. When the truth-conditions previously established for '$A \vee B$' are recalled, this sentence is seen to be true in the first three cases and false in the fourth. Similarly, '$A . B$' is seen to be true only in the first case and false in the remaining three. Again, '$A \supset B$' is false only in the second case and true in the others, while '$A \equiv B$' is true in the first and last cases, and false in the other two.

The table below, called a *truth-table* (or a truth-value table), presents compactly the truth-values of the compounds in each of the four possible cases. It is well to remark that the letters 'T' and 'F' are *not* signs in our symbolic language, but simply abbreviations for the English words "true" and "false". English itself serves here as our meta-language, i.e. the language in which we speak about the symbolic language (see **20**). Truth-tables belong to the metalanguage, not the symbolic language: they represent in tabular form what was presented in **3b** by means of English, viz. specifications of the truth-conditions of sentential compounds in our symbolic language. (Note: The '+' prefixed to certain theorems, definitions, rules, tables, etc., indicates—as here, with Tables I and II—those which are especially important.)

+TRUTH-TABLE I

	(1) A	(1) B	(2) $A \vee B$	(3) $A . B$	(4) $A \supset B$	(5) $A \equiv B$
1	T	T	T	T	T	T
2	T	F	T	F	F	F
3	F	T	T	F	T	F
4	F	F	F	F	T	T

Since a negation has only one component, only two cases are possible:

+TRUTH-TABLE II

	(1) A	(2) $\sim A$
1	T	F
2	F	T

With the help of Truth-table I and Truth-table II we can determine the truth-values of an elaborate compound involving, say, n different constituent sentences ($n=1,2,3,\ldots$) joined by our various connectives. First we set up a table whose vertical column (1) shows the 2^n possible combinations of truth-values for the n constituent sentences. Then, beginning with these constituent sentences, we determine in each case the truth-values of the successively larger compound components until we arrive at the truth-value of the original elaborate compound itself. When this has been done for all 2^n cases, the distribution of truth-values for the original compound will have been obtained. The examples below illustrate this truth-table technique.

Examples. Compounds involving just one constituent sentence. Here we deal with two compounds, '$A \vee \sim A$' and '$A . \sim A$' and display their values in Table III. Our discussion will explain how this table is built up. — *Example 1*: the sentential compound '$A \vee \sim A$'. Only one constituent sentence, viz. 'A', being involved here, we set up a truth-table whose column (1) is headed with 'A' and which contains $2^1=2$ horizontal rows. The next simplest component of '$A \vee \sim A$' is '$\sim A$'; so we head column (2) of the table with '$\sim A$', and use Table II to find the appropriate truth-value entries therein. No other components remaining, we head column (3) with the sentence '$A \vee \sim A$' itself. To find the truth-value entries in (3), we proceed as follows: '$A \vee \sim A$' is a disjunction; in the first row of our table the two components 'A' and '$\sim A$' of this disjunction have respectively (we see from columns (1) and (2)) the values T, F; in this case, as we learn from the second row of column (2) of Table I, a disjunction has the value T; we therefore enter 'T' in the first row of column (3); and proceeding similarly, the entry 'T' is made in the second row of column (3). Columns (1) and (3) of Table III thus constitute a truth-table for the sentence '$A \vee \sim A$', column (3) in particular indicating the distribution of truth-values for '$A \vee \sim A$'. — *Example 2*: the sentential compound '$A . \sim A$'. Here again we proceed as in Example 1, with the difference that we refer back to column (3) of Table I for the final values of the conjunction '$A . \sim A$'. Columns (1) and (4) of Table III thus

TRUTH-TABLE III

	(1) A	(2) $\sim A$	(3) $A \vee \sim A$	(4) $A . \sim A$
1	T	F	T	F
2	F	T	T	F

constitute a truth-table for '$A . \sim A$', with column (4) showing the actual distribution of truth-values.

Compounds involving two constituent sentences. Here we deal with three examples, the compounds, '$\sim (A \vee B)$', '$\sim A . \sim B$' and '$\sim (A \vee B) \equiv \sim A . \sim B$'. The distribution of values of '$\sim (A \vee B)$' is shown in Table IV, column (3); that of '$\sim A . \sim B$' in Table IV(6); and that of '$\sim (A \vee B) = \sim A . \sim B$' in Table IV(7). Let us now explain how the distributions of these compounds are obtained. — *Example 3*: the sentential compound '$\sim (A \vee B)$'. This negation has two constituent sentences, 'A' and 'B'; hence we construct a table of $2^2=4$ rows whose column (1) shows the possible truth-combinations for 'A' and 'B'; column (2) is headed '$A \vee B$', the only component of our negation, and the values entered under it are obtained from Table I(2); finally, column (3) is headed by the negation '$\sim (A \vee B)$' itself, and the entries thereunder obtained by reversing the corresponding

TRUTH-TABLE IV

	(1) A B	(2) $A \vee B$	(3) $\sim(A \vee B)$	(4) $\sim A$	(5) $\sim B$	(6) $\sim A . \sim B$	(7) $\sim(A \vee B) \equiv \sim A . \sim B$
1	T T	T	F	F	F	F	T
2	T F	T	F	F	T	F	T
3	F T	T	F	T	F	F	T
4	F F	F	T	T	T	T	T

values in column (2) (for we know from Table II that the negation of a sentence has a truth-value opposite that of the sentence itself). Columns (1) and (3) of Table IV thus constitute a truth-table for '$\sim(A \vee B)$', column (3) itself showing the actual distribution of truth-values. — *Example 4*: the sentential compound '$\sim A . \sim B$'. As in Example 3, so here we need for our conjunction '$\sim A . \sim B$' a table of four rows whose column (1) is that of Table IV; next, the components of the conjunction being '$\sim A$' and '$\sim B$', we want two columns so headed (these are (4) and (5) in Table IV), with entries that are opposite those in (1); finally, we make a column (it is Table IV(6)) headed with the conjunction '$\sim A . \sim B$' itself, and obtain its entries as follows: in the first row, sentences (4) and (5) have the values FF respectively, hence by Table I(3) our conjunction (6) has here the value F; and similarly we obtain the values in the other three rows of (6). Columns (1) and (6) of Table IV thus constitute a truth-table for '$\sim A . \sim B$'. — *Example 5*: the sentential compound '$\sim(A \vee B) \equiv \sim A . \sim B$'. This equivalence involves two constituent sentences, 'A' and 'B', hence calls for a table of four rows whose column (1) is that of Table IV; the two components of the equivalence are the sentences '$\sim(A \vee B)$' and '$\sim A . \sim B$' whose values are already displayed in columns (3) and (6) of Table IV; thus we need only a last column (7) headed by our equivalence; now reading (3) and (6) together, row by row, we see that the components of our equivalence furnish only two different combinations of truth-values, viz. FF and TT; hence with the help of Table I(5), rows 1 and 4, we find the value T for each entry in (7). Columns (1) and (7) of Table IV thus constitute a truth-table for '$\sim(A \vee B) \equiv \sim A . \sim B$'.

It is useful to show how the truth-table method described above can be simplified. The simplification consists in not forming separate columns for the several components of a compound, but instead listing the values directly under letters and under connective signs. E.g. Table V is such a simplification of Table IV.

TRUTH-TABLE V

$\sim$ (5)	$(A$ (1)	$\vee$ (3)	$B)$ (1)	$\equiv$ (7)	$\sim$ (4)	A (2)	$.$ (6)	$\sim$ (4)	B (2)
F	T	T	T	T	F	T	F	F	T
F	T	T	F	T	F	T	F	T	F
F	F	T	T	T	T	F	F	F	T
T	F	F	F	T	T	F	T	T	F

Example. It is evident that columns (1) and (7) of Table V furnish for '$\sim(A \vee B) \equiv \sim A . \sim B$' the same information that Table IV(1)(7) does. Let us examine the steps by which this simplified Table V is built. We number these steps (1), (2),... and label with the same number the corresponding column(s) in Table V. (1) Under the first occurrence of each different constituent letter, enter truth-values as in Table I(1). (2) Enter the same succession of values under every other occurrence of these letters. (3) Using Table I(2), enter under 'V' the values of the disjunction that correspond to the values of '*A*' and '*B*' there. (4) Under each of the two signs '$\sim$' in the right side enter the appropriate values according to Table II (these columns will then appear as Table IV(4)(5) respectively). (5) Under the first sign '$\sim$' enter values that are opposite those given under 'V' (since the values of the sentence to which this '$\sim$' applies are precisely those listed under its principal connective 'V'). (6) Using Table I(3), enter under the conjunction sign '.' the values determined for it by its components (the column resulting here is the same as Table IV(6)). (7) Finally, use Table I(5) to enter under '$\equiv$' the appropriate values, remembering here that the values of its components are listed respectively under the first '$\sim$' (i.e. in (5)) and under '.' (i.e. in (6)); the resulting column is the same as Table IV(7). In this simpler way we have determined the distribution of values for our original equivalence.

A sentence is called a *tautology*, a *contingency*, or a *contradiction* according as its distribution of truth-values shows respectively only 'T', at least one 'T' and at least one 'F', or only 'F'.

Partial truth-tables. Frequently we are interested simply in deciding whether a given sentence is a tautology. The question whether a given sentence conjectured to be a tautology actually is one can be settled by using a partial truth-table in the following way. Assign the value F to the whole sentence, and check to see if this value can be maintained when we proceed backwards step by step through the values of successively smaller components.

Example. Is the sentence '$[A \supset (\sim B \equiv C)] \supset (A . C \supset \sim B)$' a tautology? Let us apply to it the test described above. We shall explain each step of the test carefully, and show the results in Truth-table VI. (Note that the sentence to be tested has three distinct constituent sentences, '*A*' and '*B*' and '*C*', hence a full truth-table for it would require $2^3 = 8$ rows; a glance ahead at Table VI tells that our test requires only one row.) Write out the sentence being tested, and (1) enter 'F' under its principal connective '$\supset$'. (2) Since, according to Table I(4), a conditional has value F just in case its members have respectively the values T, F, we enter 'T' under the principal connective '$\supset$' of the antecedent and 'F' under the principal connective '$\supset$' of the consequent. (3) Now a conditional can take on the value T in three cases, but the value F in only one. Hence we have to examine three cases if we work with the antecedent, but only one if we work with the consequent. Therefore we proceed with the consequent. As in step (2) so here the two parts of this consequent necessarily have the values T, F respectively; in consequence, we enter under '.' the value T and under the last '$\sim$' the value F. (4) A conjunction having the value T just in case each of its components has this value, we next enter 'T' under '*A*' and under '*C*'. (5) If a negation has the value F, then the component being negated must (by Table II) have the value T; hence we enter under the last '*B*' the value T. (6) Every part of the right side of our original conditional now having a determinate value, let us give our attention to the left side. Here it is simpler to reverse our direction and proceed outwards, not inwards. So we enter under the '*A*', '*B*' and '*C*' of the left member the values T, T, T found under these same letters on the right. (7) It now follows that the entry 'F' must go under the left sign '$\sim$', and further (8) that the entry 'F' goes under the connective '$\equiv$'. (9) From this 'F' under '$\equiv$' and the 'T' already under the first '*A*', it is necessary that an 'F' be placed under the first '$\supset$'. But a 'T' has already (in step (2)) been entered under that first '$\supset$', hence this new entry is incompatible. We conclude that our initial assignment of the value F to the original sentence (done in step (1)) is impossible. Hence, the original sentence is a tautology.

PARTIAL TRUTH-TABLE VI

$[A$ (6)	$\supset$ (2)	$(\sim$ (7)	B (6)	$\equiv$ (8)	$C)]$ (6)	$\supset$ (1)	$(A$ (4)	$.$ (3)	C (4)	$\supset$ (2)	$\sim$ (3)	$B)$ (5)
T	T (9) F	F	T	F	T	F	T	T	T	F	F	T

Note that the method of partial truth-tables as used in Truth-table VI can be employed to determine whether a sentence is a contradiction or not. This method can also be employed to determine whether a sentence is a contingency or not.

Exercises. **1.** Write truth-tables like V for the following, and thus decide whether they are tautologies, contingencies, or contradictions: a) ‘$\sim(A.B)\equiv\sim A\vee\sim B$’; b) ‘$\sim(A\supset B)\supset\sim B.\sim A$’; c) ‘$(A\supset B)\vee(B\supset A)$’; d) ‘$B\equiv((A\supset B)\vee\sim B)$’; e) ‘ $A\equiv(B\equiv C)\equiv((A\equiv B)\equiv C)$’. — **2.** How can the method of Truth-table VI be used to determine whether a sentence is a contradiction? — **3.** How can the method of Truth-table VI be used to determine whether a sentence is a contingency? — **4.** Using the method of Truth-table VI on the following, decide whether a), b), c), d) are tautologies, and whether e) is a contradiction: a) ‘$(A\equiv B)\supset((C.A)\equiv(B.C))$’; b) ‘$\sim B\supset(A\equiv(\sim A.B))$’; c) ‘$((A\vee B)\supset C)\supset(A\supset C).(B\supset C)$’; d) ‘$(B\equiv(\sim A\vee B))\supset(A.\sim B)$’; e) ‘$\sim(A\supset B).\sim(A\supset\sim B)$’.

4b. Truth-conditions and meaning. What the truth-table of a connective gives is primarily a necessary and sufficient condition for the truth of a compound so connected, in terms of the truth-values of its members. Now, however, it is easy to see that the specification of such a condition amounts to the assignment of a unique meaning to the connective (and therefore that the addition of an English translation for the sign is theoretically superfluous, however helpful it may be pedagogically or psychologically). For suppose that a person knows the sense of the sentences ‘A’ and ‘B’, where perhaps ‘A’ says that it is (now, in Paris) snowing and ‘B’ says that it is raining; and suppose no translation of ‘$\vee$’ has been given him, but only the Truth-table I(2). Can the person then comprehend the meaning of the sentence ‘$A\vee B$’ so that (a) he knows when it is permissible to assert this compound on the basis of his factual information; and (b) he can extract from a communication having the form of this compound the factual information being communicated? The answer is: he can. Perceiving from the truth-table that the compound holds in the first three cases but not in the last, our subject knows precisely the conditions under which the compound may be asserted and he knows precisely what information it conveys as a communication. For on the one hand he knows the compound sentence may be asserted if his observations of the present weather in Paris indicate it is both snowing and raining (case 1), indicate it is snowing without raining (case 2), indicate it is raining without snowing (case 3); and on the other hand he knows the compound may not properly be asserted if indications are it is neither snowing nor raining (case 4). Again, were our subject to receive this compound sentence as a communication, he could gather from it (provided, of course, he believed the communicator) that one of the first three cases obtained, but certainly not the last. All this the person himself can translate into the word-language as "it is raining or it is snowing, or both", or as "it is not the case that it is neither raining nor snowing", or however he will. In any event, it is not necessary that our subject have a translation of '$\vee$'; its meaning is fully determined by the truth-table for '$\vee$'.

These remarks support a general statement: a knowledge of the truth-conditions of a sentence is identical with an understanding of its meaning.

5. L-CONCEPTS

5a. Tautologies. Suppose $\mathfrak{S}_i$ is a sentence composed out of the sentential constants 'A', 'B', etc., with the help of the sentential connectives previously discussed. (Here '$\mathfrak{S}_i$' is a sign of the metalanguage which serves to refer to sentences of the symbolic language. Cf. **20**, **21a**.) By a *value-assignment* for $\mathfrak{S}_i$ we understand any assignment of truth-values to the sentential constants occurring in $\mathfrak{S}_i$. If $\mathfrak{S}_i$ involves n distinct sentential constants, then there are 2^n possible value-assignments for $\mathfrak{S}_i$; these value-assignments are represented by the rows of the truth-table for the sentential constants. By the *range* of $\mathfrak{S}_i$ we understand the class of those possible value-assignments for $\mathfrak{S}_i$ at which $\mathfrak{S}_i$ comes out true; these particular value assignments are represented by the rows of the truth-table which have the entry 'T' in the last column. E.g. consulting Table I(2), we see that the range of '$A \vee B$' consists of the first three of the four value-assignments for '$A \vee B$' represented by the four rows of Table I(1); similarly, the range of '$A \equiv B$' consists of the first and last of these value-assignments; and similarly, the range of '$A . B$' consists of just the first of these value-assignments.

Now it is easy to see that the smaller the range of a sentence, the more the sentence says. Suppose e.g. we know the meaning of each of the two sentences 'A' and 'B'. If, then, '$A . B$' is communicated to us, we know precisely which of the four possible cases (i.e. which of the four value-assignments) actually obtains: it is the first one. On the other hand, the communication '$A \equiv B$' is indeterminate, for it does not decide between two possibilities. Again, '$A \vee B$' is even more indeterminate, for it excludes only one possibility and fails to decide between three possibilities. And if the range of a sentence is *total*, i.e. if, like '$A \vee \sim A$' (cf. Table III(3)), its range comprises all possible value-assignments, then the sentence excludes no possibility and hence says nothing. E.g. if 'A' means "it is raining here and now", then '$A \vee \sim A$' means "it is raining here and now, or it is not raining here and now"—a sentence which is true in every possible circumstance, no matter whether it is raining here now or not; if communicated to us, we could learn from it nothing whatever about actual present circumstances. Sentences which thus are true for all possible value-assignments of their constituent parts are said to be *tautologous sentences* or *tautologies*.

5b. Range and L-truth. Suppose we want to investigate a given sentence with a view towards establishing its truth-value. The procedure necessary to this end can be divided into two steps. Clearly we must, to begin with, understand the sentence; therefore, *the first step* must consist in establishing the meaning of the sentence. Here two considerations enter: on the one hand, we must attend to the meanings of the several signs that occur in the sentence (these meanings may perhaps be given by a list of meaning-rules, arranged e.g. in the form of a dictionary); and on the other, we must attend to the form of the sentence, i.e. the pattern into which the signs are assembled. *The second step* of our procedure consists in comparing what the sentence says with the actual state of the affairs to which the sentence refers. The meaning of the sentence determines what affairs are to be taken account of, i.e. what objects, what properties and relations of these objects, etc. By observation (understood in the widest sense) we settle how these affairs stand, i.e. what the facts are; then we compare these facts with what the sentence pronounces regarding them. If the facts are as the sentence says, then the sentence is true; otherwise, false.

In the usage of philosophers, the word "logical" is quite vague and ambiguous. We shall not attempt to state a general and exact definition of the word here. But we can increase somewhat the clarity of our remarks by indicating (in a non-technical way, with no claim to precision) certain situations in which we intend to use the term "logical". Our uses of this term will appear to be in reasonable agreement with those of ordinary language—complete agreement naturally cannot be demanded, considering the confused state of familiar speech. We shall call a procedure *logical* when it is grounded only in the analysis of senses (the first step of our previous paragraph) and does not require any observations of fact (the second step above); if the procedure requires the second step, we call it *non-logical*, or synthetic, or empirical. The analysis of sense we therefore term "logical analysis". Similarly, we refer to every concept which can be specified exclusively on the basis of the first step as a logical concept; concepts which depend on observation are counted as non-logical (descriptive, factual). Finally, we say a result or a statement is logical if it is based exclusively on the analysis of sense; and we say the same of a question whose answer comes about solely by analysis of sense.

Now let us introduce several concepts which are logical in the sense just indicated. We shall call them L-concepts, and shall form terms for them with the prefix "L-".

We divide all the signs of our symbolic language into two classes, the *constants* and *variables*. Every constant has a fixed specific meaning. Variables, on the other hand, serve to refer to unspecified objects, properties, etc.; they will be explained in subsequent sections. Again, we divide all our signs into *logical* and *descriptive* (or non-logical). Descriptive signs are those constants which serve to refer to objects, properties, relations, etc., in the world; they include the individual constants, the predicates, and the sentential constants. Logical signs include all the variables and the logical constants. Logical signs do not themselves refer to something in the world (the world of things has nothing like negation, disjunction, etc.); rather, they bind together the descriptive constants of a sentence and thereby contribute indirectly to the sense of a sentence. The logical constants comprise the connective signs, and such auxiliary signs as brackets, commas, etc. A compound expression is said to be descriptive if it contains at least one descriptive sign; otherwise, it is said to be logical. Thus, a logical expression is one that contains only logical signs.

We turn next to a generalization of the concepts of valuation and range. Among the *value-bearing signs* we count all the descriptive constants and certain variables. We have already taken as possible values for sentential constants the two truth-values, T and F. Later we shall lay down what other signs are to be value-bearing signs, and what their possible values are to be. The explanations which follow below will be conceived of so broadly as to apply not only to sentences, but more generally to *sentential formulas*, i.e. sentences or sentence-like expressions of other kinds to be described later. By a *value-assignment* for a given sentential formula $\mathfrak{S}_i$, we mean a coordination of values with all the value-bearing signs that occur in $\mathfrak{S}_i$. If a sign occurs in $\mathfrak{S}_i$ more than once, the same value must be coordinated with each of its occurrences. By the *evaluation* of a sentential formula at a specific value-assignment we understand the determination of the truth-value of $\mathfrak{S}_i$ for this value-assignment. When $\mathfrak{S}_i$ consists of sentential constants and connective signs, the evaluation of $\mathfrak{S}_i$ is made by means of the truth-tables. Later we will lay down additional rules of evaluation for other types of sentential formulas. In analogy with the earlier explanation, we take the *range* of the formula $\mathfrak{S}_i$ to be the class of those value-assignments at which $\mathfrak{S}_i$ comes out true. The class of all possible value-assignments for $\mathfrak{S}_i$ (i.e. for the value-bearing signs that occur in $\mathfrak{S}_i$) we call the *total range* of $\mathfrak{S}_i$; the empty class of such value-assignments we call the *null range*.

Sometimes it is said that a sentence (or a proposition, or a judgment) is logically true or logically necessary or analytic if it is true "on purely logical grounds", or if it is true independently of the accidental state of the facts, or if it holds in all possible worlds (Leibniz). It seems plausible to explicate (i.e. to conceive precisely; cf. the note on explication at the end of **1a**) this imprecise notion in the following way. We call a sentence *L-true* provided its range is the total range, hence provided it is true in every possible case. Every tautology is evidently L-true; later (**14**), we will encounter many L-true sentences that are not tautologies. Every L-true sentence is true: for since it holds in every possible case, it holds in the case actually before us. The truth of an L-true sentence is however not dependent on the facts, since it would be true whatever the disposition of the facts. Therefore it is unnecessary to institute observations in order to establish the truth of an

L-true sentence; what suffices here is logical analysis, viz. investigation of all possible value-assignments on the basis of the rules governing evaluation. L-truth is thus a logical concept in the sense previously described. The same holds for subsequent L-concepts.

We apply the notions of truth and falsity to sentences only, and not to other sentential formulas. (For these last, only the relative concepts "true (or: false) respecting this or that value-assignment" are applicable.) On the other hand, we can define L-concepts for sentential formulas in general by means of our generalized concepts of value-assignment and range. Thus, in analogy with the considerations of the last paragraph, we say that a *sentential formula* is *L-true* just in case its range is the total range, i.e. it is true for every value-assignment.

A sentential formula is said to be *L-false* (or logically false, or contradictory) in case its range is the null range, i.e. it is false for every value-assignment. Every L-false sentence is evidently false; moreover, its falsity resides entirely in the sense of the sentence and is independent of the facts.

If a sentential formula is either L-true or else L-false, we say it is *L-determinate*: otherwise (i.e. if it is neither L-true nor L-false), we say it is *L-indeterminate*. A sentential formula is L-indeterminate provided its range is neither total nor empty, i.e. when there is at least one value-assignment at which it is true, and at least one value-assignment at which it is false. Of an L-indeterminate sentence (though not of an open sentential formula) we also say that it is *factual*. This concept is intended to be an explication for the traditional notion of the synthetic judgment. Logical analysis does not suffice to ascertain the truth-value of a factual sentence; it is necessary to observe facts in order to establish whether we have before us one of the cases in which the sentence is true or one in which it is false. (As examples of factual sentences, we offer: '$Sph(moon)$', '$\sim Sph(moon)$', '$Stud(a) \vee Bro(a,b)$'.) If a factual sentence is true, we call it *F-true* (or factually true); if false, *F-false* (or factually false).

The remarks of our last four paragraphs suggest the following *classification of sentences* (this classification is not applicable to other sentential formulas):

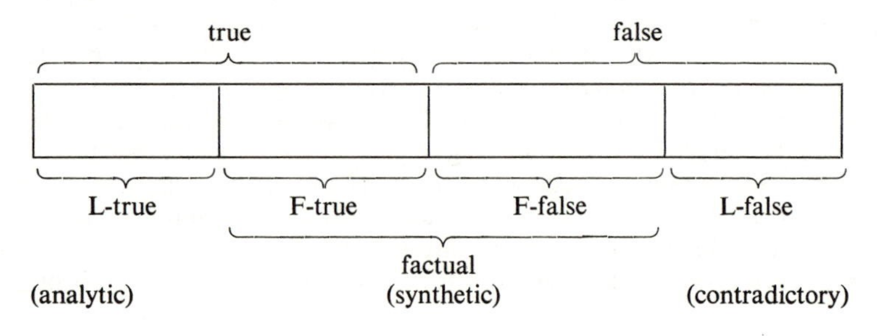

The theorems below follow from the definitions of the L-concepts and Truth-tables I and II. [We designate theorems by "T" and give each theorem two numbers, the first of which indicates the section in which the theorem is stated; this section number of the theorem is suppressed when references are made to it in the same section. (E.g. "T5-1c" refers to Theorem 1c of section **5**; if this reference is made in the text of section **5**, it is written simply "T1c".) Definitions are sometimes designated by "D", with the same sort of double numbering. As we remarked in **4a**, the sign "+" is prefixed to theorems, definitions, etc., of special importance.]

+T5-1. Ranges. **a.** Let $\mathfrak{S}_i$ be an arbitrary sentential formula, and $\sim\mathfrak{S}_i$ its negation; then the range of $\sim\mathfrak{S}_i$ is the complement of the range of $\mathfrak{S}_i$. (The *complement* of the range of $\mathfrak{S}_i$ is the class of those value-assignments in the total range of $\mathfrak{S}_i$ which do not belong to the range of $\mathfrak{S}_i$.)

b. The range of the disjunction of two or more sentential formulas is the union of the ranges of the individual sentential formulas. (The *union* of several classes is the class of all those elements which belong to at least one of the classes.)

c. The range of the conjunction of two or more sentential formulas is the intersection of the ranges of the individual sentential formulas. (The *intersection* of several classes is the class of all those elements which belong to each of the classes.)

T5-2. **a.** $\mathfrak{S}_i$ is L-false if and only if $\sim\mathfrak{S}_i$ is L-true. $\sim\mathfrak{S}_i$ is L-false if and only if $\mathfrak{S}_i$ is L-true. (From T1a.)

b. A disjunction of two or more sentential formulas is L-false if and only if each such member of the disjunction is L-false. (From T1b.)

c. A conjunction of two or more sentential formulas is L-true if and only if each such member of the conjunction is L-true. (From T1c.)

Exercise. Show that T2 follows from T1.

6. L-IMPLICATION AND L-EQUIVALENCE

6a. L-implication and L-equivalence. In this section we introduce two additional L-concepts, viz. the logical relations of L-implication and L-equivalence. First of all, we provide an illustration based on Truth-table I. Sentence '*A*' has for its range the first two cases, while the range of '$A \vee B$' comprises the first three cases. From this we see that for each case in which '*A*' is true—viz. the first and the second—'$A \vee B$' is also true. Hence we can conclude '$A \vee B$' from '*A*' without any knowledge of facts. What we do here is generalize this consideration to arbitrary sentential formulas $\mathfrak{S}_i$ and $\mathfrak{S}_j$. If $\mathfrak{S}_i$ and $\mathfrak{S}_j$ are such that the range of $\mathfrak{S}_i$ is contained in that of $\mathfrak{S}_j$ (i.e. if for the value-bearing signs of $\mathfrak{S}_i$ and $\mathfrak{S}_j$ each value-assignment at which $\mathfrak{S}_i$ is true is also one at which $\mathfrak{S}_j$ is true), then we shall say that $\mathfrak{S}_i$ *L-implies* $\mathfrak{S}_j$. L-implication is our explication (recall **1a**) for the traditional concept which is usually called "implication" or "logical implication" or "entailment", and whose inverse is ordinarly referred to by such terms as "logical consequence", "deducibility" and the like. In connection with our illustration above, we would now say that '$A \vee B$' is L-implied by 'A'.

+T6-1. **a.** A sentential formula which is L-implied by an L-true sentential formula is itself L-true.

b. A sentential formula which is tautologously (i.e. in truth-table terms) L-implied by a sentential formula that is a tautology is itself a tautology.

c. A sentential formula which L-implies an L-false sentential formula is itself L-false.

+T6-2. **a.** An L-true sentential formula is L-implied by every sentential formula.

b. An L-false sentential formula L-implies every sentential formula.

+T6-3. **a.** Every sentential formula L-implies itself.

b. Transitivity of L-implication. If $\mathfrak{S}_i$ L-implies $\mathfrak{S}_j$ and $\mathfrak{S}_j$ L-implies $\mathfrak{S}_k$, then $\mathfrak{S}_i$ L-implies $\mathfrak{S}_k$.

Now assume that two sentential formulas $\mathfrak{S}_i$ and $\mathfrak{S}_j$ are such that the conditional $\mathfrak{S}_i \supset \mathfrak{S}_j$ is L-true. Then $\mathfrak{S}_i$ L-implies $\mathfrak{S}_j$; for if there were a value-assignment at which $\mathfrak{S}_i$ is true and $\mathfrak{S}_j$ false, then by Truth-table I(4), line 2, the sentential formula $\mathfrak{S}_i \supset \mathfrak{S}_j$ would be false—which is impossible, since $\mathfrak{S}_i \supset \mathfrak{S}_j$ is presupposed to be L-true. Moreover, the converse holds: assuming $\mathfrak{S}_i$ L-implies $\mathfrak{S}_j$, it follows that the sentential formula $\mathfrak{S}_i \supset \mathfrak{S}_j$ is L-true. For otherwise there is a value-assignment at which $\mathfrak{S}_i \supset \mathfrak{S}_j$ is false, i.e. at which (by Table I(4), 2), $\mathfrak{S}_i$ is true and $\mathfrak{S}_j$ is false, which contradicts our assumption that $\mathfrak{S}_i$ L-implies $\mathfrak{S}_j$. Therefore:

+T6-4. If $\mathfrak{S}_i$ and $\mathfrak{S}_j$ are arbitrary sentential formulas, then $\mathfrak{S}_i$ L-implies $\mathfrak{S}_j$ if and only if the conditional $\mathfrak{S}_i \supset \mathfrak{S}_j$ is L-true.

Example. Take the sentences 'A' and '$A \vee B$' of our initial illustration as instances of $\mathfrak{S}_i$ and $\mathfrak{S}_j$ respectively. As we have seen, the range of 'A' is contained in that of '$A \vee B$', i.e. there is no value-assignment at which 'A' is true and '$A \vee B$' false. Hence, on the one hand 'A' L-implies '$A \vee B$'; and on the other hand the conditional '$A \supset A \vee B$', being true at every value-assignment, is itself L-true. For a conditional is false only at the values TF, and this combination cannot occur here.

T6-5. **a.** A sentential formula which L-implies $\mathfrak{S}_i$ and which also L-implies $\sim\mathfrak{S}_i$ is itself L-false. (From T5-1a.)

b. A sentential formula which L-implies its own negation is L-false. (From a and T3a.)

We call $\mathfrak{S}_i$ *L-equivalent* (or: logically equivalent) to $\mathfrak{S}_j$ just in case the range of $\mathfrak{S}_i$ is the same as the range of $\mathfrak{S}_j$.

+**T6-6.** **a.** Two sentential formulas are L-equivalent if and only if each L-implies the other.

b. Two sentential formulas are L-equivalent if and only if at each value-assignment either both are true or else both are false.

+**T6-7.** Two sentential formulas $\mathfrak{S}_i$ and $\mathfrak{S}_j$ are L-equivalent if and only if the biconditional $\mathfrak{S}_i \equiv \mathfrak{S}_j$ is L-true.

Proof. 1. Suppose $\mathfrak{S}_i$ and $\mathfrak{S}_j$ are L-equivalent. Then they both have the same range, i.e. at each of the possible value-assignments they are either both true or else both false. But by Table I(5), $\mathfrak{S}_i \equiv \mathfrak{S}_j$ is thereby true at every value-assignment; hence $\mathfrak{S}_i \equiv \mathfrak{S}_j$ is L-true. — **2.** Take $\mathfrak{S}_i \equiv \mathfrak{S}_j$ to be L-true, i.e. true at every value-assignment. Then there is no value-assignment at which $\mathfrak{S}_i$ and $\mathfrak{S}_j$ have different truth-values; thus $\mathfrak{S}_i$ and $\mathfrak{S}_j$ have the same range, and are L-equivalent.

6b. Content. A sentence says something about the world in that it excludes certain cases which are possible in themselves. In so doing, the sentence informs us that the excluded cases are not real cases. The more cases a sentence excludes, the more it says. Hence it seems plausible to define the *content* of a sentence as the class of possible cases in which it does not hold, i.e. those value-assignments which do not belong to the range of the sentence. (In the sequel we shall not make extensive use of this concept.)

The essential character of logical deduction, i.e. concluding from a sentence $\mathfrak{S}_i$ a sentence $\mathfrak{S}_j$ that is L-implied by it, consists in the fact that the content of $\mathfrak{S}_j$ is contained in the content of $\mathfrak{S}_i$ (because the range of $\mathfrak{S}_i$ is contained in that of $\mathfrak{S}_j$). We see thereby that logical deduction can never provide us with new knowledge about the world. In every deduction the range either enlarges or remains the same, which is to say the content either diminishes or remains the same. *Content can never be increased by a purely logical procedure.*

To gain factual knowledge, therefore, a non-logical procedure is always necessary. This point is also brought out by considering the sort of sentences whose truth logic is able to establish, viz. the L-true sentences: an L-true sentence excludes no possible case, and hence its content is null.

Though logic cannot lead us to anything new in the logical sense, it may well lead to something new in the psychological sense. Because of limitations on man's psychological abilities, the discovery of a sentence that is L-true or of a relation of L-implication is often an important cognition.

But this cognition is not a factual one, and is not an insight into the state of the world; rather, it is a clarification of logical relations subsisting between concepts, i.e. a clarification of relations between meanings. Suppose someone knows $\mathfrak{S}_i$ to begin with; and suppose that thereafter, by a laborious logical procedure, he finds that $\mathfrak{S}_j$ is L-implied by $\mathfrak{S}_i$. Our subject may now properly regard $\mathfrak{S}_j$ as known, but he may not count it as logically new: for the content of $\mathfrak{S}_j$, even though initially concealed, was from the beginning part of the content of $\mathfrak{S}_i$. Thus logical procedure, by disclosing $\mathfrak{S}_j$ and making it known, enables practical activities to be based on $\mathfrak{S}_j$. Again, two L-equivalent sentences have the same range and hence the same content; consequently, they are simply different formulations of this common logical content. However, the psychological content (the totality of associations) of one of these sentences may be entirely different from that of the other.

6c. Classes of sentences. We now extend to classes of sentences and other sentential formulas the concepts which up to the present have been applied to sentences. We regard a class of sentences conjunctively, i.e. we regard a class as expressing precisely what all of its sentences together express. Thus we say a class of sentences is true just in case each of its member sentences is true. Such a class is therefore false if at least one of its members is false. By the *range* of a class of sentential formulas we understand the aggregate of all value-assignments (to the value-bearing signs of all sentential formulas in the class) at which the class is true, i.e. the totality of those value-assignments at which all sentential formulas of the class come out true. L-concepts whose definitions rest on the notion of range may now be carried over unaltered. On this basis the following theorems result:

T6-8. The range of a class of sentential formulas is the intersection of the ranges of the individual sentential formulas.

From this, in view of T5-1c, follows:

+T6-9. A conjunction of two or more sentential formulas is L-equivalent to the class comprising these sentential formulas.

T6-10. A class of sentential formulas L-implies each of its sentential formulas, and each of its subclasses.

T6-11. A class of sentential formulas is L-true if and only if each of its sentential formulas is L-true.

T6-12. **a.** A sentential formula L-implies a class of sentential formulas if and only if it L-implies each sentential formula of this class.

b. A class of sentential formulas L-implies a second class if and only if it L-implies each sentential formula in the second class.

c. A sentential formula, or a class of such formulas, L-implies a conjunction with two or more components if and only if it L-implies each of these components.

T6-13. A class of sentential formulas which contains both a sentential formula and its negation is L-false.

If we say that certain sentential formulas L-imply another sentential formula or the like, we mean that the *class* of these sentential formulas L-implies the sentential formula in question, etc.

T6-14. **a.** The class comprising the sentential formulas $\mathfrak{S}_i$ and $\mathfrak{S}_i \supset \mathfrak{S}_j$ L-implies the sentential formula $\mathfrak{S}_j$. (By Truth-table I(4).)
b. If $\mathfrak{S}_i$ and $\mathfrak{S}_i \supset \mathfrak{S}_j$ are L-true, then $\mathfrak{S}_j$ is also L-true. (By a.)

T6-15. The class comprising the sentential formulas $\mathfrak{S}_i$ and $\sim\mathfrak{S}_i$ L-implies every sentential formula; and likewise the conjunction $\mathfrak{S}_i . \sim\mathfrak{S}_i$ L-implies every sentential formula. (By T13 and T2b.)

This last result is important in the treatment of deductive systems, e.g. axiom systems. If in such a system two contradictory sentences are derivable, the whole system becomes trivial inasmuch as any arbitrary sentence is thereupon derivable.

6d. Examples. **1.** From Truth-table I it is seen that the range of 'A' comprises the first two value-assignments (the first two cases), while the range of 'B' comprises the first and third value-assignments. Hence the intersection of these two ranges comprises just the first value-assignment alone. The class comprising the two sentences 'A' and 'B' therefore L-implies each of the following sentences: a) '$A . B$'; b) '$A \vee B$'; c) '$A \supset B$'; d) '$A \equiv B$'. — **2.** The part common to the ranges of 'A' and of '$A \supset B$' comprises just the first value-assignment alone; the range of 'B' comprises the first and the third value-assignments. Consequently, 'B' is L-implied by 'A' and '$A \supset B$'. (See T14a.)

Exercises. **1.** Show that T15 follows from T13 and T2b. — **2.** Determine (by means of a truth-table for 'A', 'B', 'C') the range of each of the following four classes of sentences: a) 'A', '$A . B \supset C$', '$\sim C$'; b) '$\sim B$', '$(A \supset B) \vee (\sim A \supset B)$'; c) '$C \supset A$', '$B \vee C$', '$\sim(A \vee B)$'; d) '$\sim B$'. — **3.** On the basis of your considerations in exercise 2 just above, determine which of the classes L-imply or are L-equivalent to what others. — **4.** Show if a class K of sentences L-implies a class M of sentences and every sentence in K is true, then every sentence in M is true. Do this using only the definition of "range of a class" in **6c** and the definition of "L-implies" in **6a**; do not use the theorems. — **5.** Show that, if the sentence 'A' L-implies the sentence 'B', and 'A' is true, then 'B' must be true. Hint: use the results of exercise 4. — **6.** Show that the sentence 'A' together with the sentence 'B' L-imply the sentence '$A . B$'.

7. SENTENTIAL VARIABLES

7a. Variables and sentential formulas. In mathematics variables have for centuries been used to great advantage for the purpose of representing relations between numbers exactly and concisely. Thus e.g. the formula '$x^2 = 3y + 4$' uses the number-variables 'x' and 'y' to express a relation which holds for certain pairs of numbers and not for others. Again, the formula '$x + y = y + x$' expresses a universal numerical relation, i.e. one that holds for all pairs of numbers; it is a universal or generally valid formula (often called an arithmetical law or an identity). If, when an expression is substituted for a variable of a given formula, there is produced another meaningful (but not necessarily true) formula, we say the expression is *substitutable* for the variable and call it a *substitutable expression.* The entities referred to by a variable of a formula are called the *values* of the variable. E.g. the variables 'x' and 'y' of the two formulas cited above have for their values numbers (more precisely, numbers of a certain kind—e.g, natural numbers—in accordance with the rules of the system in question), and numerical expressions (such as '6' or '6+2') are substitutable for them; thus these variables are termed "numerical variables". In mathematics the variables first used were numerical variables; later, however, use was made of variables whose values were entities of other sorts, e.g. functions, classes, operators and the like. Symbolic logic borrows the variable from mathematics, but employs it in a much more extended fashion. Symbolic logic admits as values of its variables entities of all possible kinds, e.g. things, classes, properties, relations, functions, propositions, etc. (Later a distinction shall be made between value-extensions and value-intensions, see **10b**.)

In our symbolic language system we shall use hereafter individual variables 'x', 'y', etc., for which individual constants like 'a', 'b', etc., are substitutable; and also predicate variables 'F', 'G', etc., for which predicates like 'P', 'Q', etc., are substitutable. By a *sentential formula* we shall understand an expression which is a sentence or which contains variables and becomes a sentence upon appropriate substitutions for these variables. E.g. 'Pa' is a sentence and hence a sentential formula; again 'Px', 'Fa', and 'Fx' are sentential formulas, since they go over into 'Pa' by appropriate substitutions. We make general use of the sign '$\mathfrak{S}$' for sentential formulas. Later we shall become acquainted with other kinds of formulas, e.g. numerical formulas (expressions which designate numbers, such as '6+3', or which by appropriate substitution transform into such expressions, as in the case of '$x+3$'), formulas for properties, for relations, for functions, etc. Our present concern being only with sentential formulas, we shall often write simply "formula" in place of "sentential formula".

7b. Sentential variables. Now we introduce as the first kind of variable in our language system the sentential variables (or propositional variables) 'p', 'q', 'r', etc. We agree that arbitrary sentential formulas of our language are substitutable for these sentential variables. Regarding such substitution, we understand that at every occurrence of á sentential variable in a given sentential formula the same expression is substituted. E.g. in '$p \vee q \supset q \vee p$' the same formula must be substituted at both occurrences of 'p'; similarly for 'q' (what is substituted for 'q' need not necessarily be different from what is substituted for 'p'). A sentential formula which contains at least one variable (later (**9a**) we shall say more precisely: a free variable) is called *open*; otherwise, *closed.* The closed sentential formulas are the sentences.

(In other language systems, it sometimes happens that open sentential formulas are also admitted as sentences.) Every closed sentential formula that can be derived from an open sentential formula $\mathfrak{S}_i$ by substitution is said to be a *substitution instance* (briefly: an instance) of $\mathfrak{S}_i$; if $\mathfrak{S}_i$ is a closed sentential formula, we count $\mathfrak{S}_i$ itself as its only substitution instance.

We say $\mathfrak{S}_i'$, $\mathfrak{S}_j'$, etc., are *corresponding substitution instances* of $\mathfrak{S}_i$, $\mathfrak{S}_j$, etc., if $\mathfrak{S}_i'$ is obtained from $\mathfrak{S}_i$, $\mathfrak{S}_j'$ from $\mathfrak{S}_j$, etc., by the same substitutions (i.e. for each sentential variable, the same expression is substituted at every occurrence of this variable in $\mathfrak{S}_i$, in $\mathfrak{S}_j$, etc.).

Individual constants and individual variables are called *individual signs*. A sentential formula which consists in an n-place predicate and n individual signs is said to be a *full formula of the predicate*; and further if no individual variables appear the sentential formula is said to be a *full sentence of the predicate*.

Sentential constants and sentential variables are called *sentential signs*. A sentential formula which is either a sentential sign or a full formula of a predicate is called an *atomic formula*; and if further this formula is a sentence, it is called an *atomic sentence*. A sentential formula is termed a *molecular compound* of other formulas if it is constructed from these other formulas by means of the connective signs previously considered. A sentential formula which is either an atomic formula or a molecular compound of atomic formulas is called a *molecular sentential formula*, and a *molecular sentence* if additionally it is a sentence. We say that $\mathfrak{S}_i$ *occurs molecularly* in $\mathfrak{S}_j$ in case $\mathfrak{S}_i$ and $\mathfrak{S}_j$ are such sentential formulas that $\mathfrak{S}_j$ is a molecular compound involving $\mathfrak{S}_i$ and possibly other formulas not containing $\mathfrak{S}_i$ as a part. [Example: 'Px' occurs molecularly in '$A \vee Px$', but not in '$A \vee (x)Px$'.]

The sentential variables are included among the value-bearing signs. Their possible values are the possible values of sentential constants, viz. the truth-values T and F. Suppose $\mathfrak{S}_i$ is a molecular sentence with n different sentential constants; and suppose $\mathfrak{S}_j$ is an open sentential formula obtained from $\mathfrak{S}_i$ by replacing the sentential constants by n different sentential variables. If now $\mathfrak{S}_i$ is true at a certain value-assignment to the sentential constants, then $\mathfrak{S}_j$ is evidently true at the same value-assignment to the sentential variables; and indeed, if $\mathfrak{S}_i$ is L-true, then $\mathfrak{S}_j$ is too. It is also evident that truth-tables can be applied directly to the sentential variables of a molecular formula. Thus e.g. since by Table III(3) the sentence '$A \vee \sim A$' is L-true, the open sentence '$p \vee \sim p$' is also L-true; and this result can be seen at once by a truth-table analogous to the one cited, but with 'p' in place of 'A'.

+T7-1. Substitutions. Suppose $\mathfrak{S}_i$ and $\mathfrak{S}_j$ are arbitrary sentential formulas; and suppose $\mathfrak{S}_i'$ and $\mathfrak{S}_j'$ are obtained from $\mathfrak{S}_i$ and $\mathfrak{S}_j$ respectively by the same substitutions for one or more (but not

+**T7-1** necessarily all) of the sentential variables appearing in the latter. Then it is the case that:

a. If $\mathfrak{S}_i$ is L-true, then $\mathfrak{S}_i'$ is also.

Proof. Take $\mathfrak{S}_i$ to be L-true, i.e. true at each value-assignment to the value-bearing signs that appear in $\mathfrak{S}_i$. Suppose $\mathfrak{S}_i'$ is obtained from $\mathfrak{S}_i$ by substituting the sentential formula $\mathfrak{S}_k$ at each of the occurrences of some one sentential variable (say 'p') appearing in $\mathfrak{S}_i$. The value-bearing signs of $\mathfrak{S}_k$ now appear among the value-bearing signs of $\mathfrak{S}_i'$. Suppose a value-assignment is made to the value-bearing signs of $\mathfrak{S}_i'$. This leads, in particular, to an evaluation of $\mathfrak{S}_k$ as either T or else F. But since $\mathfrak{S}_i$ is true at every value-assignment, no matter whether 'p' is assigned the value T or the value F, it must be that $\mathfrak{S}_i'$ is true at every value-assignment (which necessarily fixes the newly added value-bearing signs of $\mathfrak{S}_k$), no matter whether these assignments impart to $\mathfrak{S}_k$ the value T or the value F. Thus $\mathfrak{S}_i'$ is L-true.

b. If $\mathfrak{S}_i$ is a tautology, so also is $\mathfrak{S}_i'$. (This is a special case of a.)
c. If $\mathfrak{S}_i$ is L-false, so also is $\mathfrak{S}_i'$. (By analogy with a.)
d. If $\mathfrak{S}_i'$ is L-indeterminate, so also is $\mathfrak{S}_i$. (From a and c.)
e. If $\mathfrak{S}_i$ L-implies $\mathfrak{S}_j$, then also $\mathfrak{S}_i'$ L-implies $\mathfrak{S}_j'$. (From a and T6-4.)
f. If $\mathfrak{S}_i$ and $\mathfrak{S}_j$ are L-equivalent, so also are $\mathfrak{S}_i'$ and $\mathfrak{S}_j'$. (From a and T6-7.)

Examples related to T1b. The formula '$p \vee q \supset q \vee p$' is a tautology. Hence the formulas '$p \vee A \supset A \vee p$' and '$(p.r) \vee (A.\sim p) \supset (A.\sim p) \vee (p.r)$' are also tautologies.

8. SENTENTIAL FORMULAS THAT ARE TAUTOLOGIES

8a. Conditional formulas that are tautologies. The theorems below list sentential formulas that are tautologies. In each case, the tautological character of the formula can be established by means of a truth-table that has sentential variables 'p', 'q', etc. where formerly 'A', 'B', etc., appeared. A first reading of this book requires only that attention be given the more important formulas marked '+'.

T8-1. The following formulas are tautologies, and hence L-true:

+**a.** $p \vee \sim p$.
b. $\sim p \vee p$.
c. $\sim(p.\sim p)$.

T8-2. Let $\mathfrak{S}_i \supset \mathfrak{S}_j$ be any of the conditionals introduced below [viz. a(1) through i(2)]. Suppose $\mathfrak{S}_i' \supset \mathfrak{S}_j'$ is obtained from $\mathfrak{S}_i \supset \mathfrak{S}_j$ by arbitrary substitutions. Then each of the following holds:

A. $\mathfrak{S}_i \supset \mathfrak{S}_j$ is a tautology, and hence L-true.
B. $\mathfrak{S}_i' \supset \mathfrak{S}_j'$ is a tautology, and hence L-true. (From T7-16.)
C. $\mathfrak{S}_i$ L-implies $\mathfrak{S}_j$. (By T6-4.)
D. $\mathfrak{S}_i'$ L-implies $\mathfrak{S}_j'$. (By C, in view of T7-1d.)

T8-2 E. If $\mathfrak{S}_i$ is a conjunction (whence the whole conditional has the form $\mathfrak{S}_k . \mathfrak{S}_l \supset \mathfrak{S}_j$), then $\mathfrak{S}_j$ is L-implied by the class comprising the formulas $\mathfrak{S}_k$ and $\mathfrak{S}_l$; and similarly for formulas obtained from these three by corresponding substitutions.

a. +(1) $p \supset p \vee q$.
(2) $q \supset p \vee q$.
(3) $q \supset (p \supset q)$.
(4) $\sim p \supset (p \supset q)$.

b. +(1) $p.q \supset p$.
(2) $p.q \supset q$.

+**c.** $p.\sim p \supset q$.

d. +(1) $(p \vee q).\sim p \supset q$.
+(2) $(p \vee q).\sim q \supset p$.
+(3) $(p \supset q).p \supset q$.
(4) $p \supset [(p \supset q) \supset q]$.
(5) $(p \supset q).\sim q \supset \sim p$.

e. +(1) $(p \equiv q) \supset (p \supset q)$.
+(2) $(p \equiv q) \supset (q \supset p)$.
(3) $(p \equiv q) \supset (\sim p \supset \sim q)$.
(4) $(p \equiv q) \supset (\sim q \supset \sim p)$.
(5) $(p \equiv q).p \supset q$.
(6) $(p \equiv q).q \supset p$.
(7) $(p \equiv q).\sim p \supset \sim q$.
(8) $(p \equiv q).\sim q \supset \sim p$.

f. (1) $(p \supset q) \supset (p \vee r \supset q \vee r)$.
(2) $(p \supset q) \supset (p.r \supset q.r)$.
(3) $(p \supset q) \supset [(r \supset p) \supset (r \supset q)]$.
(4) $(p \supset q) \supset [(q \supset r) \supset (p \supset r)]$.
(5) $(p \supset q).(p \vee r) \supset q \vee r$.
+(6) $(p \supset q).(q \supset r) \supset (p \supset r)$.
(7) $(p \equiv q).(p \equiv r) \supset (q \equiv r)$.
+(8) $(p \equiv q).(q \equiv r) \supset (p \equiv r)$.

g. (1) $(p \equiv q) \supset (p \vee r \equiv q \vee r)$.
(2) $(p \equiv q) \supset (p.r \equiv q.r)$.
(3) $(p \equiv q) \supset [(p \supset r) \equiv (q \supset r)]$.
(4) $(p \equiv q) \supset [(r \supset p) \equiv (r \supset q)]$.
(5) $(p \equiv q) \supset [(p \equiv r) \equiv (q \equiv r)]$.

h. (1) $(p \supset q).(r \supset s) \supset (p \vee r \supset q \vee s)$.
(2) $(p \supset q).(r \supset s).(p \vee r) \supset q \vee s$.

i. (1) $q \supset (p \equiv p.q)$.
(2) $\sim q \supset (p \equiv p \vee q)$.

In connection with using the conditional formulas listed just above in T2, the subsidiary assertions C and D have special importance: in each case, the first member (or a substitution instance thereof) L-implies the second member (or its corresponding substitution instance). Thus it is possible in a deduction (derivation, **8d**) to infer the latter formula from the former.

From a(1) and (2), e.g., it appears we may join to a given sentential formula another arbitrary one as a member of a disjunction. From a(3) and (4): a conditional formula is L-implied by its consequent, and also by the negation of its antecedent. (Hence a conditional sentence is true if its consequent is true; and again, true if its antecedent is false; which also can be seen from Truth-table I(4).) From b(1) and (2): a conjunction L-implies each of its members. From c: a sentential formula and its negation together L-imply any arbitrary sentential formula (cf. T6-15). From d(1) and (2): a disjunction and the negation of one of its members together L-imply the other member. Regarding d(3): this supports an important type of inference, viz. from a conditional together with its antecedent to the consequent (sometimes called *modus ponens*; cf. T6-14a). Regarding d(5): this allows a similar inference from a conditional together with the negation of its consequent to the negation of the antecedent (sometimes called *modus tollens*). From e(1) and (2): a biconditional L-implies the two conditionals that can be formed from its members. From e(5) and (6): a biconditional together with one of its members L-implies the other member. From e(7) and (8): a biconditional together with the negation of one of its members L-implies the negation of the other member. From f(1) and (2): in a given conditional it is possible to join to each member the same formula as a member of a disjunction, or as a member of a conjunction; and from f(3) and (4) likewise, this added formula may be joined as the antecedent of a conditional, or as the consequent (in this event, the original members exchange position). From f(6): conditional is transitive. From g(1) to (5): in a given biconditional the same formula may be joined to both members either as member of a disjunction or of a conjunction, or as first or second member of conditionals, or as first or second members of biconditionals. From i(1): an arbitrary true sentence can be conjoined to a given sentence without changing its truth-value; and the conjunctive addition of an L-true sentence does not change the content of the original, i.e. the result is L-equivalent to the original sentence. Finally, i(2) permits an analogous claim for the disjunctive addition of a false (or L-false) sentence.

8b. Interchangeability. We say an expression $\mathfrak{A}_i$ is *interchangeable* with an expression $\mathfrak{A}_j$ just in case the following holds for arbitrary sentential formulas $\mathfrak{S}_i$ and $\mathfrak{S}_j$: if $\mathfrak{S}_i$ contains $\mathfrak{A}_i$ and $\mathfrak{S}_j$ is obtained from $\mathfrak{S}_i$ by replacing $\mathfrak{A}_i$ by $\mathfrak{A}_j$ at one or more (but not necessarily all) occurrences of $\mathfrak{A}_i$ in $\mathfrak{S}_i$, then $\mathfrak{S}_i \equiv \mathfrak{S}_j$ is true. We say $\mathfrak{A}_i$ is *L-interchangeable* with $\mathfrak{A}_j$ if additionally $\mathfrak{S}_i \equiv \mathfrak{S}_j$ is always L-true, i.e. $\mathfrak{S}_i$ and $\mathfrak{S}_j$ are always L-equivalent.

The truth-value of a sentence involving just one of our connective signs is uniquely determined by the truth-values of its components, with the aid of the truth-table for the connective. (It is for this reason that our connectives are also called "truth-functions".) Therefore the truth-value of an arbitrarily compounded molecular sentence is also uniquely determined by the truth-values of the atomic sentences occurring in it. Suppose $\mathfrak{S}_i$ is a molecular sentence in which $\mathfrak{S}_j$ occurs as a component ($\mathfrak{S}_j$ may be an atomic sentence or a compound molecular sentence). If now this $\mathfrak{S}_j$ in $\mathfrak{S}_i$ is interchanged with any other sentence $\mathfrak{S}_k$ whose truth-value is the same as that of $\mathfrak{S}_j$, then from our previous remarks it is clear that the truth-value of $\mathfrak{S}_i$ remains unaltered. In effect: a sentential formula is translated into one L-equivalent to it when any component formula of the original is interchanged with any formula L-equivalent to that component. This important result is proved more exactly in the following theorems.

T8-3. Suppose '...p...' is one of the following formulas: '$\sim p$', '$p \vee r$', '$r \vee p$', '$p . r$', '$r . p$', '$p \supset r$', '$r \supset p$', '$p \equiv r$', '$r \equiv p$'. Suppose '...q...', '...A...' and '...B...' are corresponding formulas, with 'q' (or 'A', or 'B' respectively) standing in place of 'p'. Then the following hold:

a. '$(p \equiv q) \supset [(...p...) \equiv (...q...)]$' is L-true.
b. '$p \equiv q$' L-implies '$(...p...) \equiv (...q...)$'.
c. '$(p \equiv q) . (...p...) \supset (...q...)$' is L-true.
d. '$p \equiv q$' and '...p...' together L-imply '...q...'.
e. '$(A \equiv B) \supset [(...A...) \equiv (...B...)]$' is L-true.
f. '$A \equiv B$' L-implies '$(...A...) \equiv (...B...)$'.
g. '$(A \equiv B) . (...A...) \supset (...B...)$' is L-true.
h. '$A \equiv B$' and '...A...' together L-imply '...B...'.

Proof. We state a proof for the formula '$p \vee r$'; proofs for the other formulas are analogous. — (a). From T2g (1), or from the truth-table. — (b). From (a), by T6-4. — (c). '$(p \equiv q) . (p \vee r) \supset q \vee r$' is a tautology. — (d). From (c), by T6-4 and T6-9. — (e) through (h) follow from (a) through (d), by T7-1.

[It is to be noted that assertions analogous to (a) and (b), with '$\supset$' instead of '$\equiv$' in both places, do *not* hold except in certain cases—of which several are specified in T2f (1) (2) (3).]

T8-4. Suppose '...p...' is a molecular sentential formula containing 'p'. Suppose '...q...', '...A...' and '...B...' result from '...p...' by the introduction of 'q' or 'A' or 'B' respectively in place of 'p'. Then assertions (a) through (h) of T3 hold.

Proof. A proof of (b) results from applying T3 first of all to the smallest component formula of '...p...' that contains 'p', and then to successively larger component formulas until '...p...' itself is reached. These successive stages make use of the following tautologies:

(α) $(p \equiv q) \supset [(r \equiv s) \supset (p \vee r \equiv q \vee s)]$.
(β) $(p \equiv q) \supset [(r \equiv s) \supset (p . r \equiv q . s)]$.
(γ) $(p \equiv q) \supset [(r \equiv s) \supset ((p \supset r) \equiv (q \supset s))]$.
(δ) $(p \equiv q) \supset [(r \equiv s) \supset ((p \equiv r) \equiv (q \equiv s))]$.

[A proof, with the formula '$(r.\sim p) \vee (p.s)$' taken for '...p...', will illustrate these considerations. First, beginning with '$(p \equiv q)$', we see by T3b that: (1) '$p \equiv q$' L-implies '$\sim p \equiv \sim q$'. Next: (2) '$p \equiv q$' L-implies '$r.p \equiv r.q$'. This last yields, by substitution: (3) '$\sim p \equiv \sim q$' L-implies '$r.\sim p \equiv r.\sim q$'. From (1) and (3), by T6-3b: (4) '$p \equiv q$' L-implies '$r.\sim p \equiv r.\sim q$'. Again, by T3b: (5) '$p \equiv q$' L-implies '$p.s \equiv q.s$'. By (α), with substitution: (6) '$r.\sim p \equiv r.\sim q$' and '$p.s \equiv q.s$' together L-imply '$(r.\sim p) \vee (p.s) \equiv (r.\sim q) \vee (q.s)$'. From this last, in view of (4) and (5), we have: '$p \equiv q$' L-implies '$(r.\sim p) \vee (p.s) \equiv (r.\sim q) \vee (q.s)$', the biconditional desired.] To finish the original proof, we need only note that (a) follows from (b) by T6-4, and that other parts of the theorem follow in analogy with T3.

+T8-5. Suppose $\mathfrak{S}_i$ and $\mathfrak{S}_j$ are L-equivalent; and suppose $\mathfrak{S}_i$ occurs in $\mathfrak{S}_k$ one or several times, but only molecularly. Now let $\mathfrak{S}_l$ be obtained from $\mathfrak{S}_k$ by interchanging $\mathfrak{S}_i$ with $\mathfrak{S}_j$ at one or more (but not necessarily all) of the occurrences of $\mathfrak{S}_i$ in $\mathfrak{S}_k$. Then $\mathfrak{S}_k$ and $\mathfrak{S}_l$ are L-equivalent.

Proof. To begin, $\mathfrak{S}_i \equiv \mathfrak{S}_j$ is L-true. This formula L-implies $\mathfrak{S}_k \equiv \mathfrak{S}_l$, by T4b. Hence $\mathfrak{S}_k \equiv \mathfrak{S}_l$ is also L-true, in view of T6-1a. By T6-7, therefore, $\mathfrak{S}_k$ and $\mathfrak{S}_l$ are L-equivalent.

T5 tells us that L-equivalent sentential formulas are L-interchangeable in places where they occur molecularly. Later we shall state a more general theorem on L-interchangeability (it is T15-3) that has T5 as a special case.

8c. Biconditional formulas that are tautologies.

T8-6. Let $\mathfrak{S}_i \equiv \mathfrak{S}_j$ be any one of the biconditional formulas (a) through (q)(5) introduced below. Suppose $\mathfrak{S}_i' \equiv \mathfrak{S}_j'$ is obtained from $\mathfrak{S}_i \equiv \mathfrak{S}_j$ by arbitrary substitutions. Then the following hold:

A. $\mathfrak{S}_i \equiv \mathfrak{S}_j$ is a tautology, and hence L-true.

B. $\mathfrak{S}_i' \equiv \mathfrak{S}_j'$ is a tautology, and hence L-true. (By T7-1b.)

C. $\mathfrak{S}_i$ and $\mathfrak{S}_j$ are L-equivalent. (From (A), by T6-7.)

D. $\mathfrak{S}_i'$ and $\mathfrak{S}_j'$ are L-equivalent. (From (B), by T6-7.)

E. $\mathfrak{S}_i$ and $\mathfrak{S}_j$ are mutually L-interchangeable in molecular compounds. (From (C), by T5.)

F. $\mathfrak{S}_i'$ and $\mathfrak{S}_j'$ are mutually L-interchangeable in molecular compounds. (From (D), by T5.)

a. $p \equiv p$.

+b. $p \equiv \sim\sim p$.

c. $p \equiv p \vee p$.

d. $p \equiv p.p$.

e. Commutative laws.

+(1) $p \vee q \equiv q \vee p$.

+(2) $p.q \equiv q.p$.

+(3) $(p \equiv q) \equiv (q \equiv p)$.

T8-6 **f.** +(1) $(p \equiv q) \equiv (p \supset q).(q \supset p)$.
(2) $(p \equiv q) \equiv [(p \equiv r) \equiv (q \equiv r)]$.
(3) $(p \equiv q) \equiv ({\sim}p \vee q).(p \vee {\sim}q)$.
(4) $(p \equiv q) \equiv (p.q) \vee ({\sim}p.{\sim}q)$.

g. Duality laws.
+(1) ${\sim}(p \vee q) \equiv {\sim}p.{\sim}q$.
(2) ${\sim}(p_1 \vee p_2 \vee \ldots \vee p_n) \equiv {\sim}p_1.{\sim}p_2.\ldots.{\sim}p_n$.
+(3) ${\sim}(p.q) \equiv {\sim}p \vee {\sim}q$.
(4) ${\sim}(p_1.p_2.\ldots.p_n) \equiv {\sim}p_1 \vee {\sim}p_2 \vee \ldots \vee {\sim}p_n$.
(5) $p \vee q \equiv {\sim}({\sim}p.{\sim}q)$.
(6) $p.q \equiv {\sim}({\sim}p \vee {\sim}q)$.

h. Negation laws.
+(1) ${\sim}(p \supset q) \equiv p.{\sim}q$.
(2) ${\sim}(p.{\sim}q) \equiv (p \supset q)$.
+(3) ${\sim}(p \equiv q) \equiv (p \equiv {\sim}q)$.
(4) ${\sim}(p \equiv q) \equiv ({\sim}p \equiv q)$.
(5) ${\sim}(p \equiv q) \equiv (p \supset {\sim}q).({\sim}q \supset p)$.
(6) ${\sim}(p \equiv q) \equiv ({\sim}p \supset q).(q \supset {\sim}p)$.
(7) ${\sim}(p \equiv q) \equiv (p.{\sim}q) \vee ({\sim}p.q)$.
(8) ${\sim}(p \equiv q) \equiv (p \vee q).({\sim}p \vee {\sim}q)$.

i. Transposition laws.
+(1) $(p \supset q) \equiv ({\sim}q \supset {\sim}p)$.
(2) $({\sim}p \supset q) \equiv ({\sim}q \supset p)$.
(3) $(p \supset {\sim}q) \equiv (q \supset {\sim}p)$.
+(4) $(p \equiv q) \equiv ({\sim}p \equiv {\sim}q)$.
(5) $(p \equiv {\sim}q) \equiv ({\sim}p \equiv q)$.
(6) $(p.q \supset r) \equiv (p.{\sim}r \supset {\sim}q)$.
(7) $(p \supset q \vee r) \equiv (p.{\sim}q \supset r)$.
(8) $(p \supset {\sim}q \vee r) \equiv (p.q \supset r)$.

j. Transformations of the conditional.
(1) $(p \supset q) \equiv {\sim}p \vee q$.
(2) $(p \supset q) \equiv (p \supset p.q)$.
(3) $(p \supset q) \equiv (p \equiv p.q)$.
(4) $(p \supset q) \equiv (p \vee q \supset q)$.
(5) $(p \supset q) \equiv (p \vee q \equiv q)$.

k. (1) $p \equiv (p \vee q).(p \vee {\sim}q)$.
(2) $p \equiv (p.q) \vee (p.{\sim}q)$.

l. (1) $\big(p \supset (q \supset r)\big) \equiv (p.q \supset r)$.
(2) $\big(p \supset (q \supset r)\big) \equiv \big(q \supset (p \supset r)\big)$.

m. Associative laws.
+(1) $(p \vee q) \vee r \equiv p \vee (q \vee r)$.
+(2) $(p.q).r \equiv p.(q.r)$.

T8-6 **n.** Distributive laws.

+(1) $p.(q \vee r) \equiv (p.q) \vee (p.r)$.

(2) $p.(q_1 \vee q_2 \vee \ldots \vee q_n) \equiv (p.q_1) \vee (p.q_2) \vee \ldots \vee (p.q_n)$.

(3) $(p_1 \vee p_2 \vee \ldots \vee p_m).(q_1 \vee q_2 \vee \ldots \vee q_n) \equiv (p_1.q_1) \vee (p_1.q_2) \vee \ldots \vee (p_1.q_n) \vee (p_2.q_1) \vee \ldots \vee (p_m.q_1) \vee (p_m.q_2) \vee \ldots \vee (p_m.q_n)$, where the conjunctions on the right represent possible pairs comprising one p-variable and one q-variable.

+(4) $p \vee (q.r) \equiv (p \vee q).(p \vee r)$.

(5) $p \vee (q_1.q_2.\ldots.q_n) \equiv (p \vee q_1).(p \vee q_2).\ldots.(p \vee q_n)$.

(6) $(p_1.p_2.\ldots.p_m) \vee (q_1.q_2.\ldots.q_n) \equiv (p_1 \vee q_1).(p_1 \vee q_2).\ldots.(p_1 \vee q_n).(p_2 \vee q_1).\ldots.(p_m \vee q_1).(p_m \vee q_2).\ldots.(p_m \vee q_n)$, in analogy with (3).

(7) $p \vee (q \equiv r) \equiv (p \vee q \equiv p \vee r)$.

(8) $(p \supset q.r) \equiv (p \supset q).(p \supset r)$.

(9) $(p \supset q_1.q_2.\ldots.q_n) \equiv (p \supset q_1).(p \supset q_2).\ldots.(p \supset q_n)$.

(10) $(p \supset q \vee r) \equiv (p \supset q) \vee (p \supset r)$.

(11) $(p \supset q_1 \vee q_2 \vee \ldots \vee q_n) \equiv (p \supset q_1) \vee (p \supset q_2) \vee \ldots \vee (p \supset q_n)$.

(12) $p \supset (q \supset r) \equiv (p \supset q) \supset (p \supset r)$.

(13) $p \supset (q \equiv r) \equiv ((p \supset q) \equiv (p \supset r))$.

o. (1) $(p.q \supset r) \equiv (p \supset r) \vee (q \supset r)$.

(2) $(p_1.p_2.\ldots.p_n \supset r) \equiv (p_1 \supset r) \vee (p_2 \supset r) \vee \ldots \vee (p_n \supset r)$.

(3) $(p \vee q \supset r) \equiv (p \supset r).(q \supset r)$.

(4) $(p_1 \vee p_2 \vee \ldots \vee p_n \supset r) \equiv (p_1 \supset r).(p_2 \supset r).\ldots.(p_n \supset r)$.

p. $(p \supset (q \equiv r)) \equiv (p.q \equiv p.r)$.

q. (1) $p \equiv p \vee (p.q)$.

(2) $p \equiv p.(p \vee q)$.

(3) $p \vee q \equiv p \vee (q.\sim p)$.

(4) $p.q \equiv p.(q \vee \sim p)$.

(5) $p.q \equiv p.(p \supset q)$.

Our application of the tautological biconditionals listed just above depends heavily on two features, viz. the two main components are L-equivalent, and these two components are mutually L-interchangeable in molecular compounds. In particular, (b) permits the suppression of double negation signs. Again, (e)(1) to (3) permit the commutation of the components of a disjunction, of a conjunction, and of a biconditional. The laws (g)—these are sometimes called *De Morgan's* laws—and the laws (h) show how the negations of certain compounds are transformed. The laws (i) allow what is called transposition (or contraposition); in particular, (i)(1) says that the components of a conditional are exchanged and negated. The biconditional (j)(1) states the interpretation of the implication sign given earlier. The laws (m) state that disjunction and conjunction are associative: when a disjunction (or conjunction) has three components, the way they are put together may be altered arbitrarily. Thus, in these cases parentheses may be omitted and e.g. expressions written simply '$A \vee B \vee C$' or '$A . B . C$'; cf. **3c**, rule (3) for omission of parentheses. [The same remarks hold true when the disjunction (or conjunction) has more than three components.] Finally, (n)(1) and (4) permit distribution through parentheses. These two laws are analogous to the arithmetical theorem "$x \cdot (y+z) = x \cdot y + x \cdot z$"; however, there is this difference: while arithmetic permits a multiplying-out (as in the theorem just cited) and not a similar adding-out, here both (1) and (4) hold. [In (1), conjunction corresponds to multiplication; in (4), disjunction corresponds to multiplication.]

8d. Derivations. The L-implications set forth above can be utilized in deducing from certain assumptions (the "premisses") a result (the "conclusion"). By a *derivation* with given premisses we will understand a sequence of sentential formulas which begins with the premisses and which continues through other sentential formulas one at a time, each step being a formula that is L-implied by the ones preceding it.

Example. Suppose we know (or assume) that '$A . B \supset C$' is true, that 'A' is true, and that 'C' is false. What, then, can be said about the truth-value of 'B'? This question can be answered either by a truth-table (cf. **6d**, exercise 3) or by a derivation. We give below an illustrative derivation. (To the left of a line in a derivation we sometimes note which of its preceding formulas were used, and what theorems were applied, to produce that line of the derivation.)

Derivation.		Premisses:		
		1)	$A . B \supset C$	(1)
		2)	A	(2)
		3)	$\sim C$	(3)
(1)	T6 l (1)		$A \supset (B \supset C)$	(4)
(2)(4)	T6-14a		$B \supset C$	(5)
(5)	T6i(1)		$\sim C \supset \sim B$	(6)
(3)(6)	T6-14a		$\sim B$	(7)

Hence, '$\sim B$' is L-implied by the premisses, i.e. on the basis of our original assumptions 'B' is false.

Exercises. Transform each of the following two sentences into an L-equivalent sentence which has no negation sign before a parenthesis (hint: use theorems T6b, g(1),(3) and h(1),(3)): **1.** '$\sim [A.(B \supset C)]$'. — **2.** '$\sim [(A \equiv B) \vee (C. \sim D)]$'. — **3.** Suppose that '$(A \supset B.C) \equiv D$' and 'B' are true, and 'D' is false; make a derivation to determine the truth-values for 'A' and 'C' from these assumptions. — **4.** According to T4 the sentence '$(D \equiv \sim B.C) \supset E$' is L-implied by '$A \equiv B$' and '$(D \equiv \sim A.C) \supset E$'. Show this by a derivation which uses only T3, and not T4. (This L-implication can also be established by means of a truth-table; how many lines must that table contain?) — **5.** Give a derivation for each of the following cases of L-implication: a) '$\sim D \vee B$', '$B \supset C$' and '$A \supset D$' L-imply '$\sim A \vee C$' (hint: use T6j(1),T2f(4)); b) '$A \vee (B.C)$' and '$\sim B$' L-imply 'A' (use, among others, T6n(4), T2d(2)); c) '$B \supset A$' L-implies '$\sim\sim B \supset A$' (use T3); d) '$A \supset \sim A$' L-implies '$\sim A$'; e) '$\sim (A \supset C)$' and 'C' L-imply 'D'.

9. UNIVERSAL AND EXISTENTIAL SENTENCES

9a. Individual variables and quantifiers. As was previously indicated, we use '*a*', '*b*', etc. as individual constants, and '*P*', '*Q*', etc. as predicates. Further, from atomic sentences (e.g. '*Pa*', '*Rbc*') and the familiar connectives we form compound molecular sentences (e.g. '$Pa \vee \sim Rbc$'). Now suppose we have a sentence dealing with an individual *a*, i.e. a sentence '...*a*...*a*...' in which '*a*' occurs one or more times (e.g. '$Pa \vee Rab$'). Suppose further we wish to state that what this sentence says about *a* does in fact hold for *every* individual in the domain of individuals to which *a* belongs. Then we say "for every *x*, ...*x*...*x*..." and write '(*x*)(...*x*...*x*...)'. [For the particular sentence cited above, we write '$(x)(Px \vee Rxb)$'. Thus, the sentence '$(x)(Px \vee Rxb)$' means "for every individual *x*, *x* has property *P* or *x* bears relation *R* to *b*".] Instead of '*x*', any of the letters '*u*', '*v*', '*w*', '*y*', '*z*' can be used as well. We term '*x*', '*u*', '*v*', '*w*', '*y*' and '*z*' *individual variables*. Individual constants and individual variables are called *individual signs*. The whole sentence '(*x*)(...*x*...*x*...)' is known as a *universal sentence*. The expression '(*x*)' at the head of a universal sentence is called a *universal quantifier*, and the parenthetical expression following it is called the *operand* of this quantifier. [E.g. the *operand* of the universal quantifier '(*x*)' in the universal sentence '$(x)(Px \vee Rxb)$' is '$Px \vee Rxb$'.]

If we wish rather to state that what the sentence '...*a*...*a*...' says about *a* does in fact hold for *at least one* individual of the domain (leaving open the question whether *a* is that individual), we again employ a variable, e.g. '*x*', saying "for at least one *x*, ...*x*...*x*..." and writing '($\exists x$)(...*x*...*x*...)'. Other readings for '($\exists x$)(...*x*...*x*...)' are: "for some *x*, ...*x*...*x*...", and "there is an *x* such that ...*x*...*x*...". The whole sentence '($\exists x$)(...*x*...*x*...)' is called an *existential sentence*. The expression '($\exists x$)' at the head of an existential sentence is called an *existential quantifier*, and the parenthetical expression following it is called the *operand* of this quantifier.

Our explanations above of universal and existential sentences indicate that the sense of these sentences depends on what is taken as the *domain of individuals*. In connection with any application of the symbolic language, it must be established what this domain is. The domain can be fixed at will; and in particular, it may be finite or infinite. However, it is customary to assume that the domain is not empty, i.e. there is at least one individual in the domain. Another frequent presupposition is that the domain is so chosen as to have a specified number of individuals in it.

A sentential formula having the structure of either of the two special sentence forms just described is called a *universal formula* or an *existential formula*, as the case may require. Formulas of these two types can, of course, appear as components in compound formulas. In this connection again, it is important to have rules which permit the *omission of parentheses*. We give two such rules below, and regard them as continuing the list begun in **3c** with rules (1), (2), (3). The first of these new rules, rule 4, actually applies to certain other sorts of formulas besides the universal and existential ones. Hence it is convenient to phrase this rule in a more general fashion. To this end we use the word "operator", understanding by it one of our two quantifiers or one of certain other expressions to be explained later (in **33** and **35**).

It is considered permissible to omit the parentheses that enclose a component formula $\mathfrak{S}_i$ of a given formula provided one of the following conditions is satisfied:

4. $\mathfrak{S}_i$ consists of an operator (of any kind) together with its operand. [E.g. $\mathfrak{S}_i$ may be a component of a compound, as in '$\sim(\exists x)(Px \vee Qx)$' or '$A.(x)(Px \vee Qx)$'; again, $\mathfrak{S}_i$ may be the operand of an earlier operator, as in '$(\exists y)(x)(Rxy)$'.]

5. $\mathfrak{S}_i$ is the operand of a universal or existential quantifier and is the smallest sentential formula following that quantifier. [E.g. '$(x)Px$', '$(x)\sim Px$', '$(\exists x)\sim(y)\sim(\exists z)Txyz$'; in the last of these three formulas, rule (5) permitted omission of three pairs of parentheses and rule (4), two.]

One should note the difference between the sentence '$\sim(x)Px$' (read: "not every individual has property P") and the sentence '$(x)\sim Px$' (read: "every individual has property not-P", i.e. "no individual has property P").

We say that an occurrence of a variable (either an individual variable or a variable of the other kinds to be discussed later) is bound by a quantifier, and for short call the variable a *bound variable*, provided it is in a quantifier or is in the operand of a quantifier that contains the same variable. A variable which at a certain occurrence is not bound is said to be *free* at this occurrence. An expression with no free variables (i.e. an expression which contains no variables or else only bound variables) is called *closed*. An expression with at least one free variable is called *open*. An open sentential formula with n different free variables is said to be *n-place*, or of *degree n*. The closed sentential formulas are the *sentences* of language A.

9b. Multiple quantification. The sentence '$(x)(Px \vee Rxb)$' says something about the individual b, viz. it ascribes to b a certain property (in the broad sense of the word "property" adopted in this book). To assert that every individual of the domain has this property, we employ a second variable and a second quantifier with this variable, and write '$(y)[(x)(Px \vee Rxy)]$'. To assert that this property attaches to at least one individual of the domain, we proceed similarly and write: '$(\exists y)[(x)(Px \vee Rxy)]$'. It should be recognized that rule (4) of **9a** permits the omission of the square brackets in these two formulas.

The sentences '$\sim(x)Px$' and '$(\exists x)\sim Px$' say the same thing: for if not every individual has property P, there must be at least one which fails to have it, and conversely. Again, the sentences '$\sim(\exists x)Px$' and '$(x)\sim Px$' say the same thing: for if not at least one individual has property P, then every individual fails to have it (i.e. no individual has property P), and conversely. We will see later that the two pairs of sentences mentioned here are pairs of L-equivalent sentences.

9c. Universal conditionals. Of special importance for the language of science are universal sentences with operands in the form of a conditional. Such sentences are called *universal conditionals.* E.g. '$(x)(Px \supset Qx)$' has this form. Since '$(x)(Px \supset Qx)$' and '$(x)(\sim Px \vee Qx)$' say the same thing, the sentence '$(x)(Px \supset Qx)$' is true provided that for every individual, at least one of the following conditions holds: 1. the individual is not P (i.e. does not have property P); 2. the individual is Q (i.e. has property Q). It may happen that a certain individual c is not P; in this event, so far as the truth of '$(x)(Px \supset Qx)$' is concerned, it is a matter of indifference whether c is Q or not. However, if any individual is P, then it must also be Q if the sentence '$(x)(Px \supset Qx)$' is to be true. For if individual c, say, were P but not Q, then neither condition (1) nor condition (2) would hold for c; thus '$\sim Pc \vee Qc$' would be false and so, consequently, would the all-sentence under discussion. Hence this all-sentence '$(x)(Px \supset Qx)$' states: "For every x, if x is P then x is Q". Notice here that the if-then translation is well suited to the universal conditional, even though it is not always adequate for the simple conditional '$A \supset B$' (cf. **3b**). Another reading for '$(x)(Px \supset Qx)$' is: "All P is Q". Most of the laws of science—physics, biology, even psychology and social science—can be phrased as conditionals. E.g. a physical law that runs something like "if such-and-such a condition obtains or such-and-such a process occurs, then so-and-so follows" can be rephrased as "for every physical system, if such-and-such conditions obtain, then so-and-so obtains".

If a sentence of the form "all...are..." is to be translated into the symbolic language, notice should be taken of the following remarks. Generally, such a sentence is to receive the symbolic formulation '$(x)(Px \supset Qx)$'. However, if the first predicate of the sentence (i.e. the one following right after the "all" and receiving the symbol 'P') serves merely to characterize the domain of individuals in view—so that it necessarily attaches to each individual—, then we can suppress this first predicate and formulate the translation simply as '$(x)Qx$'. Predicates of this kind—called "universal words"—are necessary in the word-language to fix the domain in respect to which the word "all" (or such words as "each", "a", and the like) is to function (cf. [Syntax] §76). Such predicates are not needed in a symbolic language, where it is presupposed that each variable employed has a determinate domain of values; for individual variables, this domain is the domain of individuals of the language in question. Examples (cf. the list of predicates in **2c**): 1. The domain: things (characterized by the universal word "thing"). The sentence "All books are blue" is translated '$(x)(Book(x) \supset Blue(x))$'; on the other hand, "All things are blue" is rendered '$(x)Blue(x)$'. — 2. The domain: natural numbers. A sentence running "For each prime number there is..." becomes '$(x)[Prime(x) \supset (\exists y)(...)]$', whereas "For each natural number there is a greater" is written simply '$(x)(\exists y)Gr(y,x)$'.

Exercises. Translate the following sentences into the word-language: **1.** '$Ml(a) \vee Fl(a)$'. — **2.** '$(x)(Ml(x) \vee Fl(x))$'. — **3.** '$Gr(5,3).Gr(5,2)$'. — **4.** '$Gr(5,3) \supset Gr(5,2)$'. — **5.** '$(x)(Gr(x,3) \supset Gr(x,2))$'. — **6.** '$Prime(3).Gr(3,2).\sim Even(3)$'. — **7.** '$(x)[Prime(x).Gr(x2) \supset \sim Even(x)]$'. — **8.** '$(\exists x)(Prime(x).Gr(x,3))$'. — **9.** '$Sq(9,3)$'. — **10.** '$(\exists x)Sq(x,3)$'. — **11.** '$\sim(\exists x)Sq(3,x)$': "there is (in the domain of natural numbers) no square root of 3". — **12.** '$(x)(\sim Sq(3,x))$'. — **13.** '$(x)[(\exists y)Hus(y,x) \supset Fl(x)]$'.

Translate the following sentences into the symbolic language. (The words "thing", "number", "man" in parentheses—being universal words; see above—are not carried over in the translation.)—**14.** "Every (thing) is blue". — **15.** "There is a blue (thing)". — **16.** "Every (number) is either even or not even". — **17.** "There is a blue book" (conjunction). — **18.** "Every book is blue". — **19.** "There are (numbers) x and y such that x is the square of y". — **20.** "There is no (number) which is the immediate predecessor of zero" (use an existential quantifier). — **21.** "No (number) is the immediate predecessor of zero". — **22.** "a is a father" (i.e. "a is the father of someone", or "there is a (man) such that a is his father"). — **23.** "Fathers are male" (i.e. "for every x, if x is the father of someone, then x is male"). — **24.** "For each square number there is a greater" (use one universal quantifier and two existential quantifiers).

9d. Translation from the word-language. In connection with translations into the symbolic language, it is to be noted that universality is not always expressed in the word-language by terms like "each", "all", etc; sometimes universality is also expressed simply by the definite or indefinite articles ("the", "an"), though these words do not ordinarily have this significance. When articles are so used, it can only be gathered from the context that universality is intended. E.g. the phrase "the lion" has a universal sense in the sentence "the lion is a beast of prey", but not in the sentence "the lion is now fed". The first sentence here means "all lions are beasts of prey" and hence is to be translated into a symbolic sentence like '$(x)(Px \supset Qx)$'. The second sentence means "this object a is a lion, and a is now fed"—symbolically, '$Pa.Sa$'. Again, "a lion" expresses universality in the sentence "a lion is a beast of prey", but just existence in the sentence "Charles is shooting a lion". The first of these two sentences means "the (or: every) lion is a beast of prey", and hence is rendered '$(x)(Px \supset Qx)$'. The second sentence states "there is an x such that: x is a lion and Charles is shooting x", and so receives a symbolic translation like '$(\exists x)(Px.Rax)$'. There are still other words, e.g. "anything" and "anyone", which have this dual use—serving to express universality in some cases, and existence in others. To produce a correct symbolic translation of a sentence containing words like "a", "the", "something", "anyone", "nothing", etc., it is best first to expose the sense of the sentence by paraphrasing it so that expressions such as "for every x" and "there is an x" appear in place of the words mentioned.

Exercises. Translate the following sentences into the symbolic language. Besides the signs specified in **2c**, use the following: **1.** Individual constants. For "Charles", use 'a'; for "the table", use 'b'. **2.** One-place predicates. For "is at home", use 'H'.

3. Two-place predicates. For "sees", use '*S*'; for "lies on", use '*L*'; and for "belongs to", use '*B*'. Each of the given sentences involves at least one quantifier. — **1.** "Charles sees something". — **2.** "Charles sees a blue book". — **3.** "Something is lying on the table". — **4.** "If something is lying on the table, it belongs to Charles". — **5.** "If something is lying on the table, Charles is at home". [Note the difference between (4) and (5), which the word-language discloses only by the "it" in the second clause of (4); because of this "it", the operand of the quantifier in (4) must include the whole sentence, whereas that of the quantifier in (5) comprises just the first clause of (5).] — **6.** "If any (number) is smaller than 4, it is (also) smaller than 5" (use "for every x"). — **7.** "If any (number) is greater than c and smaller than d, c is smaller than d" (use "there is..."; note the difference from (6), which has an "it" in the second clause). — **8.** "If one (number) is the predecessor of another, it is smaller than the other". — **9.** "If one number is the predecessor of another, then it or the other is even". — **10.** "a is a friend of a brother of e" (i.e. "there is a (third man) such that..."). — **11.** "9 is a square number", (i.e. "9 is the square of some (number)"). — **12.** "Zero is not greater than any (number)".

10. PREDICATE VARIABLES

10a. Predicate variables. According to our treatment of the universal quantifier and the existential quantifier, a sentence of the form '$(x)(...x...)$' is true if and only if the sentential formula '$...x...$' holds for every individual; and a sentence of the form '$(\exists x)(...x...)$' is true if and only if the formula '$...x...$' holds for at least one individual.

Now, it is easy to see that the sentence '$(x)Px \supset Pa$' (i.e. '$(x)(Px) \supset Pa$') is true in every possible case, no matter what the facts are regarding the individual a and the property P. Only two cases need to be distinguished. Case (1): the individual a has property P. In this case, 'Pa' is true; hence (by Truth-table I(4)) the whole sentence is true. Case (2): a fails to have property P. In this case, the sentence '$(x)Px$' is false because it asserts that all individuals have property P; hence (again by the truth-table) the whole sentence is true. The sentence in question is thus necessarily true, regardless of the facts. We may also see this immediately from the word-language version of '$(x)Px \supset Pa$'. "If all individuals are P, then a is P". Indeed, the sentence '$(x)Px \supset Pa$' can be included among sentences that are L-true in our technical sense, provided we extend in a suitable way the rules governing value-assignments. Let us make that extension now.

Let us agree that free variables and descriptive signs count as value-bearing signs. [Thus, in '$(x)Px \supset Pa$' only 'P' and 'a' are value-bearing.] As values of individual signs, let us take all individuals of the domain in question; and as values of one-place predicates, let us take all classes of these individuals (i.e. all subclasses of the domain in question).

Let us agree to regard a one-place atomic formula as true at a given value-assignment if and only if the individual (assigned as the value of the individual sign) belongs to the class (assigned as the value of the predicate). Further, we agree to regard a universal sentence (say, '$(x)Px$') as true at a given value-assignment provided the operand of this sentence (here 'Px') is true at each value-assignment to 'x', in view of the assignment already given to the remaining value-bearing signs (here, only 'P').

In view of the above, it is readily seen that the sentence '$(x)Px \supset Pa$' is true at every value-assignment to the value-bearing signs 'P' and 'a', and hence is L-true. [The argument is essentially the same as that given at the beginning of this section; we repeat it here because the formulation must now be phrased in terms of value-assignments. Case (1): the value-assignment to 'P' and 'a' is such that the individual assigned to 'a' does in fact belong to the class assigned to 'P'. At this value-assignment, 'Pa' is true; and hence the whole sentence '$(x)Px \supset Pa$' is true. Case (2): the value-assignment to 'P' and 'a' is such that the individual assigned to 'a' does not belong to the class assigned to 'P'. At this value-assignment, '$(x)Px$' is false since 'Px' is not true at every value-assignment to 'x' (in particular, 'Px' is not true if we assign to 'x' the individual presently assigned to 'a'); hence at this value-assignment the whole sentence '$(x)Px \supset Pa$' is again true. Consequently the sentence is true at every value-assignment.] Similarly, the open formula '$(x)Px \supset Py$' is L-true; for the possible value-assignments to the free variable 'y' are identical with those to 'a'.

It is further evident that any other sentence with the same form as '$(x)Px \supset Pa$', but with a different predicate in place of 'P', is true just as '$(x)Px \supset Pa$' is. E.g. '$(x)Qx \supset Qa$' is true. Now we saw earlier that sentential variables are useful because they facilitate the creation of open L-true formulas from which L-true sentences can be obtained by arbitrary substitutions. Here, analogously, it is useful to introduce *predicate variables.* Let us agree to use 'F', 'G', 'H', 'K' (and other letters, as occasion demands) for predicate variables, and to count as expressions substitutable for these variables either predicate constants or other predicate variables. In making value-assignments for a sentential formula, we assign classes of individuals to one-place predicate variables and also to one-place predicate constants. Thus e.g. the open formula '$(x)Fx \supset Fa$' is L-true, since in fact the possible value-assignments to 'F' are the same as those originally possible for 'P'. From this L-true formula our earlier L-true sentences can then be obtained by substituting 'P' for 'F', or else 'Q' for 'F'. The open formula '$(x)Fx \supset Fy$' with both 'F' and 'y' as free variables is also L-true, and is in fact the most general formula of the form considered here; it has as substitution instances the previous L-true formulas of this section. '$(x)Fx \supset Fy$' is a purely logical formula, devoid of descriptive constants.

10b. Intensions and extensions. Our practice has been to define L-concepts on the basis of value-assignment. Now let us take up several questions regarding the sorts of values we have used in such assignment. Why do we take the values of sentential variables to be truth-values and not propositions? Of course, it is simpler to work with just two truth-values than with indefinitely many propositions. But the question is: Is this simplification justifiable? A similar question occurs in connection with one-place predicate variables: Is it justifiable to take as values of these predicate variables just classes of individuals, rather than properties?

In order to resolve these questions we introduce here the semantic concepts of intension and extension. (A reader concerned chiefly to master the technique of the symbolic language, and having less interest in semantic and philosophic matters, may omit this section (**10b**).)

A one-place predicate designates a property. (E.g. 'Book' designates the property of being a book; 'Blue' designates the colour blue, a property of certain things.) We shall call this property the *intension* of the predicate. By the *extension* of a predicate we shall understand the class of individuals having the property designated by the predicate. (E.g. the extension of '*Book*' is the class of books; and the extension of '*Blue*' is the class of blue things.) Analogously, we consider the intension of a two-place predicate to be the two-place relation designated by the predicate, and the extension of a two-place predicate to be the class of ordered pairs of individuals for which the predicate holds (i.e. the class of ordered pairs that satisfy the relation designated by the predicate). (E.g. the intension of the predicate '*Fa*' is the relation of fatherhood, and the extension of this predicate is the class of pairs comprising a father and one of his children.) In general, for any natural number n, $n \geq 2$, we take the intension of an n-place predicate to be the n-place relation designated by that predicate, and its extension to be the class of ordered n-tuples for which the predicate holds.

We agree that the intension of a sentence shall be the proposition designated by this sentence, and that its extension shall be its truth-value. The last part of this agreement reflects the fact that the truth-value of a sentence has a role similar to that of the class of individuals corresponding to a predicate.

While not customary, it is useful to make analogous distinctions for individual constants (or, more generally, for closed individual expressions). Suppose the father of Peter Brown is also mayor of Lexington. Then the two phrases "the father of Peter Brown" and "the mayor of Lexington" (more precisely, the individual expressions in our symbolic language that correspond to these two phrases; such expressions are introduced later (**35**) as "descriptions") refer to the same individual. Of these two phrases, therefore, we say that they have the same extension, viz. this particular individual. On the other hand, it is evident that the two phrases have different senses. By the intension of an individual expression we understand its sense. This is a concept similar to property or relation, but of a different type for which there is no established designation; we agree to use for it the term "individual concept". We will become acquainted later with still other such concepts, among them functions like e.g. the arithmetical sum-function designated '+'. By the intension of such a function sign (or: functor) we understand the function designated by the sign; by its extension we understand the value-distribution of the function (a notion to be explicated later).

Next, suppose that the symbolic language contains variables whose substitutable expressions include the constants and closed compound expressions of some fixed kind. Following out the distinction between the intension and the extension of a constant, it is possible here to set up a similar distinction between the *value-intensions* and the *value-extensions* of a variable. Expressions substitutable for a variable have both intensions and extensions; we count all such intensions among the value-intensions of the variable, and similarly all such extensions among the value-extensions of the variable. When we think of the "values" of a variable, we usually have its value-intensions in mind. However, in examining the L-truth of logical formulas constructed in a language with so simple a structure as the symbolic languages treated in this book, it is quite sufficient to consider the values of a variable as its value-extensions. E.g. the values (regarded as value-intensions) of the sentential variables 'p', 'q', etc., are propositions; as we have seen, however, the tautological character of (say) the formula '$p \vee \sim p$' can be ascertained without considering numerous (in some circumstances, infinitely numerous) propositions, but simply the two truth-values which are the value-extensions of the variable 'p'.

So far as the truth-value of a sentential compound is concerned, it is sufficient to consider just the value-extensions (the truth-values) of constituent sentential variables because the truth-value of this compound is uniquely determined by the truth-values of its components; i.e. the sentential connectives used in such compounds are themselves *extensional*. Again, the truth-value of an atomic sentence obviously depends only on the extension of its predicate and the extensions of its individual constants; hence, an atomic sentence is also extensional. Continuing, the truth-value of a universal sentence depends only on the extension of the property determined by the operand of the quantifier (i.e. on whether this property attaches to all individuals, or not); thus a universal sentence is also extensional. The same remark applies to an existential sentence. Indeed, each of our symbolic languages—the present language A, and the languages B and C to be introduced later—is an *extensional language* in the following sense: a sentence in any one of these languages does not change its truth-value if any expression in the sentence is replaced by another with the same extension. Consequently, it suffices for the evaluation of any formula to consider simply the possible extensions of the formula's descriptive constants and the value-extensions of the formula's variables.

A symbolic language which, in contrast to the one treated here, also contains symbols for the so-called *logical modalities*—i.e. such concepts as necessity, possibility, impossibility, contingency and the like—is not extensional. [For suppose it is not raining here now. Then the sentence "it is raining" is false, and so has the same extension (or truth-value) as the L-false sentence "it is raining and it is not raining". Now let this last sentence be a component in a larger modal sentence; when "it is raining and it is not raining" is replaced by "it is raining", the truth-value of the whole modal sentence does not always remain unchanged. E.g. the modal sentence "it is impossible that it is raining and it is not raining" is true, whereas the sentence "it is impossible that it is raining" (produced therefrom by the indicated replacement) is false—for while it is not the case that it is raining here now, this case is nevertheless logically possible. Thus symbolic languages with modality symbols are generally not extensional.] In such non-extensional symbolic languages, one must consider intensions as well as extensions as values of descriptive constants and variables.

Most systems of symbolic logic employ an extensional language, for the reason that such a language has radically simpler structure and hence simpler constitutive rules. However, it cannot justly be said that this procedure compels a neglect of the logical modalities: these can be expressed in another way, viz. in the metalanguage, with the aid of L-concepts. Instead of saying a certain proposition (or state of affairs) is necessary—or impossible, or possible, or contingent—, we say that a corresponding sentence (i.e. one that designates the proposition in question) is L-true—or L-false, or not L-false, or L-indeterminate, respectively. E.g. let 'A' designate the proposition (the possible state of affairs) that it is raining here now; '$A \vee \sim A$' then designates the proposition that it is raining or it is not raining. Within a modal language containing words one would say "it is necessary that it is raining or it is not raining"; in a symbolic modal language having the symbol 'N' for "necessary", the sentence would appear as '$N(A \vee \sim A)$'. By contrast, this sentence cannot be stated in our object language because this language is extensional; however, we can formulate the corresponding sentence "the sentence '$A \vee \sim A$' is L-true" in our metalanguage.

Intensions and Extensions of the Chief Types of Expressions

Expression	Intension	Extension
Sentence	Proposition	Truth-value
Individual constant	Individual concept	Individual
One-place predicate	Property	Class of individuals
n-place predicate ($n > 1$)	n-place relation	Class of ordered n-tuples of individuals
Functor	Function	Value-distribution

11. VALUE-ASSIGNMENTS

On the basis of the preceding considerations we now undertake to clarify generally the two concepts of value-assignment and evaluation. Earlier (in **5**) we applied these concepts just to sentential constants and sentential variables; this application will now be extended to other kinds of signs.

We count as *value-bearing signs* in a given sentential formula $\mathfrak{S}_i$ all descriptive signs and all free variables in $\mathfrak{S}_i$. A *value-assignment* for $\mathfrak{S}_i$ consists in associating with each value-bearing sign in $\mathfrak{S}_i$ a possible extension of that sign. Let us use the sign '$\mathfrak{V}$' of the metalanguage for value-assignments. Values of the following kinds are then associated with those value-bearing signs we have already introduced into the symbolic language:

(1) with a sentential sign: a truth-value;
(2) with an individual sign: an individual (of the given domain of individuals);
(3) with a one-place (descriptive) predicate: a class of individuals;
(4) with an n-place (descriptive) predicate, or an n-place predicate variable ($n>1$): a class of ordered n-tuples of individuals.

If a certain value-bearing sign occurs more than once in the given formula, the same extension is associated with it at each of its occurrences (in case the sign is a variable, the same extension is associated with it at each of its *free* occurrences).

Now suppose we are given a sentential formula $\mathfrak{S}_i$ and an arbitrary value-assignment $\mathfrak{V}_k$ for the value-bearing signs in this $\mathfrak{S}_i$. Then the *evaluation* of $\mathfrak{S}_i$ at $\mathfrak{V}_k$, i.e. the establishment of the truth-value of $\mathfrak{S}_i$ relative to $\mathfrak{V}_k$, is made in accordance with the following *rules of evaluation*. In stating each rule, we employ "T: ..." as short-hand for "the following is a necessary and sufficient condition that formula $\mathfrak{S}_i$ have the truth-value T relative to $\mathfrak{V}_k$: ...". In other words, "T: ..." means "if ..., then $\mathfrak{S}_i$ is true relative to $\mathfrak{V}_k$; and if not ..., then $\mathfrak{S}_i$ is false relative to $\mathfrak{V}_k$".

+R11-1. *Rules of evaluation* for a sentential formula $\mathfrak{S}_i$ at a value-assignment $\mathfrak{V}_k$:

a. Suppose $\mathfrak{S}_i$ is a one-place atomic formula. Then $\mathfrak{V}_k$ comprises a class of individuals (as value for the predicate), and a single individual (as value for the individual sign).
T: the single individual belongs to the class.

b. Suppose $\mathfrak{S}_i$ is an n-place atomic formula ($n>1$). Then $\mathfrak{V}_k$ comprises a class of ordered n-tuples of individuals, and a single such n-tuple.
T: the single n-tuple belongs to the class.

c. Suppose $\mathfrak{S}_i$ is $\sim\mathfrak{S}_j$.
T: the value of $\mathfrak{S}_j$ at $\mathfrak{V}_k$ is F.

d. Suppose $\mathfrak{S}_i$ is $\mathfrak{S}_j \vee \mathfrak{S}_k$.
T: at least one of $\mathfrak{S}_j$ and $\mathfrak{S}_k$ has at $\mathfrak{V}_k$ the value T.

e. Suppose $\mathfrak{S}_i$ is $\mathfrak{S}_j . \mathfrak{S}_m$.
T: Each of $\mathfrak{S}_j$ and $\mathfrak{S}_m$ has at $\mathfrak{V}_k$ the value T.

+**R11-1** **f.** Suppose $\mathfrak{S}_i$ is $\mathfrak{S}_j \supset \mathfrak{S}_m$.
T: at $\mathfrak{B}_k$, $\mathfrak{S}_j$ has the value F or $\mathfrak{S}_m$ the value T or both.

g. Suppose $\mathfrak{S}_i$ consists of a universal quantifier whose operand is the formula $\mathfrak{S}_j$.
T: $\mathfrak{S}_j$ at $\mathfrak{B}_k$ is true for every value-assignment to the variable occurring in the universal quantifier.
[In case the variable of the universal quantifier does not occur free in $\mathfrak{S}_j$, then: T: $\mathfrak{S}_j$ at $\mathfrak{B}_k$ is true.]

h. Suppose $\mathfrak{S}_i$ consists of an existential quantifier whose operand is the formula $\mathfrak{S}_j$.
T: $\mathfrak{S}_j$ at $\mathfrak{B}_k$ is true for at least one value-assignment to the variable occurring in the existential quantifier.
[In case the variable of the existential quantifier does not occur free in $\mathfrak{S}_j$, then: T: $\mathfrak{S}_j$ at $\mathfrak{B}_k$ is true.]

i. Suppose $\mathfrak{S}_i$ is an identity formula (**17a**), with the sign '=' of identity standing between two individual expressions.
T: the two individual expressions have at $\mathfrak{B}_k$ the same individual as value.

A given value-assignment for a sentential formula fixes initially the values to be associated with the value-bearing signs of this formula. Thereafter, application of the rules of evaluation—first to atomic formulas, then step by step to increasingly comprehensive component formulas—eventuates in the truth-value of the entire formula at the given value-assignment.

If formula $\mathfrak{S}_i$ is true at the value-assignment $\mathfrak{B}_k$, we also say that $\mathfrak{B}_k$ *satisfies* formula $\mathfrak{S}_i$, i.e. the values associated through $\mathfrak{B}_k$ satisfy $\mathfrak{S}_i$.

By the *range* of a sentential formula $\mathfrak{S}_i$ we understand the class of those value-assignments at which $\mathfrak{S}_i$ comes out true. While this definition is phrased in the same way as the earlier one (given in **5**), it should be borne in mind that 'value-assignment' now refers to the extended concept introduced at the beginning of this section. The definitions of the various *L-concepts* can thus be carried over unchanged; we shall not repeat them here. Note, however, that in the present context these L-concepts apply to additional sorts of forms, in particular forms containing individual variables and predicate variables.

We say a formula is *descriptive* when it contains at least one descriptive sign; otherwise, *logical*. A logical formula, therefore, contains only variables and logical constants. In connection with open logical formulas (formulas whose only value-bearing signs are free variables), the following terminology is often employed: such a formula is called *universally valid* when it is satisfied by every value-assignment; *satisfiable* when there is at least one value-assignment that satisfies it; and *unsatisfiable* when there is no value-assignment that satisfies it. Instead of these terms we use for the most part the L-terms 'L-true', 'not L-false' and 'L-false' respectively; these last have the advantage of supplying a single terminology suitable at once to open logical formulas, to open descriptive formulas and to closed formulas (or sentences).

12. SUBSTITUTIONS

12a. Substitutions for sentential variables. We have already noted that the sentential formula obtained from a given L-true sentential formula with a free sentential variable by an arbitrary substitution for this variable is again an L-true formula. Similar remarks may now be made respecting other variables. Therefore we will soon present lists of purely logical L-true sentential formulas with free individual variables and predicate variables (the latter employed initially simply as free variables), and make the observation that any formula obtained from a listed one by substitutions is again an L-true formula. Before presenting these lists, however, we must state more exactly how substitutions are made for free variables of different kinds. One general remark can be made at the outset: every substitution for a variable in a given formula requires that the *same* expression be substituted at each free occurrence of the variable in the formula (these occurrences are termed the "substitution positions"); an exception to this regulation is formula-substitution for a predicate variable—a type of substitution that will be described below (**12c**).

Up to the present it has been permissible to substitute an arbitrary sentential formula for a free sentential variable. But our system now includes bound variables; hence we must limit substitution in the following way. For a sentential variable in a given sentential formula $\mathfrak{S}_i$ an arbitrary sentential formula $\mathfrak{S}_j$ may be substituted, provided no individual variable which occurs free in $\mathfrak{S}_j$ becomes bound at one of the substitution positions in $\mathfrak{S}_i$. E.g. in '$(x)(p \supset Fx) \equiv [p \supset (x)Fx]$' no formula in which '$x$' is free may be substituted for 'p' because 'x' would become bound at the first substitution position. [This example suggests a ready explanation for the limitation we have placed on substitution. The formula given is L-true. Substitution of 'Px' for 'p' in this formula produces the following formula $\mathfrak{S}_k$: '$(x)(Px \supset Fx) \equiv [Px \supset (x)Fx]$'. Now formula $\mathfrak{S}_k$ can be seen to be not L-true. For suppose we obtain from $\mathfrak{S}_k$ the substitution instance $\mathfrak{S}_l$: '$(x)(Px \supset Px) \equiv [Pa \supset (x)Px]$' by substituting '$P$' for '$F$' and '$a$' for '$x$' (note that only the fourth occurrence of 'x' is free and so open to the substitution of 'a'). Suppose further we take $\mathfrak{V}_l$ to be a value-assignment that associates with 'a' a certain individual and that associates with 'P' a class containing the individual associated with 'a' but not all individuals. At this value-assignment $\mathfrak{V}_l$ it is clear that 'Pa' is true and '$(x)Px$' is false, i.e. the right member of the biconditional $\mathfrak{S}_l$ is false; and similarly it is clear that the left member of $\mathfrak{S}_l$ is always true. Hence $\mathfrak{S}_l$ is false at $\mathfrak{V}_l$. Consequently $\mathfrak{S}_l$ is not L-true, and so $\mathfrak{S}_k$ is not L-true either.]

12b. Substitutions for individual variables. For an individual variable there may be substituted an arbitrary individual constant or an individual variable, provided the following limitation is observed: no individual variable is to be substituted which becomes bound at one of the substitution positions. E.g. in '$(x)Ryx \vee (\exists z)Szy$' there may be substituted for 'y' any individual constant, and any individual variable except 'x' and 'z'—for 'x' would become bound at the first substitution position, and 'z' would become bound at the second. [The following example from the domain of natural numbers suggests the reason for the limitation we have placed on substitutions for individual variables. The formula '$(\exists x)Gr(x,y)$' holds for every y, since it says simply "there is a number x which is greater than y". If now in this formula we were to allow the substitution of 'x' for the free variable 'y' (in violation of our restriction, since 'x' clearly becomes bound at the substitution position), we would obtain the sentence '$(\exists x)Gr(x,x)$'. This sentence says "there is a number which is greater than itself", and is evidently false.

12c. Substitutions for predicate variables. Here we must distinguish between two different kinds of substitutions. One kind, *simple substitution*, has already been mentioned: for an n-place predicate variable there may be substituted an arbitrary n-place predicate or an arbitrary n-place predicate variable, with no limitations whatever. [Later, when bound predicate variables are used (**16a**), the following limitation holds: no predicate variable is to be substituted which becomes bound at one of the substitution positions.]

There is, however, another kind of substitution for a predicate variable, which we shall call *formula-substitution*. Let us lead into a discussion of formula-substitution by way of an example.

Suppose $\mathfrak{S}_i$ is the sentential formula '$(x)Fx \supset Fa$'. It has been brought out above that $\mathfrak{S}_i$ is L-true, hence $\mathfrak{S}_i$ holds for every property F. Now it is easy to state that what $\mathfrak{S}_i$ claims for all properties holds in particular for the properties P,Q, etc.; we merely use simple substitution and produce the substitution instances '$(x)Px \supset Pa$', '$(x)Qx \supset Qa$', etc. However, we must note an important fact, viz. not all properties expressible in our symbolic language are designated by predicates like 'P', 'Q', etc. Indeed, every arbitrary sentential formula with an individual variable as its sole free variable expresses a property of individuals. If e.g. $\mathfrak{S}_k$ is '$Qx \vee Rxb$', then $\mathfrak{S}_k$ is such a formula (the individual variable 'x' is its only free variable); and the property of x expressed by $\mathfrak{S}_k$ is the property of being Q or bearing the relation R to b. Moreover, what $\mathfrak{S}_i$ asserts about all properties must in particular be true of the property expressed by $\mathfrak{S}_k$—a claim conveyed by the sentence $\mathfrak{S}_l$: '$(x)(Qx \vee Rxb) \supset (Qa \vee Rab)$'. It is our intention to count the sentence $\mathfrak{S}_l$ as still another substitution instance of $\mathfrak{S}_i$. But we must recognize this sort of substitution as not another version of simple substitution; we are not simply substituting a predicate for 'F', but rather substituting first the compound $\mathfrak{S}_k$ for the *full formula* '*Fx*' and then the corresponding compound for '*Fa*' in accordance with the following scheme:

'*Fx*', '$Qx \vee Rxb$';
Fa', '$Qa \vee Rab$'.

This scheme is constructed as follows: In the first line, write an open full formula of '*F*' (called the *nominal formula*) and follow it by that formula (having the same free variable) which we have selected to be the *substitutum* (the substitutum expresses the property in terms of which we wish to form a special case of the given formula $\mathfrak{S}_i$). Since $\mathfrak{S}_i$ involves '*Fa*' as well as '*Fx*', we add to our scheme a second line that begins with '*Fa*' and follows it with a formula obtained from the second formula of the first line by the same substitution as that leading from '*Fx*' to '*Fa*', viz. the substitution of '*a*' for '*x*'. Had it happened that our original formula $\mathfrak{S}_i$ also involved, say, '*Fu*' and '*Fb*', we would continue our scheme with two more lines of formula-pairs, thus:

'*Fu*', '$Qu \vee Rub$';
'*Fb*', '$Qb \vee Rbb$'.

Each of these pairs of formulas is obtained from the formula-pair in the first line of the scheme by a uniform substitution for the individual variables that occur in the nominal formula. The substitutions are so chosen that the first formula of the resulting pair is one which appears in a determinate place in the original formula. Conceived in its entirety as a single act of substitution, we see that our procedure consists in substituting into the original given formula simultaneously for all full formulas of '*F*'. What is substituted for a particular full formula of '*F*' is the substitutum that stands alongside this full formula in our scheme. The first line in the scheme (i.e. the first formula-pair) represents the substitution we have chosen; thereupon, in all subsequent lines (i.e. all subsequent formula-pairs), the second formula or substitutum is uniquely determined.

The example treated above introduces us to the type of substitution we call *formula-substitution*. The scheme developed in connection with formula-substitution serves mainly to guide the substitution. As we have said, the first formula-pair in the scheme represents the substitution chosen, and subsequent pairs of the scheme follow systematically from the first in accordance with the demands of the original formula. What we take for our first formula-pair (i.e. for our substitution) is to a large extent arbitrary, but is not entirely without restrictions. These limitations are suggested in the next paragraph, where we state general rules governing formula-substitution.

Let the formula $\mathfrak{S}_i$ be given, and suppose $\mathfrak{S}_i$ contains an *n*-place predicate variable for which substitution is to be made. Let $\mathfrak{S}_j$ be the nominal formula, and $\mathfrak{S}_k$ the substitutum chosen for $\mathfrak{S}_j$. Formula-substitution may then proceed, subject to the following rules:

1. The nominal formula $\mathfrak{S}_j$ consists of the predicate variable in question, together with *n* arbitrary different individual variables;

2. The substitutum $\mathfrak{S}_k$ for $\mathfrak{S}_j$ is any sentential formula such that:

- **a.** the variables of $\mathfrak{S}_j$ do not occur in the quantifiers (or other operators) that appear in $\mathfrak{S}_k$ (these variables usually occur free in $\mathfrak{S}_k$, but this is not necessary);
- **b.** the variables which occur in $\mathfrak{S}_i$ but not in $\mathfrak{S}_j$ do not occur in $\mathfrak{S}_k$ (variables which occur neither in $\mathfrak{S}_i$ nor in $\mathfrak{S}_j$ may occur arbitrarily in $\mathfrak{S}_k$, free or bound);

3. From the formula-pair $\mathfrak{S}_j$, $\mathfrak{S}_k$ other formula-pairs are obtained by the same substitutions for the variables occurring in $\mathfrak{S}_j$;

4. The substitution of $\mathfrak{S}_k$ for $\mathfrak{S}_j$ in $\mathfrak{S}_i$ proceeds as follows: each full formula in $\mathfrak{S}_i$ with a (free) occurrence of the predicate variable in question is replaced by the substitutum which is paired with this full formula in accordance with rule (3).

12d. Theorems on substitutions.

+T12-1. Suppose $\mathfrak{S}_i$ and $\mathfrak{S}_j$ are arbitrary sentential formulas. Suppose $\mathfrak{S}_i'$ and $\mathfrak{S}_j'$ are obtained from $\mathfrak{S}_i$ and $\mathfrak{S}_j$ respectively by the same substitutions of the following four kinds for one or more (but not necessarily all) of the free variables: (1) substitution for a sentential variable; (2) substitution for an individual variable; (3) simple substitution for a predicate variable; and (4) formula-substitution for a predicate variable. Then the following hold:

- **a.** If $\mathfrak{S}_i$ is L-true, then $\mathfrak{S}_i'$ is also L-true.
- **b.** If $\mathfrak{S}_i$ is L-false, then $\mathfrak{S}_i'$ is also L-false. (By (a) and T5-2a.)
- **c.** If $\mathfrak{S}_i'$ is L-indeterminate, then $\mathfrak{S}_i$ is also L-indeterminate. (By (a) and (b).)
- **d.** If $\mathfrak{S}_i$ L-implies $\mathfrak{S}_j$, then $\mathfrak{S}_i'$ L-implies $\mathfrak{S}_j'$. (By (a) and T6-4.)
- **e.** If $\mathfrak{S}_i$ and $\mathfrak{S}_j$ are L-equivalent, then $\mathfrak{S}_i'$ and $\mathfrak{S}_j'$ are also L-equivalent. (By (a) and T6-7.)

Proof of (a). Assertion (a) was proved earlier for sentential variables; cf. T7-1a. Similar considerations obtain for the other kinds of substitutions. For $\mathfrak{S}_i$ is satisfied by every value-assignment to the variables concerned (and the other value-bearing signs), hence in particular $\mathfrak{S}_i$ must be satisfied by the value-assignments which result from arbitrary value-assignments to the value-bearing signs that occur in the substituted expressions.—These remarks apply particularly to formula-substitution for a predicate variable, where the situation—though somewhat more complicated—is still essentially the same. We illustrate this fact by means of an example similar to an earlier one, viz. the substitution of the formula $\mathfrak{S}_k$: '$Qx \vee Rxb$' for 'Fx' in an L-true sentential formula $\mathfrak{S}_i$ where 'F' occurs by way of the following atomic formulas: 'Fa'; 'Fx' (in the compound '$(x)Fx$'); 'Fb'; and 'Fu' (where 'u' is a free variable in $\mathfrak{S}_i$). Formula $\mathfrak{S}_i'$ is obtained from

$\mathfrak{S}_i$ by systematically replacing '*Fx*' by '*Qx*∨*Rxb*', '*Fa*' by '*Qa*∨*Rab*', '*Fb*' by '*Qb*∨*Rbb*', and '*Fu*' by '*Qu*∨*Rub*'. Now let $\mathfrak{B}_j$ be an arbitrary value-assignment to whatever value-bearing signs besides '*F*' happen to occur in $\mathfrak{S}_i$ (these include '*a*', '*b*', '*u*', and possibly other signs). In $\mathfrak{S}_k$, besides '*x*' and '*b*', there appear two new value-bearing signs, '*Q*' and '*R*'; let $\mathfrak{B}_i'$ be an arbitrary value-assignment to these latter two signs. On the basis of $\mathfrak{B}_i'$ and $\mathfrak{B}_j$ (which last assigns a value to '*b*'), the formula $\mathfrak{S}_k$ determines a certain class *K*. (The class *K* is the class of all those individuals which, when treated as value-assignments to '*x*', render $\mathfrak{S}_k$ true; i.e. *K* is the class of those individuals which either belong to the class chosen for '*Q*' or bear the relation chosen for '*R*' to the individual chosen for '*b*'.) Let $\mathfrak{B}_i$ be the value-assignment which associates this class *K* with '*F*'. Now it is easily seen that the truth-value of $\mathfrak{S}_i$ at the value-assignment $\mathfrak{B}_j+\mathfrak{B}_i$ is the same as the truth-value of $\mathfrak{S}_i'$ at the value-assignment $\mathfrak{B}_j+\mathfrak{B}_i'$. For each of '*Fa*' and '*Qa*∨*Rab*' is true if and only if the individual assigned to '*a*' by $\mathfrak{B}_j$ belongs to the class *K*. And further, with an arbitrary value assigned to '*a*', each of '*Fx*' and '*Qx*∨*Rxb*' is true just in case the individual assigned to '*x*' actually belongs to *K*; whence it follows that each of '(*x*)*Fx*' and '(*x*)(*Qx*∨*Rxb*)' is true if and only if *K* is the domain of all individuals. And continuing further, similar remarks are seen to apply to the atomic formulas '*Fu*' and '*Fb*' and their substituta. The argument in respect to our illustrative example is now completed as follows: Since $\mathfrak{S}_i$ is L-true, it is true at every value-assignment, and so in particular at the assignment $\mathfrak{B}_j+\mathfrak{B}_i$. Therefore $\mathfrak{S}_i'$ is true at the assignment $\mathfrak{B}_j+\mathfrak{B}_i'$. It being the case that $\mathfrak{B}_j$ and $\mathfrak{B}_i$ are arbitrary assignments, we see that $\mathfrak{S}_i'$ is true at every value-assignment. Consequently, $\mathfrak{S}_i'$ is L-true.

The content of T1a, viz. that *L-truth persists under arbitrary substitutions*, is of great importance. E.g., an instance of this importance is the matter of proof. Recall (from **8d**) our understanding of a *derivation* as a sequence of sentential formulas which begins with given formulas (premisses) and which proceeds through other sentential formulas one at a time, each step being a formula that is L-implied by the ones preceding it. Now, by a *proof* we shall understand a derivation whose premisses are L-true. The object in setting up a proof is to show that its last formula is L-true. In this connection, two remarks are pertinent. According to T6-1a, every formula of a proof is L-true. And according to the results of the present section, arbitrary substitutions are allowable in obtaining steps in proofs. [In this last respect, proofs are in sharp contrast to derivations. Generally, a derivation cannot admit a step which depends on substitutions because an initial formula $\mathfrak{S}_i$ generally does not L-imply a substitution variant $\mathfrak{S}_i'$ of itself. We return to this matter in the next section.] Another reason for the importance of T1a lies in the practical utility it imparts to lists of L-true formulas, e.g. the lists presented in **14**.

+T12-2. Suppose $\mathfrak{S}_i$ is formula in which '*x*' occurs as a free variable, but '*y*' does not occur. Suppose $\mathfrak{S}_i'$ results from $\mathfrak{S}_i$ by the substitution of '*y*' for '*x*'. Then the following hold (and analogous assertions hold for other arbitrary individual variables):

a. If $\mathfrak{S}_j$ consists of an all-operator '(*x*)' with $\mathfrak{S}_i$ as operand, and $\mathfrak{S}_j'$ similarly consists of '(*y*)' with $\mathfrak{S}_i'$ as operand, then: $\mathfrak{S}_j$ and $\mathfrak{S}_j'$ are L-equivalent.

+**T12-2** **b.** If $\mathfrak{S}_j$ consists of '$(\exists x)$' with $\mathfrak{S}_i$ as operand, and $\mathfrak{S}_j'$ of '$(\exists y)$' with $\mathfrak{S}_i'$ as operand, then: $\mathfrak{S}_j$ and $\mathfrak{S}_j'$ are L-equivalent. ((b) is an analog of (a).)

Proof of (a). Suppose $\mathfrak{V}_j$ is any value-assignment that makes $\mathfrak{S}_j$ true. Then $\mathfrak{V}_j$, together with an arbitrary value-assignment $\mathfrak{V}_x$ to 'x', satisfies $\mathfrak{S}_i$ (in view of our rule R11-1g of evaluation). Hence $\mathfrak{V}_j$, together with an arbitrary value-assignment to 'y', satisfies $\mathfrak{S}_i'$. Therefore $\mathfrak{V}_j$ satisfies $\mathfrak{S}_j'$, as was to be shown. The converse is argued analogously.

This theorem countenances an operation that is called *revising* (or rewriting) *a bound variable*: Given a universal formula or an existential formula, the variable occurring in the operator may be replaced at this occurrence by any other variable that does not occur in the operand, provided only that the same replacement is made at every free occurrence of the original variable in the operand. The new formula which results is L-equivalent to the original formula. This revision of bound variables is, on its face, an entirely plausible operation; it is evident, e.g. that '$(x)Px$' and '$(y)Py$' say exactly the same thing, viz. every individual is P. Later we will establish a theorem on *interchangeability* (it is T15-3) which permits revision of bound variables in a formula that is a component of another formula.

12e. Example. The formula '$(x)(Fx)\supset Fx$' is L-true (cf. **10a**), hence any substitution instance of it made in accordance with the rules should also be L-true (cf. T1a). Suppose we take as the nominal formula 'Fx' and as the substitutum 'Fxy'. To check restrictions 2a and 2b (**12c**) we notice that: (a) the variables of 'Fx' (viz. 'x') do not occur in operators that appear in 'Fxy' (since there are no operators there), and (b) variables which occur in '$(x)Fx\supset Fx$' (viz. 'x') but not in 'Fx' (since 'x' also occurs here there are none) do not occur in 'Fxy'. The substitution now proceeds and yields '$(x)Fxy\supset Fxy$' as L-true. Now since 'x' is free in its last occurrence in this result, we may substitute any variable or constant which does not become bound at the substitution position. Let us substitute 'y'. We thus establish that '$(x)Fxy\supset Fyy$' is L-true. (How?) To show why certain restrictions on substitution are necessary we shall consider a substitution which violates restriction 2a (**12c**). Beginning with the L-true formula '$(x)Fxy\supset Fyy$', choose 'Fxy' as the nominal formula and '$(\exists y)Rxy$' as substitutum. This substitutum violates restriction 2a since 'y' (which is a variable of 'Fxy') occurs in an operator that appears in '$(\exists y)Rxy$'. Note though, that the chosen substitutum violates no other restrictions. Now write:

$$\text{'}Fxy\text{'},\quad \text{'}(\exists y)Rxy\text{'};$$
$$\text{'}Fyy\text{'},\quad \text{'}(\exists y)Ryy\text{'}.$$

Completion of the substitution yields '$(x)(\exists y)Rxy\supset(\exists y)Ryy$', which is not L-true (cf. exercise 1 below). Thus if restriction 2a were dropped, substitution would no longer have the L-truth preserving characteristic we want of it.

Exercises. **1.** Show that '$(x)(\exists y)Rxy\supset(\exists y)Ryy$' is not L-true by finding an interpretation for 'R' which makes the formula false. (Hint: use the domain of natural numbers; see sec. **2c**, (3).) — **2.** Show that restriction 2b cannot be dropped and substitution still have its L-truth preserving characters. — **3.** Not every substitution which violates one of the restrictions fails to preserve L-truth. Show that this is true by constructing a substitution instance of the L-true formula '$(x)Fx\supset Fy$' which violates restriction 2b, but preserves L-truth. — **4.** Decide whether each of the following can be obtained directly (regardless of restrictions 2a and 2b) by substitution from '$Fy\supset(\exists x)Fx$'. If it can be so obtained, give the nominal formula, the substitutum, and the other formula pair used. Also indicate whether restrictions 2a or 2b were violated. The first case is solved as an example.

a) '$(\exists z)(Hyz \vee Hzx) \supset (\exists x)(\exists z)(Hxz \vee Hzx)$' is obtainable using '$Fy$', '$(\exists z)(Hyz \vee Hzx)$'; '$Fx$', '$(\exists z)(Hxz \vee Hzx)$'. The substitution violates restriction 2b.

b) '$Hyy \supset (\exists x)Hxy$'; c) '$Hyy \supset (\exists x)Hxx$'; d) '$Hxy \supset (\exists x)Hxx$'; e) '$Hyy \supset (\exists x)Hzx$'; f) '$Hyz \supset (\exists x)Hxz$'; g) '$Gx \vee Hxx \supset (\exists x)(Gx \vee Hxx)$'; h) '$Hyz \supset (\exists x)Hyx$'; i) '$(\exists z) \sim Hyz \supset (\exists x)(\exists z) \sim Hxz$'; j) '$(y)Hyy \supset (\exists x)(y)Hyx$'; k) '$(\exists z)(Mzyy \vee Hzx) \supset (\exists x)(\exists z)(Mzxy \vee Hzy)$'. — **5.** Show that (4b) and (4g) are obtainable without violating any restrictions by combining formula substitution with individual variable substitution.

13. THEOREMS ON QUANTIFIERS

In this section we establish theorems on quantifiers, mainly universal quantifiers, with special attention to theorems that deal with transformations affecting universal quantifiers. It is important in this connection to distinguish clearly between transformations of this type which can be employed in any derivation and transformations which can be used only in proofs. The fundamental distinction is as follows: If a theorem asserts that a formula $\mathfrak{S}_i$ L-implies another formula $\mathfrak{S}_j$, then the step from $\mathfrak{S}_i$ to $\mathfrak{S}_j$ is admissible in any derivation. [Whence, of course, the step from $\mathfrak{S}_i$ to $\mathfrak{S}_j$ is admissible in any proof; for by T6-1a, if $\mathfrak{S}_i$ is L-true, so also is $\mathfrak{S}_j$.] When, however, a theorem asserts only the weaker claim that if $\mathfrak{S}_i$ is L-true, then $\mathfrak{S}_j$ is also L-true, the step from $\mathfrak{S}_i$ to $\mathfrak{S}_j$ is admissible in any proof, but is not generally admissible in derivations.

T13-1. Suppose $\mathfrak{S}_x$ is an arbitrary sentential formula in which 'x' occurs free. Suppose the formulas $\mathfrak{A}(\mathfrak{S}_x)$ and $\mathfrak{E}(\mathfrak{S}_x)$ are obtained from $\mathfrak{S}_x$ by prefixing to $\mathfrak{S}_x$ the quantifiers '(x)' and '$(\exists x)$' respectively. Finally, suppose $\mathfrak{S}_a$ results from $\mathfrak{S}_x$ by substituting 'a' for 'x' in $\mathfrak{S}_x$. Then the following hold (as well as analogous results phrased in terms of other individual variables and individual constants):

a. $\mathfrak{A}(\mathfrak{S}) \supset \mathfrak{S}_x$ is L-true. (Cf. the discussion at the outset of **10a.**)
+**b.** $\mathfrak{A}(\mathfrak{S}_x)$ L-implies $\mathfrak{S}_x$. (By (a) and T6-4.)
c. $\mathfrak{A}(\mathfrak{S}_x) \supset \mathfrak{S}_a$ is L-true. (By (a) and T12-1a.)
+**d.** $\mathfrak{A}(\mathfrak{S}_x)$ L-implies $\mathfrak{S}_a$. (By (c) and T6-4.)
+**e.** If $\mathfrak{S}_x$ is L-true, so also is $\mathfrak{A}(\mathfrak{S}_x)$. (By rule R11-1g.)
f. If $\mathfrak{S}_x$ is L-true, so also is $\mathfrak{S}_a$. (By T12-1a.)
g. If $\mathfrak{S}_a$ is L-true, and 'a' does not occur in $\mathfrak{S}_x$, then $\mathfrak{S}_x$ is also L-true. (A proof of this assertion appears below.)
h. If $\mathfrak{S}_a$ is L-true and 'a' does not occur in $\mathfrak{S}_x$, then $\mathfrak{A}(\mathfrak{S}_x)$ is also L-true. (By (g) and (e).)
+**i.** $\mathfrak{S}_x$ L-implies $\mathfrak{E}(\mathfrak{S}_x)$. (By rule R11-1b.)

T13-1 **j.** $\mathfrak{S}_x \supset \mathfrak{E}(\mathfrak{S}_x)$ is L-true. (By (i) and T6-4.)
k. $\mathfrak{S}_a \supset \mathfrak{E}(\mathfrak{S}_x)$ is L-true. (By (j) and T12-1a.)
+l. $\mathfrak{S}_a$ L-implies $\mathfrak{E}(\mathfrak{S}_x)$. (From (k).)

Proof of (g). Suppose that $\mathfrak{S}_a$ is L-true and that 'a' does not occur in $\mathfrak{S}_x$. Then every value-assignment to 'a' (together with arbitrary value-assignments to the remaining value-bearing signs) makes $\mathfrak{S}_a$ true. Thus every value-assignment to 'x' makes $\mathfrak{S}_x$ true, because 'a' occurs in $\mathfrak{S}_a$ at precisely those places in which 'x' occurs free in $\mathfrak{S}_x$. Hence $\mathfrak{S}_x$ is L-true. — [The requirement that 'a' does not appear in $\mathfrak{S}_x$ cannot be relaxed, as the following counterexample shows. Take $\mathfrak{S}_x$ to be '$Px \supset Pa$'. Then $\mathfrak{S}_a$ is the L-true formula '$Pa \supset Pa$'. But '$Px \supset Pa$' is not L-true; e.g. it is false at any value-assignment for which x belongs to the class P and a does not.]

It is to be emphasized that in general $\mathfrak{S}_x \supset \mathfrak{S}_a$ is not L-true, and that in general $\mathfrak{S}_x$ does not L-imply $\mathfrak{S}_a$. E.g. taking 'Px' for $\mathfrak{S}_x$, we see from the remark immediately following the proof of (g) just above that '$Px \supset Pa$' is not L-true.

T13-1 tells us that the following transformations are permissible steps in *derivations*, and therefore in proofs: omission of a universal quantifier (b); omission of a universal quantifier together with substitution in the operand (d)—this transformation is known as "specialization"; prefixing an existential quantifier (i); changing an individual constant into a variable and prefixing the appropriate existential quantifier (l)—a transformation known as "existential inference". On the other hand, T13-1 tells us that the following transformations are permissible steps in *proofs*, but not generally in derivations: prefixing a universal quantifier (e); substitution (f); changing an individual constant at each of its occurrences into one and the same variable (g).

T13-2. Vacuous operator. Suppose $\mathfrak{S}_j$ consists of a universal quantifier or an existential quantifier, together with an operand $\mathfrak{S}_i$. Suppose further that the variable in the quantifier does not occur free in $\mathfrak{S}_i$. Then $\mathfrak{S}_i$ and $\mathfrak{S}_j$ are L-equivalent. [This follows from the parenthetical additions to rules R11-1g and R11-1h.]

According to T13-2, a vacuous operator may at will be prefixed to, or removed from, a formula.

T13-3. Let $\mathfrak{S}_i$ and $\mathfrak{S}_j$ be arbitrary formulas. Let $\mathfrak{A}_k$ be a universal quantifier or a sequence of such quantifiers. Then the following hold:

+a. $\mathfrak{A}_k(\mathfrak{S}_i . \mathfrak{S}_j)$ is L-equivalent to $\mathfrak{A}_k(\mathfrak{S}_i) . \mathfrak{A}_k(\mathfrak{S}_j)$. (Proved below.)
b. Lemma. $\mathfrak{A}_k[(\mathfrak{S}_i \supset \mathfrak{S}_j) . \mathfrak{S}_i]$ L-implies $\mathfrak{A}_k(\mathfrak{S}_j)$. (Proved below.)
c. $\mathfrak{A}_k(\mathfrak{S}_i \supset \mathfrak{S}_j) \supset [\mathfrak{A}_k(\mathfrak{S}_i) \supset \mathfrak{A}_k(\mathfrak{S}_j)]$ is L-true. (Proved below.)
+d. $\mathfrak{A}_k(\mathfrak{S}_i \supset \mathfrak{S}_j)$ L-implies $\mathfrak{A}_k(\mathfrak{S}_i) \supset \mathfrak{A}_k(\mathfrak{S}_j)$. (From (c).)
e. If $\mathfrak{S}_i$ L-implies $\mathfrak{S}_j$, then $\mathfrak{A}_k(\mathfrak{S}_i)$ L-implies $\mathfrak{A}_k(\mathfrak{S}_j)$. (Proved below.)

T13-3 **f.** If $\mathfrak{S}_i$ and $\mathfrak{S}_j$ are L-equivalent, then $\mathfrak{A}_k(\mathfrak{S}_i)$ and $\mathfrak{A}_k(\mathfrak{S}_j)$ are also L-equivalent. (By (e) and T6-6a.)

+g. $\mathfrak{A}_k(\mathfrak{S}_i \equiv \mathfrak{S}_j)$ L-implies $\mathfrak{A}_k(\mathfrak{S}_i) \equiv \mathfrak{A}_k(\mathfrak{S}_j)$. (Proved below.)

Before taking up the proofs of these assertions, let us note that T3a, d and g are distribution laws, i.e. assertions which indicate respectively that a universal quantifier (or a series of such) distributes over the components of a conjunction, of a conditional and of a biconditional.

Proof of (a). Suppose $\mathfrak{B}_{ij}$ is a value-assignment to the value-bearing signs in $\mathfrak{A}_k(\mathfrak{S}_i \,.\, \mathfrak{S}_j)$, and suppose further that $\mathfrak{A}_k(\mathfrak{S}_i \,.\, \mathfrak{S}_j)$ is true at $\mathfrak{B}_{ij}$. Let $\mathfrak{B}_i$ be the part of $\mathfrak{B}_{ij}$ which pertains to the value-bearing signs of $\mathfrak{S}_i$; and similarly, $\mathfrak{B}_j$ the part of $\mathfrak{B}_{ij}$ pertaining to $\mathfrak{S}_j$ ($\mathfrak{B}_i$ and $\mathfrak{B}_j$ may overlap); then $\mathfrak{B}_{ij}$ is the union $\mathfrak{B}_i + \mathfrak{B}_j$. Now by rule R11-1g, $\mathfrak{S}_i \,.\, \mathfrak{S}_j$ is true at $\mathfrak{B}_{ij}$ for any value-assignment to the variables in $\mathfrak{A}_k$; whence by R11-1e both $\mathfrak{S}_i$ and $\mathfrak{S}_j$ separately are true there. Thus, $\mathfrak{S}_i$ alone is true at $\mathfrak{B}_i$ for any value-assignment to the variables of $\mathfrak{A}_k$, which (in view of R11-1g) tells us that $\mathfrak{A}_k(\mathfrak{S}_i)$ is true at $\mathfrak{B}_i$. And similarly, $\mathfrak{S}_j$ alone is true at $\mathfrak{B}_j$ for any value-assignment to the variables of $\mathfrak{A}_k$, whence $\mathfrak{A}_k(\mathfrak{S}_j)$ is true at $\mathfrak{B}_j$. Therefore $\mathfrak{A}_k(\mathfrak{S}_i) \,.\, \mathfrak{A}_k(\mathfrak{S}_j)$ is true at $\mathfrak{B}_{ij}$. An analogous argument establishes the converse: If $\mathfrak{A}_k(\mathfrak{S}_i) \,.\, \mathfrak{A}_k(\mathfrak{S}_j)$ is true at $\mathfrak{B}_{ij}$, then $\mathfrak{A}_k(\mathfrak{S}_i \,.\, \mathfrak{S}_j)$ is also true there.

Proof of (b). Suppose $\mathfrak{A}_k[(\mathfrak{S}_i \supset \mathfrak{S}_j) \,.\, \mathfrak{S}_i]$ is true at the value-assignment $\mathfrak{B}_i$. Then, by R11-1g, $(\mathfrak{S}_i \supset \mathfrak{S}_j) \,.\, \mathfrak{S}_i$ is true at $\mathfrak{B}_i$ for any value-assignment to the variables in $\mathfrak{A}_k$. Now $(\mathfrak{S}_i \supset \mathfrak{S}_j) \,.\, \mathfrak{S}_i$ L-implies $\mathfrak{S}_j$; hence $\mathfrak{S}_j$ itself is true at $\mathfrak{B}_i$ for any value-assignment to the variables in $\mathfrak{A}_k$. Consequently, $\mathfrak{A}_k(\mathfrak{S}_j)$ is true at $\mathfrak{B}_i$.

Proof of (c). By (a), $\mathfrak{A}_k(\mathfrak{S}_i \supset \mathfrak{S}_j) \,.\, \mathfrak{A}_k(\mathfrak{S}_i)$ is L-equivalent to $\mathfrak{A}_k[(\mathfrak{S}_i \supset \mathfrak{S}_j) \,.\, \mathfrak{S}_i]$; hence, by (b), $\mathfrak{A}_k(\mathfrak{S}_i \supset \mathfrak{S}_j) \,.\, \mathfrak{A}_k(\mathfrak{S}_i)$ L-implies $\mathfrak{A}_k(\mathfrak{S}_j)$. T6-4 now tells us that $[\mathfrak{A}_k(\mathfrak{S}_i \supset \mathfrak{S}_j) \,.\, \mathfrak{A}_k(\mathfrak{S}_i)] \supset \mathfrak{A}_k(\mathfrak{S}_j)$ is L-true. Assertion (c) follows from this by an application of T8-61(1).

Proof of (e). If $\mathfrak{S}_i$ L-implies $\mathfrak{S}_j$, then by T6-4 the formula $\mathfrak{S}_i \supset \mathfrak{S}_j$ is L-true. Hence by T1e $\mathfrak{A}_k(\mathfrak{S}_i \supset \mathfrak{S}_j)$ is L-true, and by (d) so also is $\mathfrak{A}_k(\mathfrak{S}_i) \supset \mathfrak{A}_k(\mathfrak{S}_j)$. An application of T6-4 to this last result yields assertion (e) as desired.

Proof of (g). Since T8-6f(1) guarantees $\mathfrak{S}_i \equiv \mathfrak{S}_j$ is L-equivalent to $(\mathfrak{S}_i \supset \mathfrak{S}_j) \,.\, (\mathfrak{S}_j \supset \mathfrak{S}_i)$, we may use (f) to see $\mathfrak{A}_k(\mathfrak{S}_i \equiv \mathfrak{S}_j)$ is L-equivalent to $\mathfrak{A}_k[(\mathfrak{S}_i \supset \mathfrak{S}_j) \,.\, (\mathfrak{S}_j \supset \mathfrak{S}_i)]$. Applying (a) to this last formula, it appears $\mathfrak{A}_k(\mathfrak{S}_i \equiv \mathfrak{S}_j)$ is L-equivalent to $\mathfrak{A}_k(\mathfrak{S}_i \supset \mathfrak{S}_j) \,.\, \mathfrak{A}_k(\mathfrak{S}_j \supset \mathfrak{S}_i)$. Thus, $\mathfrak{A}_k(\mathfrak{S}_i \equiv \mathfrak{S}_j)$ L-implies the conjunction $\mathfrak{A}_k(\mathfrak{S}_i \supset \mathfrak{S}_j) \,.\, \mathfrak{A}_k(\mathfrak{S}_j \supset \mathfrak{S}_i)$, hence by T6-12c the formula $\mathfrak{A}_k(\mathfrak{S}_i \equiv \mathfrak{S}_j)$ L-implies each component of the conjunction separately. Recalling (d), we now see that $\mathfrak{A}_k(\mathfrak{S}_i \equiv \mathfrak{S}_j)$ L-implies each of $\mathfrak{A}_k(\mathfrak{S}_i) \supset \mathfrak{A}_k(\mathfrak{S}_j)$ and $\mathfrak{A}_k(\mathfrak{S}_j) \supset \mathfrak{A}_k(\mathfrak{S}_i)$ separately, hence (by T6-12c again) the conjunction of these last two formulas, and finally (by another appeal to T8-6f(1)) the biconditional $\mathfrak{A}_k(\mathfrak{S}_i) \equiv \mathfrak{A}_k(\mathfrak{S}_j)$, as was to be proved.

T13-4. Suppose $\mathfrak{S}_i$, $\mathfrak{S}_j$, and $\mathfrak{S}_m$ are arbitrary sentential formulas. Let $\mathfrak{A}_k$ be a universal quantifier or a sequence of such quantifiers. Then the following hold:

a. $\mathfrak{A}_k(\mathfrak{S}_i)$ is L-true if and only if $\mathfrak{S}_i$ is L-true. (By T1b,e.)

b. If $\mathfrak{S}_i \supset \mathfrak{S}_j$ is L-true, then $\mathfrak{A}_k(\mathfrak{S}_i)$ L-implies $\mathfrak{A}_k(\mathfrak{S}_j)$. (Proved below.)

c. If $\mathfrak{S}_i \,.\, \mathfrak{S}_j \supset \mathfrak{S}_m$ is L-true, then $\mathfrak{A}_k(\mathfrak{S}_i)$ and $\mathfrak{A}_k(\mathfrak{S}_j)$ together L-imply $\mathfrak{A}_k(\mathfrak{S}_m)$. (Proved below.)

d. If $\mathfrak{S}_i \equiv \mathfrak{S}_j$ is L-true, then $\mathfrak{A}_k(\mathfrak{S}_i)$ and $\mathfrak{A}_k(\mathfrak{S}_j)$ are L-equivalent. (Proved below.)

Proofs. Proof of (b): supposing $\mathfrak{S}_i \supset \mathfrak{S}_j$ L-true, by (a) it appears $\mathfrak{A}_k(\mathfrak{S}_i \supset \mathfrak{S}_j)$ is L-true, whence $\mathfrak{A}_k(\mathfrak{S}_i) \supset \mathfrak{A}_k(\mathfrak{S}_j)$ is L-true by T3d, and an application of T6-4 yields (b). — Proof of (c): supposing $\mathfrak{S}_i . \mathfrak{S}_j \supset \mathfrak{S}_m$ L-true, we see that $\mathfrak{A}_k(\mathfrak{S}_i . \mathfrak{S}_j \supset \mathfrak{S}_m)$ is L-true by (a), that then $\mathfrak{A}_k(\mathfrak{S}_i . \mathfrak{S}_j) \supset \mathfrak{A}_k(\mathfrak{S}_m)$ is L-true by (b), and finally, T3a that $\mathfrak{A}_k(\mathfrak{S}_i) . \mathfrak{A}_k(\mathfrak{S}_j) \supset \mathfrak{A}_k(\mathfrak{S}_m)$ is L-true, whence assertion (c). — Proof of (d): supposing $\mathfrak{S}_i \equiv \mathfrak{S}_j$ L-true, $\mathfrak{A}_k(\mathfrak{S}_i \equiv \mathfrak{S}_j)$ is L-true and so $\mathfrak{A}_k(\mathfrak{S}_i) \equiv \mathfrak{A}_k(\mathfrak{S}_j)$ by T3g, hence (d).

T4 frequently proves useful in connection with formulas that are tautologies. E.g. since '$p.q \supset p$' is a tautology, the following formulas are L-true: '$Fx.Gx \supset Fx$', '$(x)(Fx.Gx \supset Fx)$', and '$(x)(Fx.Gx) \supset (x)Fx$'. Whence we see that '$(x)(Fx.Gx)$' L-implies '$(x)Fx$'.

T13-5. **a.** '$\sim(x)Fx$' is L-equivalent to '$(\exists x)(\sim Fx)$'.
b. '$(x)(p \vee Fx)$' is L-equivalent to '$p \vee (x)Fx$'.

Proof of (a). The only value-bearing sign in the two formulas of (a) is 'F'. So take $\mathfrak{B}_k$ to be any value-assignment to 'F' at which '$\sim(x)Fx$' is true. By R11-1c, '$(x)Fx$' is false at $\mathfrak{B}_k$. Hence, in view of R11-1g, it is not the case that, for every value-assignment to 'x', 'Fx' comes out true at $\mathfrak{B}_k$. Thus there is a value-assignment $\mathfrak{B}_x$ to 'x' such that 'Fx' is false at $\mathfrak{B}_k + \mathfrak{B}_x$. By R11-1c, therefore, '$\sim Fx$' is true at $\mathfrak{B}_k + \mathfrak{B}_x$. Consequently '$(\exists x)(\sim Fx)$' is true at $\mathfrak{B}_k$, by R11-1h. A similar argument establishes the converse.

Proof of (b). The value-bearing signs in the formulas in (b) being 'p' and 'F', let $\mathfrak{B}_k$ be any value-assignment to 'p' and 'F' at which '$(x)(p \vee Fx)$' comes out true. Now two cases are possible, viz. 'p' is true at $\mathfrak{B}_k$, or else false. (i) Suppose 'p' is true at $\mathfrak{B}_k$. In this case, we see at once by R11-1d that '$p \vee (x)Fx$' is true at $\mathfrak{B}_k$. (ii) Suppose 'p' is false at $\mathfrak{B}_k$. Since '$(x)(p \vee Fx)$' is true at $\mathfrak{B}_k$, we know from R11-1g that '$p \vee Fx$' at $\mathfrak{B}_k$ is true for any value-assignment to the variable 'x'; thus (by R11-1d) we have that 'Fx' is true at any value-assignment to 'x'. Then it follows from R11-1g that '$(x)Fx$' is true at $\mathfrak{B}_k$, whence by R11-1d we see that '$p \vee (x)Fx$' is true at $\mathfrak{B}_k$. Consequently, at every value-assignment for which '$(x)(p \vee Fx)$' is true, '$p \vee (x)Fx$' is also true. The converse may be established similarly.

We learn from T5a that the negation of a universal sentence may be transformed into an existential sentence whose operand is the negation of the original operand. The force of T5b becomes more apparent when we recall (from earlier remarks in **12a**) that in the formulas of T5b there may be substituted for 'p' any sentential formula in which 'x' is not free. Thus, T5b says that a universal sentence whose operand is a disjunction with one component devoid of free occurrences of the quantifier-variable may be transformed by shifting this component out of the operand.

14. L-TRUE FORMULAS WITH QUANTIFIERS

14a. L-true conditionals. We set down here lists of L-true formulas with quantifiers—first, in T1, a list of conditionals which includes results on L-implications; and second, in T2, a list of biconditionals which includes results on L-equivalence. The lists are chiefly for reference, but assertions marked with '+' deserve special attention because of their frequent use in practical work. The role these lists will have is rather like that of the lists of tautologies given earlier (in T8-2 and T8-6).

T14-1. Suppose $\mathfrak{S}_i \supset \mathfrak{S}_j$ is any one of the conditionals (a)(1) through (k) mentioned below. Suppose $\mathfrak{S}_i' \supset \mathfrak{S}_j'$ is obtained from $\mathfrak{S}_i \supset \mathfrak{S}_j$ by arbitrary substitutions. Finally, let $\mathfrak{A}_k$ be a universal quantifier or a sequence of such quantifiers. Then the following hold:

A. $\mathfrak{S}_i \supset \mathfrak{S}_j$ is L-true.
B. $\mathfrak{S}_i$ L-implies $\mathfrak{S}_j$. (By (A).)
C. $\mathfrak{S}_i' \supset \mathfrak{S}_j'$ is L-true. (By (A), in view of T12-1a.)
D. $\mathfrak{S}_i'$ L-implies $\mathfrak{S}_j'$. (By (C).)
E. $\mathfrak{A}_k(\mathfrak{S}_i \supset \mathfrak{S}_j)$ is L-true. (By (A), and T13-4a.)
F. $\mathfrak{A}_k(\mathfrak{S}_i) \supset \mathfrak{A}_k(\mathfrak{S}_j)$ is L-true. (By (A), and T13-4b.)
G. $\mathfrak{A}_k(\mathfrak{S}_i)$ L-implies $\mathfrak{A}_k(\mathfrak{S}_j)$. (By (F).)
H. Bound variables that occur may be revised arbitrarily into other variables. (See remark following T12-2.)

Some of the formulas below have the form $\mathfrak{S}_m . \mathfrak{S}_n \supset \mathfrak{S}_j$. For conditionals of this type the following additional assertions hold, in view of T13-4c. (We understand $\mathfrak{S}_m' . \mathfrak{S}_n' \supset \mathfrak{S}_j'$ to be formed from $\mathfrak{S}_m . \mathfrak{S}_n \supset \mathfrak{S}_j$ by arbitrary substitutions.)

I. The class comprising formulas $\mathfrak{S}_m$ and $\mathfrak{S}_n$ L-implies $\mathfrak{S}_j$.
J. The class comprising formulas $\mathfrak{S}_m'$ and $\mathfrak{S}_n'$ L-implies $\mathfrak{S}_j'$.
K. The class comprising formulas $\mathfrak{A}_k(\mathfrak{S}_m)$ and $\mathfrak{A}_k(\mathfrak{S}_n)$ L-implies $\mathfrak{A}_k(\mathfrak{S}_j)$.
L. The class comprising formulas $\mathfrak{A}_k(\mathfrak{S}_m')$ and $\mathfrak{A}_k(\mathfrak{S}_n')$ L-implies $\mathfrak{A}_k(\mathfrak{S}_j')$.

+**a.** Law of specialization (or instantiation).
(1) $(x)(Fx) \supset Fx$. (By T13-1a.)
(2) $(x)(Fx) \supset Fy$. (By (1), and T12-1a.)

b. Law of existential inference (or existential generalization).
(1) $Fx \supset (\exists x)Fx$. (By T13-1j.)
(2) $Fy \supset (\exists x)Fx$. (By (1), and T12-1a.

c. $(x)Fx \supset (\exists x)Fx$. (By (a)(1), (b)(1), and T8-2f(6).)

d. +(1) $(x)(Fx \supset Gx) \supset [(x)Fx \supset (x)Gx]$. (By T13-3c.)
+(2) $(x)(Fx \supset Gx) . (x)Fx \supset (x)Gx$. (By (1) and T8-61(1).)
(3) $(x)Fx \supset [(x)(Fx \supset Gx) \supset (x)Gx]$. (By (1) and T8-61(2).)
+(4) $(x)(Fx \supset Gx) . (\exists x)Fx \supset (\exists x)Gx$.
(5) $(x)(Fx \supset Gx) \supset [(\exists x)Fx \supset (\exists x)Gx]$. (By (4) and T8-61(1).)
(6) $(\exists x)Fx \supset [(x)(Fx \supset Gx) \supset (\exists x)Gx]$. (By (5) and T8-61(2).)

T14-1

e. $+(1)$ $(x)(Fx \equiv Gx) \supset (x)(Fx \supset Gx)$. (By T8-2e(1) and T13-4b.)

(2) $(x)(Fx \equiv Gx) \supset (x)(Gx \supset Fx)$. (By T8-2e(2) and T13-4b.)

(3) $(x)(Fx \equiv Gx) \supset [(x)Fx \supset (x)Gx]$. (By (1) and (d)(1).)

$+(4)$ $(x)(Fx \equiv Gx).(x)Fx \supset (x)Gx$. (By (3) and T8-61(1).)

(5) $(x)Fx \supset [(x)(Fx \equiv Gx) \supset (x)Gx]$. (By (3) and T8-61(1).)

(6) $(x)(Fx \equiv Gx) \supset [(x)Gx \supset (x)Fx]$. (By (2) and (d)(1).)

(7) $(x)(Fx \equiv Gx).(x)Gx \supset (x)Fx$. (By (6) and T8-61(1).)

(8) $(x)Gx \supset [(x)(Fx \equiv Gx) \supset (x)Fx]$. (By (6) and T8-61(2).)

$+(9)$ $(x)(Fx \equiv Gx) \supset [(x)Fx \equiv (x)Gx]$. (By T13-3g.)

f. (1) $(x)(Fx \equiv Gx) \supset [(\exists x)Fx \supset (\exists x)Gx]$. (By (e)(1) and (d)(5).)

(2) $(x)(Fx \equiv Gx).(\exists x)Fx \supset (\exists x)Gx$. (By (1) and T8-61(1).)

(3) $(\exists x)Fx \supset [(x)(Fx \equiv Gx) \supset (\exists x)Gx]$. (By (1) and T8-61(2).)

(4) $(x)(Fx \equiv Gx) \supset [(\exists x)Gx \supset (\exists x)Fx]$. (By (e)(2) and (d)(5).)

(5) $(x)(Fx \equiv Gx) \supset [(\exists x)Fx \equiv (\exists x)Gx]$. (By (1), (4), and T8-6f(1).)

g. $+(1)$ $(\exists x)(Fx.Gx) \supset (\exists x)Fx.(\exists x)Gx$.

$+(2)$ $(\exists x)(Fx.Gx) \supset (\exists x)Fx$. (By (1).)

(3) $(\exists x)(Fx.Gx) \supset (\exists x)Gx$. (By (1).)

h. (1) $(x)Fx.(\exists x)Gx \supset (\exists x)(Fx.Gx)$.

(2) $(x)Fx \supset [(\exists x)Gx \supset (\exists x)(Fx.Gx)]$. (By (1) and T8-61(1).)

i. (1) $(x)Fx \vee (x)Gx \supset (x)(Fx \vee Gx)$.

(2) $(x)(Fx \vee Gx) \supset (\exists x)Fx \vee (x)Gx$.

(3) $(x)(Fx \vee Gx) \supset (x)Fx \vee (\exists x)Gx$. (By (2).)

j. Syllogism.

$+(1)$ $[(x)(Fx \supset Gx).(x)(Gx \supset Hx)] \supset (x)(Fx \supset Hx)$.

(2) $(x)(Fx \supset Gx) \supset [(x)(Gx \supset Hx) \supset (x)(Fx \supset Hx)]$. (By (1) and T8-61(1).)

T14-1 j. (3) $(x)(Gx \supset Hx) \supset [(x)(Fx \supset Gx) \supset (x)(Fx \supset Hx)]$.
(By (2) and T8-61(2).)

+(4) $[(x)(Fx \supset Gx) . (\exists x)(Fx . Hx)] \supset (\exists x)(Gx . Hx)$.

(5) $(x)(Fx \supset Gx) \supset [(\exists x)(Fx . Hx) \supset (\exists x)(Gx . Hx)]$.
(By (4) and T8-61(1).)

(6) $(\exists x)(Fx . Hx) \supset [(x)(Fx \supset Gx) \supset (\exists x)(Gx . Hx)]$.
(By (5) and T8-61(2).)

+**k.** Interchange of two dissimilar quantifiers.
$(\exists x)(y)Kxy \supset (y)(\exists x)Kxy$.

Most of the formulas listed above are accompanied by references to previous formulas and theorems in earlier sections, by means of which the formula in question can be established. Formulas which carry no such reference can be proved easily in a similar way, with the help of rules R11-1g,h.

Remarks on the formulas in T1. Regarding the use of formulas (a) and (b), reference may be made to our earlier comments on T13-1. — From (c) we learn it is permissible to pass from a universal sentence to the corresponding existential sentence; such a step is possible in our present system because this system admits only non-empty domains of individuals (a customary restriction). — Formula (d)(1) countenances the *distribution* of a universal quantifier over the components of a conditional. — From (d)(4) we see that if some individual satisfies the first component of the operand in a universal conditional (e.g. a law of nature), then some (the same) individual satisfies the second component. — Formula (e)(1) says that a universal equivalence implies the corresponding universal conditional. — By (e)(9) we see that a universal quantifier may be distributed over the components of a biconditional; note further that from (e)(9) follow the two conditionals (e)(3) and (e)(6). — From (f)(2) we see that if the first component of the operand in a universal biconditional is satisfied, so also is the second. — Formula (g)(1) countenances the distribution of an existential quantifier over the components of a conjunction. (The result here holds in one direction only; in the case of disjunction, however, a similar result holds in both directions. Cf. T2c(2) below.) — Note that formulas (j) involve three predicate variables. We recognize (j)(1) as the well known inference called "Barbara" in traditional logic. From (j)(4) we have: if all F are G and some individual has both F and H, then some (the same) individual has both G and H.

Finally, some observations regarding formula (k). A sentence of the form '$(\exists x)(y)Kxy$' is an *absolute* existential sentence. This sentence says: "there is at least one individual x such that for each individual y, x bears the relation K to y". On the other hand, a sentence of the form '$(y)(\exists x)Kxy$' is a *relative* existential sentence. A relative existential sentence is weaker than (i.e. says less than; cf. **6b**) the corresponding absolute one. The sentence '$(y)(\exists x)Kxy$' says: "for each individual y there is at least one individual x such that x bears the relation K to y". Formula (k) tells us it is permissible to pass from an absolute existential sentence to the corresponding relative one. Which is entirely plausible, since if there is an individual (say b) which bears the relation K to every individual, then obviously for each individual there is one (viz. b) that bears the relation K to it. Contrariwise, however, it is generally not possible to pass from a relative existential sentence to its absolute counterpart. For the relative sentence affirms only that for each y there is an x which bears the relation K to y; nothing is said to prevent different individuals y from associating with different individuals x, i.e. nothing is said that requires some x to bear the relation K to every y. E.g. in the domain of natural numbers, the relative existential sentence '$(y)(\exists x)Gr(x,y)$' is true because for each number there is a greater; however, the corresponding absolute existential sentence '$(\exists x)(y)Gr(x,y)$' is clearly false, since it claims there is a number greater than all numbers. (It will be seen in T2g below that the interchange of two *similar* quantifiers leads to an L-equivalent formula.)

14b. L-true biconditionals.

T14-2. Suppose $\mathfrak{S}_i \equiv \mathfrak{S}_j$ is any one of the biconditionals (a)(1) through (h)(2) mentioned below. Suppose $\mathfrak{S}_i' \equiv \mathfrak{S}_j'$ is obtained from $\mathfrak{S}_i \equiv \mathfrak{S}_j$ by arbitrary substitutions. Finally, let $\mathfrak{A}_k$ be a universal quantifier or a sequence of such quantifiers. Then the following hold:

A. $\mathfrak{S}_i \equiv \mathfrak{S}_j$ is L-true.
B. $\mathfrak{S}_i$ and $\mathfrak{S}_j$ are L-equivalent. (By (A).)
C. $\mathfrak{S}_i' \equiv \mathfrak{S}_j'$ is L-true. (By (A), in view of T12-1a.)
D. $\mathfrak{S}_i'$ and $\mathfrak{S}_j'$ are L-equivalent. (By (C).)
E. $\mathfrak{S}_i$ and $\mathfrak{S}_j$ are mutually L-interchangeable. (This follows from (B), as will be shown later; cf. T15-3g.)
F. $\mathfrak{S}_i'$ and $\mathfrak{S}_j'$ are mutually L-interchangeable. (By (D).)
G. $\mathfrak{A}_k(\mathfrak{S}_i \equiv \mathfrak{S}_j)$ and $\mathfrak{A}_k(\mathfrak{S}_i' \equiv \mathfrak{S}_j')$ are both L-true. (By (A) and T13-4a.)
H. $\mathfrak{A}_k(\mathfrak{S}_i) \equiv \mathfrak{A}_k(\mathfrak{S}_j)$ and $\mathfrak{A}_k(\mathfrak{S}_i') \equiv \mathfrak{A}_k(\mathfrak{S}_j')$ are both L-true. (By (G) and T13-3g.)
I. $\mathfrak{A}_k(\mathfrak{S}_i)$ and $\mathfrak{A}_k(\mathfrak{S}_j)$ are L-equivalent, and so are $\mathfrak{A}_k(\mathfrak{S}_i')$ and $\mathfrak{A}_k(\mathfrak{S}_j')$. (By (H).)
K. Bound variables that occur may be revised arbitrarily into other variables. (In view of T12-2.)

a. Laws of negation.
+(1) $\sim(x)Fx \equiv (\exists x)\sim Fx$. (By T13-5a.)
+(2) $\sim(\exists x)Fx \equiv (x)\sim Fx$. (By (1) and T8-6i(5), substituting '$\sim Fx$' for 'Fx'.)

T14-2 **a.** +(3) $(x)Fx \equiv \sim(\exists x)\sim Fx$. (By (1) and T8-6i(5).)
+(4) $(\exists x)Fx \equiv \sim(x)\sim Fx$. (By (1), substituting '$\sim Fx$' for 'Fx'.)
(5) $\sim(x)(Fx \supset Gx) \equiv (\exists x)(Fx . \sim Gx)$. (By (1) and T8-6h(1).)
(6) $\sim(x)(Fx \supset \sim Gx) \equiv (\exists x)(Fx . Gx)$. (By (5).)
(7) $\sim(\exists x)(Fx . Gx) \equiv (x)(Fx \supset \sim Gx)$. (By (6) and T8-6i(5).)
(8) $\sim(\exists x)(Fx . \sim Gx) \equiv (x)(Fx \supset Gx)$. (By (5) and T8-6i(5).)

b. Laws of negation for several similar quantifiers. (Each of the following four formulas—which are analogous to (a)(1)-(4)—contains a sequence of n universal quantifiers ($n \geq 2$) indicated by '$(x)...(z)$', a corresponding sequence of n existential quantifiers indicated by '$(\exists x)...(\exists z)$' with the same variable in corresponding quantifiers, and an n-place predicate variable 'K' followed by a sequence of n individual variables indicated by '$x...z$'.)
(1) $\sim(x)...(z)(Kx...z) \equiv (\exists x)...(\exists z)(\sim Kx...z)$.
(2) $\sim(\exists x)...(\exists z)(Kx...z) \equiv (x)...(z)(\sim Kx...z)$.
(3) $(x)...(z)(Kx...z) \equiv \sim(\exists x)...(\exists z)(\sim Kx...z)$.
(4) $(\exists x)...(\exists z)(Kx...z) \equiv \sim(x)...(z)(\sim Kx...z)$.

c. Distribution laws.
+(1) $(x)(Fx . Gx) \equiv (x)Fx . (x)Gx$. (By T13-3a.)
+(2) $(\exists x)(Fx \vee Gx) \equiv (\exists x)Fx \vee (\exists x)Gx$. (By (1), (a)(4), T8-6g(1), (3).)

d. Shifting a universal quantifier.
(Recall from previous explanations (**12a**) that in the formulas listed under (d), (e) and (f) below it is permissible to substitute for 'p' any sentential formula in which 'x' does not occur free.)
(1) $(x)(p \vee Fx) \equiv p \vee (x)Fx$. (By T13-5b.)
(2) $(x)(Fx \vee p) \equiv (x)(Fx) \vee p$. (By (1).)
(3) $(x)(p . Fx) \equiv p . (x)Fx$. (An analog of T13-5b.)
(4) $(x)(Fx . p) \equiv (x)(Fx) . p$. (By (3).)
(5) $(x)(p \supset Fx) \equiv [p \supset (x)Fx]$. (By (1) and T8-6j(1).)

e. Shifting an existential quantifier.
(1) $(\exists x)(p \vee Fx) \equiv p \vee (\exists x)Fx$. (An analog of T13-5b.)
(2) $(\exists x)(Fx \vee p) \equiv (\exists x)(Fx) \vee p$. (By (1).)
(3) $(\exists x)(p . Fx) \equiv p . (\exists x)Fx$. (An analog of T13-5b.)
(4) $(\exists x)(Fx . p) \equiv (\exists x)(Fx) . p$. (By (3).)
(5) $(\exists x)(p \supset Fx) \equiv [p \supset (\exists x)Fx]$. (By (1) and T8-6j(1).)

T14-2 **f.** Shifting and altering a quantifier.

(1) $(x)(Fx \supset p) \equiv [(\exists x)(Fx) \supset p]$. (By T8-6j(1), d(2) and a(2).)

(2) $(\exists x)(Fx \supset p) \equiv [(x)(Fx) \supset p]$. (An analog of (1).)

g. Interchange of two similar quantifiers.

+(1) $(x)(y)Kxy \equiv (y)(x)Kxy$. (By R11-1g.)

+(2) $(\exists x)(\exists y)Kxy \equiv (\exists y)(\exists x)Kxy$. (By R11-1h.)

h. Permutation of n similar quantifiers ($n > 2$).

(What was said in (b) above applies here regarding the notations '$(x)...(z)$', '$(\exists x)...(\exists z)$', '$K$', and '$x...z$'. By '$..(z)..(x)..$' is meant an arbitrary permutation of the quantifiers in the sequence '$(x)...(z)$' and similarly for '$..(\exists z)..(\exists x)..$'.)

(1) $(x)...(z)(Kx...z) \equiv ..(z)..(x)..(Kx...z)$. (By R11-1g.)

(2) $(\exists x)...(\exists z)(Kx...z) \equiv ..(\exists z)..(\exists x)..(Kx...z)$. (By R11-1h.)

Remarks on the formulas in T2. Formulas (a)(1) and (a)(2) tell us how to transform the negation of a universal formula or of an existential formula: the negation sign is moved over the quantifier onto the operand, and the quantifier itself converted to one of the opposite sort. These transformations are entirely plausible (cf. **9b**). If the domain of individuals is finite, (a)(1) and (a)(2) correspond to *De Morgan's* laws (T8-6g). We can see this as follows: Suppose the rules of a certain language system show that the domain of individuals comprises a fixed finite number n of individuals; let these n individuals be denoted by the individual constants 'a_1', 'a_2', ..., 'a_n'. Now in this system the universal sentence '$(x)Px$' is synonymous with the n-tuple conjunction '$Pa_1 . Pa_2 . \ldots . Pa_n$', and the existential sentence '$(\exists x)Px$' with the n-tuple disjunction '$Pa_1 \vee Pa_2 \vee \ldots \vee Pa_n$'. Here, therefore, '$\sim(x)Px$' is '$\sim(Pa_1 . Pa_2 . \ldots . Pa_n)$' which by T8-6g(4) is L-equivalent to '$\sim Pa_1 \vee \sim Pa_2 \vee \ldots \vee \sim Pa_n$', and this last in turn is '$(\exists x)\sim Px$'. The same applies to '$\sim(\exists x)Px$'. — Formula (a)(3) indicates the possibility of defining the universal quantifier in terms of the existential quantifier; and (a)(4), the possibility of defining the latter in terms of the former. — From (a)(8) we see that a universal conditional, e.g. a law of nature, is synonymous with a certain negated existential sentence: "all crows are black" has the same meaning as "there is no non-black crow". — Formulas (b) are similar to (a)(1)-(4): a continuous sequence of two or more similar quantifiers may be treated as a single such quantifier.

From (c)(1) we see that a universal quantifier distributes over the components of a conjunction, and from (c)(2) that an existential quantifier likewise distributes over the components of a disjunction. [Note that both these formulas give rise to L-equivalences, i.e. each direction of a formula gives an allowable transformation. By contrast, the distribution of a universal quantifier over the components of a conditional or of a biconditional is permissible only in one direction (cf. T1-d(1), e(9)).]

The formulas (d) indicate certain cases in which the universal quantifier may be relocated: a sentential formula with no free occurrence of the quantifier-variable may at will be inserted into (or removed from) the operand, provided this formula is one of the components of a conjunction or of a disjunction, or the antecedent of a conditional. — Formulas (e) indicate that the existential quantifier may be similarly relocated in similar cases. — In contrast to (d)(5) and (e)(5), formulas (f) assert that if the sentential formula in question is the consequent of a conditional, the quantifier is not simply to be relocated but must also be converted into one of the opposite kind. E.g. (f)(1) says in effect that L-equivalence holds between the two formulas which correspond respectively to the following two word-language sentences about the inhabitants of Sodom (who here constitute the domain of individuals): 'For each Sodomite it is the case that if he is righteous, then Sodom will be spared", and "If at least one Sodomite is righteous, then Sodom will be spared". — Formula f(2) is seldom used; the operand of an existential quantifier is usually a conjunction, and only rarely a conditional. — Finally, (g) and (h) indicate that the members of a sequence of two or more similar quantifiers may be reordered at will.

14c. Exercises. Translate each of the following sentences into our symbolic language; more specifically, give each sentence two symbolic translations (which by T2a(1),(2) are L-equivalent to each other), viz. one with a universal quantifier, the other with an existential quantifier. — **1.** "No (thing) is spherical." ((a) "There is nothing ..."; (b) "Each (thing) ... not..."). — **2.** "0 is not greater than any (number)." — **3.** "Not every (number) is greater than 0." — **4.** "There is a (number) such that no (number) is smaller than it." — **5.** "For every (number) x it is the case that no (number) is both greater than and smaller than x."—Translate each of sentences (6) and (7) below into our symbolic language; then use T13-1(1) to obtain from each of these translations a corresponding existential sentence; and finally, translate each of these existential sentences back into the word language. — **6.** "The moon is spherical." — **7.** "2 is a prime number and even."—In the exercise which follows "taut" indicates the application of a tautological formula (e.g. one of the formulas listed in T8-1 or in T8-6). Part (a) of the exercise is worked out as an example. — **8.** Give a derivation for each of the following cases of L-implication:

a) '$\sim(x)(Fx \supset Gx)$' and '$(x)(Hx \supset Gx)$' L-imply '$(\exists x)\sim Hx$'.

		$(x)(Hx \supset Gx)$	1.
		$\sim(x)(Fx \supset Gx)$	2.
(2)	T2a(5)	$(\exists x)(Fx . \sim Gx)$	3.
(3)	T1g(1)	$(\exists x)Fx . (\exists x)\sim Gx$	4.
(4)	taut.	$(\exists x)\sim Gx$	5.
(1)	T1d(1)	$(x)Hx \supset (x)Gx$	6.
(5)	T2a(1)	$\sim(x)Gx$	7.
(6) (7)	taut.	$\sim(x)Hx$	8.
(8)	T2a(1)	$(\exists x)\sim Hx$	9.

b) '$(x)(Fx \supset p)$' and '$\sim p$' L-imply '$(x)\sim Fx$'. (Hint: use T2f(1).)

c) '$(x)(Hxz \supset Hax)$' and 'Haz' L-imply 'Haa'.

d) '$(x)(Fx \equiv Gx)$' and 'Ga' L-imply '$(\exists y)Fy$'.
e) '$(x)(Fx \equiv Gx)$' and '$\sim(\exists y)\sim Gy$' L-imply 'Fb'.
f) '$(y)((\exists z)(Hzy) \supset Gyy)$' and '$(\exists y)((\exists z)(Hzy).Gyx)$' L-imply '$(\exists w)(Gwx.Gww)$'.
g) '$(\exists x)Hxx$' L-implies '$(x)(\exists x)Hxx$'. (See the remark on R11-1g.)
h) '$Gb \vee Fb$' and '$(x)\sim Fx$' L-imply '$(\exists x)Gx$'.
i) '$(y)(x)(Mxyx \supset Hxy)$' and '$(z)(Maza)$' L-imply 'Haa'.
j) '$(\exists x)(y)Hxy$' L-implies '$(\exists z)Hzz$'.
k) '$(x)(Fx.Gx)$' L-implies '$(\exists y)(Fy \vee Gy)$'.
l) '$(x)(\exists y)Hxy$', '$(x)(y)(z)(Hxy.Hyz \supset Hxz)$', and '$(x)(y)(Hxy \supset Hyz)$' L-imply '$(x)Hxx$'.

Note: l) is of special interest in the study of the theory of relations. The second premise says that H is transitive, the third premise that H is symmetric, and the conclusion that H is totally reflexive (cf. **16c** and T31-1).

15. DEFINITIONS

15a. Interchangeability. We are now in a position to state theorems on interchangeability which are more general than those of **8b.** The source of this increased generality is in the fact that here the component formula subject to interchange can occur not simply as a component of a sentential connective, but as an operand as well.

Suppose $\mathfrak{S}_i$, $\mathfrak{S}_j$, $\mathfrak{S}_k$, and $\mathfrak{S}_m$ are sentential formulas. We shall say that *$\mathfrak{S}_k$ and $\mathfrak{S}_m$ are equivalent respecting $\mathfrak{S}_i$ and $\mathfrak{S}_j$* provided $(\,)(\mathfrak{S}_i \equiv \mathfrak{S}_j)$ L-implies $(\,)(\mathfrak{S}_k \equiv \mathfrak{S}_m)$, where '( )' stands for a sequence of universal quantifiers —one for each of the variables (except sentential variables) that occur free in the operand in question.

The notion just introduced is used in the following three theorems.

T15-1. Let $\mathfrak{S}_i$, $\mathfrak{S}_j$ and $\mathfrak{S}_k$ be arbitrary sentential formulas, $\mathfrak{A}$ an arbitrary universal quantifier, and $\mathfrak{E}$ an existential quantifier. For each of the following pairs of sentential formulas it is then the case that the two given formulas are equivalent with respect to $\mathfrak{S}_i$ and $\mathfrak{S}_j$:

a. $\sim\mathfrak{S}_i$ and $\sim\mathfrak{S}_j$. (Each of (a) through (i) follows from T8-3b and T13-3e.)
b. $\mathfrak{S}_i \vee \mathfrak{S}_k$ and $\mathfrak{S}_j \vee \mathfrak{S}_k$.
c. $\mathfrak{S}_k \vee \mathfrak{S}_i$ and $\mathfrak{S}_k \vee \mathfrak{S}_j$.
d. $\mathfrak{S}_i.\mathfrak{S}_k$ and $\mathfrak{S}_j.\mathfrak{S}_k$.
e. $\mathfrak{S}_k.\mathfrak{S}_i$ and $\mathfrak{S}_k.\mathfrak{S}_j$.
f. $\mathfrak{S}_i \supset \mathfrak{S}_k$ and $\mathfrak{S}_j \supset \mathfrak{S}_k$.
g. $\mathfrak{S}_k \supset \mathfrak{S}_i$ and $\mathfrak{S}_k \supset \mathfrak{S}_j$.
h. $\mathfrak{S}_i \equiv \mathfrak{S}_k$ and $\mathfrak{S}_j \equiv \mathfrak{S}_k$.
i. $\mathfrak{S}_k \equiv \mathfrak{S}_i$ and $\mathfrak{S}_k \equiv \mathfrak{S}_j$.
j. $\mathfrak{A}(\mathfrak{S}_i)$ and $\mathfrak{A}(\mathfrak{S}_j)$. (By T14-1e(9).)
k. $\mathfrak{E}(\mathfrak{S}_i)$ and $\mathfrak{E}(\mathfrak{S}_j)$. (By T14-1f(5).)

T15-2. If two sentential formulas are equivalent respecting a second pair, and again these latter two are equivalent respecting a third pair, then the two formulas of the first pair are equivalent respecting the two formulas of the third pair.

T15-3. *Interchangeability.* Suppose $\mathfrak{S}_i$ and $\mathfrak{S}_j$ are arbitrary sentential formulas. Suppose $\mathfrak{S}_i'$ is constructed from $\mathfrak{S}_i$ and possibly other arbitrary formulas by means of connectives and quantifiers. And suppose, finally, that $\mathfrak{S}_j'$ is obtained from $\mathfrak{S}_i'$ through the replacement of $\mathfrak{S}_i$ by $\mathfrak{S}_j$. Then the following hold:

a. $\mathfrak{S}_i'$ and $\mathfrak{S}_j'$ are equivalent respecting $\mathfrak{S}_i$ and $\mathfrak{S}_j$, i.e. $(\,)(\mathfrak{S}_i \equiv \mathfrak{S}_j)$ L-implies $(\,)(\mathfrak{S}_i' \equiv \mathfrak{S}_j')$. (Proved below.)
b. $(\,)(\mathfrak{S}_i \equiv \mathfrak{S}_j) \supset (\,)(\mathfrak{S}_i' \equiv \mathfrak{S}_j')$ is L-true. (By (a).)
c. $(\,)(\mathfrak{S}_i \equiv \mathfrak{S}_j)$ L-implies $\mathfrak{S}_i' \equiv \mathfrak{S}_j'$. (By (a) and T13-1b.)
d. $(\,)(\mathfrak{S}_i \equiv \mathfrak{S}_j) \supset (\mathfrak{S}_i' \equiv \mathfrak{S}_j')$ is L-true. (By (c).)
e. $(\,)(\mathfrak{S}_i \equiv \mathfrak{S}_j) . \mathfrak{S}_i' \supset \mathfrak{S}_j'$ is L-true. (By (d) and T8-61(1).)
f. $(\,)(\mathfrak{S}_i \equiv \mathfrak{S}_j)$ and $\mathfrak{S}_i'$ together L-imply $\mathfrak{S}_j'$. (By (e).)
+g. If $\mathfrak{S}_i$ and $\mathfrak{S}_j$ are L-equivalent, then $\mathfrak{S}_i'$ and $\mathfrak{S}_j'$ are also L-equivalent. (Proved below.)

A comment, before proving T3a and T3g above. From T3g we see that L-equivalent sentential formulas are L-interchangeable not only in molecular, but in general formulas as well; and further, that here it is a matter of indifference whether the variables occurring free in $\mathfrak{S}_i$ are bound or free in $\mathfrak{S}_i'$.

Proof of T3a: In view of T2, (a) follows by application of appropriate parts of T1 — first, to the smallest component formula of $\mathfrak{S}_i'$ in which $\mathfrak{S}_i$ occurs in the place in question as an operand or as a truth-functional component; and then, step by step, to more inclusive component formulas until finally $\mathfrak{S}_i'$ itself is reached.

Proof of T3g: If $\mathfrak{S}_i$ and $\mathfrak{S}_j$ are L-equivalent, then $\mathfrak{S}_i \equiv \mathfrak{S}_j$ is L-true and so also (by T13-1e) is $(\,)(\mathfrak{S}_i \equiv \mathfrak{S}_j)$. Thus, by (c), $\mathfrak{S}_i' \equiv \mathfrak{S}_j'$ is L-true, whence we see that $\mathfrak{S}_i'$ and $\mathfrak{S}_j'$ are L-equivalent.

Examples. 1. If we know '$(x)(Rxa \equiv Sbx)$' to begin with, then in '$(\exists x)(Px \vee Rxa)$' we may interchange '*Rxa*' with '*Sbx*'; the result of this interchange is '$(\exists x)(Px \vee Sbx)$'. [I.e. by T3f, this last sentence is L-implied by the first two.] — **2.** According to T8-6i(1), '$Px \supset Rxy$' and '$\sim Rxy \supset \sim Px$' are L-equivalent. If, now, it happens that the factual sentence '$(x)(\exists y)[(Px \supset Rxy) \vee Qy]$' is given, then by T3g this factual sentence can be transformed into the L-equivalent one '$(x)(\exists y)[(\sim Rxy \supset \sim Px) \vee Qy]$'.

15b. Definitions. To *define* a new sign on the basis of previous signs is to introduce this new sign in such a way that its meaning is specified in terms of the older signs. A definition must enable us to eliminate the new sign for any given sentence containing it, i.e. to transform the given sentence into an L-equivalent one that no longer contains the new sign. (This transformation must be possible at least for sentences of certain simple forms, though not necessarily for all sentences in general.)

It is often the case that a new sign is taken to be synonymous with an expression composed exclusively of previous signs, e.g. the new sign 'A' might be introduced as an abbreviation for the sentence '$Pa \vee (x)Qx$'. Such cases are not the only ones, however. Suppose we want to introduce the designation 'Q' for the property affirmed of the individual a by the sentence '$Pa \vee Rab$'. Here there is no expression composed of old signs which is synonymous with 'Q'. What we need in this case is e.g. a convention which, formulated in words, runs as follows: "The sentence 'Qa' is an abbreviation for '$Pa \vee Rab$', and similarly for other full sentences obtainable from 'Q'." But T3g enables us to state this convention simply and directly. Let us take the sentential formula '$Qx \equiv Px \vee Rxb$' as a *definition*. In so doing, we impart to the predicate 'Q' such a meaning that the definitional formula (and thus every substitution instance of it) is true—true, moreover, not on factual but on logical grounds, i.e. strictly on the basis of meaning. Naturally, therefore, we want to extend our use of L-terminology so that the definitional formula and all its substitution instances count as L-true. Suppose that this is done. Then the biconditional (the definition) is taken as L-true; and in consequence the two components of the biconditional are L-equivalent, and hence L-interchangeable, and the same holds for any substitution instance of the biconditional. Thus e.g. 'Qa' can always be transformed into '$Pa \vee Rab$', and conversely (not only if one of these sentences occurs independently, but also if it occurs as a component part of a larger sentence); and further, in any context 'Qx' can be substituted for '$Px \vee Rxb$' (or conversely), no matter whether 'x' is bound or free in that context.

The mode of definition suggested in the previous paragraph applies equally well to the definition of a many-place predicate. E.g. a definition of the two-place predicate 'R' can have the form '$Rxy \equiv ..x..y..$', where the right component of this biconditional is a sentential formula in which at most the variables 'x' and 'y' occur free.

Every definitional formula has two components, one containing the new sign and the other not. The component containing the new sign is called the *definiendum* (e.g. 'Qx' and 'Rxy' above are definienda); we shall follow the practice of writing the definiendum as the first, or left, component of the definition. The other component of a definition contains only earlier signs; it is called the *definiens*. All variables that occur free in the definiens must likewise occur free in the definiendum, and indeed each such variable must have precisely one occurrence in the definiendum. (More exact characterizations are given in **21e**.) The definition of a sentential constant, say 'A', has the simple form '$A \equiv ...$', where the definiens '...' must be closed. (In introducing an abbreviation for an open sentential formula, we can use as definiendum not a sentential constant but only a new predicate with appropriate arguments, e.g. 'Qx' for '$Rax \vee Px$'.)

15c. Examples.

I. Domain of individuals: human beings. Primitive signs already at hand: '*Par*' ("parent") and '*Ml*' ("male"). Definitions:

1. ("Human being") $Hu(x) \equiv (\exists y)(Par(x,y) \vee Par(y,x))$.
2. ("Female") $Fl(x) \equiv Hu(x) . \sim Ml(x)$.
3. ("Father") $Fa(x,y) \equiv Par(x,y) . Ml(x)$.
4. ("Child") $Ch(x,y) \equiv Par(y,x)$.
5. ("Son") $Son(x,y) \equiv Ch(x,y) . Ml(x)$.
6. ("Grandparent") $GrPar(x,y) \equiv (\exists z)(Par(x,z) . Par(z,y))$.

Other concepts, e.g. "Brother", will be defined later (**17b**). It should be remarked that some of these definitions can be formulated in an essentially simpler way in language C (cf. **30c.**).

II. Domain of individuals: natural numbers. Suppose the predicates '*E*' (two-place) and '*Prod*' (three-place) are already at hand, i.e. are either primitive signs or previously defined signs (let '*E(a,b)*' mean "*a* is equal to *b*", and '*Prod(a,b,c)*' mean "*a* is the product of *b* and *c*"). How, then, can we introduce by definition the (two-place) predicate '*Div*' and the (one-place) predicate '*Prim*', where '*Div(a,b)*' stands for "*a* is divisible by *b*" and '*Prim(x)*' means "*x* is a prime number"? These definitions may be phrased as follows:

7. $Div(x,y) \equiv (\exists z) Prod(x,y,z)$.
8. $Prim(x) \equiv (y)[Div(x,y) \supset E(y,1) \vee E(y,x)]$.

Exercises. I. Continuing the list of definitions given in I above, define the following predicates: 1. '*Mo*' ("Mother"). — 2. '*Dau*' ("Daughter"). — 3. '*GrFa*' ("Grandfather"). — 4. '*GrMo*' ("Grandmother"). — 5. '*GrCh*' ("Grandchild"). — 6. '*GrSon*' ("Grandson"). — 7. '*GrDau*' ("Grand-daughter"). In defining the following predicates, use '*Hus*' ("Husband") as a third primitive sign. — 8. '*Wif*' ("Wife"). — 9. '*FaL*' ("Father-in-law"). — 10. '*MoL*' ("Mother-in-law"). — 11. '*SonL*' ("Son-in-law"). — 12. '*DauL*' ("Daughter-in-law"). II. Domain of individuals: natural numbers. In exercises 13 and 14 below, use the predicates '*E*' and '*Prod*' (as in Example II above), the predicate '*Sum*' (where '*Sum(a,b,c)*' is read "$a=b+c$"), and the individual constants '1' and '2' in their usual sense. — 13. Define the following predicates (cf. **2c**(3)):

a) '*Even(x)*' (x is even).
b) '*S(xy)*' (x is the immediate successor of y).
c) '*Gr(xy)*' (x is greater than y).
d) '*Sm(xy)*' (x is smaller than y).
e) '*Sq(xy)*' (x is the square of y).
f) '*Dif(xyz)*' ($x=y-z$).
g) '*Pred(xy)*' (x is the immediate predecessor of y).

— 14. Formulate the following, using the signs indicated above under exercise II (but not those defined in (13)):

a) $x+y=y+x$.
b) $x \cdot (y \cdot z) = (x \cdot y) \cdot z$.
c) The square of a prime number greater than 2 is not even.
d) If y is the successor of x, then the difference between y^2 and x^2 is $x+y$.

16. PREDICATES OF HIGHER LEVELS

16a. Predicates and predicate variables of different levels. Suppose a certain theory, formulated in our symbolic language, asserts a complicated sentence $\mathfrak{S}_1$ having one or more occurrences of the predicate 'P_1'; let '$..P_1..P_1..$' represent the sentence $\mathfrak{S}_1$. Suppose, further, that this theory asserts similar sentences $\mathfrak{S}_2$ and $\mathfrak{S}_3$ phrased respectively in terms of predicates 'P_2' and 'P_3'; i.e. $\mathfrak{S}_2$: '$..P_2..P_2..$' results from $\mathfrak{S}_1$ by writing 'P_2' in place of 'P_1', and likewise for $\mathfrak{S}_3$: '$..P_3..P_3..$'. And suppose, finally, that regarding other properties P_4 and P_5 our theory asserts in sentences $\mathfrak{S}_4$ and $\mathfrak{S}_5$ the opposite of what $\mathfrak{S}_1$ asserts for P_1; thus $\mathfrak{S}_4$: '$\sim(..P_4..P_4..)$' and $\mathfrak{S}_5$: '$\sim(..P_5..P_5..)$'. [The dots stand for the other symbols in the sentence; according to our presupposition, these symbols are the same in each of $\mathfrak{S}_1$, $\mathfrak{S}_2$, $\mathfrak{S}_3$, $\mathfrak{S}_4$, and $\mathfrak{S}_5$.] Now, it is useful to avoid writing out these long sentences in full each time. To this end, therefore, we naturally introduce abbreviations. E.g. we could introduce '$M_1(P_1)$' as an abbreviation for $\mathfrak{S}_1$. Here 'P_1' appears as an argument-expression, and 'M_1' as a sign of a new sort—a predicate differing from the predicates used heretofore in that its argument-expression is not an individual sign, but again a predicate. Following out the parallels between $\mathfrak{S}_1$ and $\mathfrak{S}_2$, $\mathfrak{S}_3$, $\mathfrak{S}_4$, $\mathfrak{S}_5$, we would now use similar abbreviations for these last four sentences, viz. '$M_1(P_2)$', '$M_1(P_3)$', '$\sim M_1(P_4)$', '$\sim M_1(P_5)$'.

Predicates whose argument-expressions are individual signs (and this is the case for all predicates considered to date) are called *predicates of the first level* (or order). A predicate whose argument-expression is a predicate of the first level (as in the case e.g. with the predicate 'M_1' introduced just above) is called a *predicate of the second level.* When, in turn, predicates of the second order are taken as argument-expressions, we arrive at predicates of the *third level.* Individual signs are said to be of *zero level* in this context. Also, we wish to admit many-place predicates of various levels, i.e. sentences of the form '$M_2(P,Q)$', '$M_3(P,Q,R)$', etc. The argument-expressions in the different places of such a predicate do not themselves all need to be of the same level. E.g. we can legitimately abbreviate a sentence '$..a..P..$' by, say, '$M_4(a,P)$', in which argument-expressions at the first place of 'M_4' are of zero level, while those at the second place are of the first level. If level n is the highest of the levels of the argument-expressions for a predicate, then the predicate itself is said to be a *predicate of the* $(n+1)$th *level.* E.g. the predicate 'M_4' just mentioned is of the second level.

Earlier we used individual variables along with individual constants to make possible the assertion of universality or existence respecting the objects of some domain. Here, we wish to make similar use of *predicate variables* (of any desired level) along with predicate constants. We shall admit such predicate variables not only as free variables, but also as variables in universal and existential quantifiers. (To date we have used predicate variables of the first level only, and used them simply as free variables. Cf. **10**.) In so doing, we make it possible to assert universality or existence respecting some domain of attributes (properties or relations).

As predicate variables of the first level we shall continue to use 'F', 'G', 'H', 'K'. Now, given a sentence $\mathfrak{S}_1$: '$..P..P..$' containing a first-level predicate P, we can state that what $\mathfrak{S}_1$ says about property P does in fact hold for every property (of individuals comprising the domain in question) by writing: '$(F)(..F..F..)$' (read: "For every F, $..F..F..$"). We can also state that what $\mathfrak{S}_1$ says about property P does in fact hold for at least one property of individuals (leaving open the question whether P is that property) by writing: '$(\exists F)(..F..F..)$' (read: "For at least one F, ..."; or "For some F, ..."; or "There is an F such that ...").

These remarks about first-level predicates apply without change to higher-level predicates. Further, what was said in **10b** regarding the intensions and extensions of (first-level) predicates may by analogy be carried over to predicates of higher levels: the intensions of higher-level predicates are attributes (properties or relations) of higher levels, and their extensions are classes of higher levels. And as in **10** and **11**, so here the only values we need to consider in making value-assignments to predicates of higher levels are the extensional values, i.e. the classes of higher levels. The definitions of L-concepts may be brought up unchanged from **5** and **6**, the notion of value assignment now being understood to include the assignment of values to higher-level descriptive predicates and predicate variables. (In our subsequent application of these L-concepts, however, we shall usually find it simpler to forego the technical method of value-assignments. Thus, in showing a certain formula to be L-true we shall ordinarily be content to make intuitively clear that this formula holds "in all possible cases".)

16b. Raising levels. Consider any L-true sentential formula containing as individual signs and predicate signs only variables, not constants; e.g. $\mathfrak{S}_1$: '$(x)(Fx) \supset Fy$'. Write down a corresponding sentence $\mathfrak{S}_2$ with first-level predicate variables where $\mathfrak{S}_1$ had individual variables and second-level predicate variables where $\mathfrak{S}_1$ had first-level predicate variables; e.g. $\mathfrak{S}_2$: '$(F)(N(F)) \supset N(G)$', where 'N' is a predicate variable of the second level. Now $\mathfrak{S}_2$ is evidently L-true also. For if every first-level property has the second-level property N, then certainly P has property N; hence '$(F)(N(F)) \supset N(P)$' is L-true. The same claim can be made for any other first-level property instead of P. Thus $\mathfrak{S}_2$ is also L-true. Similar considerations and results would obtain had we employed in the same way predicate variables of consecutive, but still higher, levels. Further, for every other sentential formula previously specified as L-true and containing no descriptive constants, it can be shown that the corresponding formula appropriately phrased with higher-level variables is likewise L-true. Thus we have the following theorem (the technical proof of this theorem appears to be unduly complicated, and so will not be given here):

T16-1. Suppose $\mathfrak{S}_i$ is any one of the sentential formulas specified as L-true in T14-1 or T14-2. Suppose the sentential formula $\mathfrak{S}_j$ is obtained from $\mathfrak{S}_i$ by replacing the individual variables of $\mathfrak{S}_i$ with nth-level predicate variables and the (first-level) predicate variables of $\mathfrak{S}_i$ with $(n+1)$th-level predicate variables. Then $\mathfrak{S}_j$ is also L-true.

Substitutions for higher-level predicate variables—both simple substitutions and formula-substitutions—are accomplished in exactly the same fashion as they are for first-level predicate variables. Theorems T12-1 and T12-2 hold here by analogy, as do the theorems in **13** and **14** for quantifiers with predicate variables of arbitrary levels.

Note. Our T1 above validates raising levels only in certain L-true sentential formulas. This practice is also valid for every other L-true sentential formula considered to date, provided the formula has variables—not constants—for its sentential signs, its individual signs, and its predicate signs. However, the practice is not generally applicable to arbitrary L-true sentential formulas of this sort, but only to those formulas that are L-true respecting any (non-empty) domain of individuals, regardless of the number of individuals therein. The technique of raising levels cannot generally be used in connection with sentential formulas whose validity depends on the number of individuals in the domain (cf. the different forms possible for P12 in **22a**, **b** and **37c**, and other sentences related to such a primitive sentence).

16c. Examples. Domain of natural numbers.

The following two assertions hold for natural numbers:

(1) $(x)(y)(z)(Sm(x,y) . Sm(y,z) \supset Sm(x,z))$.

(2) $(x)(y)(z)(Gr(x,y) . Gr(y,z) \supset Gr(x,z))$.

Since sentences of this form occur frequently, it is worth while to introduce an abbreviation for them. Relations which satisfy the condition expressed in (1) and (2) are said to be *transitive* relations. Thus (1) says that *Sm* is transitive; and (2), that *Gr* is transitive. Being a property of relations, not individuals, transitivity is to be expressed in our symbolic language by a second-level predicate, say '*Trans*'. We introduce this predicate by the following definition:

(3) $Trans(H) \equiv (x)(y)(z)(H(x,y) . H(y,z) \supset H(x,z))$.

Substituting for the free (first-level predicate) variable '*H*' the constant '*Sm*' e.g., we obtain

(4) $Trans(Sm) \equiv (x)(y)(z)(Sm(x,y) . Sm(y,z) \supset Sm(x,z))$.

Now, in view of (4) and the interchangeability theorem T15-3, we can always replace the original sentence (1) with the abbreviation '*Trans*(*Sm*)', even if (1) occurs as a component part of another sentence; and conversely, any occurrence of this abbreviation can be replaced by sentence (1). Similar remarks apply to (2) and its abbreviation '*Trans*(*Gr*)'. (Later, in **31c**, simplified definitions will be given for '*Trans*' and for the predicates '*Sym*', '*Refl*' and '*Reflex*' explained in the exercises just below.)

Exercises. 1. By analogy with '*Trans*', define the second-level predicate '*Sym*', where '*Sym*(*R*)' means "(The relation) *R* is symmetric". We say that *R* is symmetric just in case: for any (individuals), if one bears the relation *R* to a second, then the second also bears the relation *R* to the first. The constant '*R*' should not appear in the definition, but rather some corresponding predicate variable, e.g. '*H*'. — **2.** Define the second-level predicate '*Refl*', where '*Refl*(*R*)' means "the relation *R* is reflexive". We say that *R* is reflexive just in case: for any individual, if it bears the relation *R* to some individual or if some individual bears the relation *R* to it, then it bears the relation *R* to itself. — **3.** Define the second-level predicate '*Reflex*(*R*)', where '*Reflex*(*R*)' means "the relation *R* is totally reflexive". We say that *R* is totally reflexive if every individual (in the domain) bears the relation *R* to itself. — **4.** Define the predicate '*NSm*', where '*NSm*(*R*,*a*)' means "the relation *R* is not symmetric with respect to the individual *a*". We shall say that *R* is not symmetric with respect to the individual *a* if either *a* bears the relation *R* to some individual which does not bear the relation *R* to *a*, or some individual bears the relation *R* to *a* and *a* does not bear the relation *R* to this individual. What is the level of '*NSm*'?

17. IDENTITY. CARDINAL NUMBERS

17a. Identity. The sentence '$a=b$' is taken to mean that *a* and *b* are identical, i.e. *a* is the same individual as *b*. The sign '=' is called the *identity sign*. In our present symbolic language A we shall use the identity sign only between individual expressions. (Regarding other uses of '=', cf. **29a**.) Clearly, all substitution instances of '$x=x$', e.g. '$a=a$', hold—and in fact are L-true, by R11-1(i). Consequently, '$x=x$' is also L-true, and so likewise is '$(x)(x=x)$'. The sign '≠' is used for "not-identical".

When a sentential formula involving '=' or '≠' occurs as part of a larger context, the parentheses enclosing this formula may be omitted (see **3c**, Rule (1)). If *a* is the same individual as *b*, everything that can correctly be said about *a* must also hold for *b*; i.e. '$a=b$' ($\mathfrak{S}_1$) L-implies '$(F)(Fa \supset Fb)$' ($\mathfrak{S}_2$). Sentence $\mathfrak{S}_2$ says in effect that whatever property *a* has, *b* has also. It is an important fact that $\mathfrak{S}_2$ also L-implies $\mathfrak{S}_1$. For one of the properties *a* has is that of being identical with *a*; and hence, by $\mathfrak{S}_2$, *b* must have this property, too. In technical terms, the derivation of $\mathfrak{S}_1$ from $\mathfrak{S}_2$ is as follows: By analogy with T13-1d, sentence $\mathfrak{S}_2$ L-implies every substitution-instance of '$Fa \supset Fb$' obtained by substituting for the free variable '*F*'. Following the procedures of formula-substitution, in **12c**, let us substitute '$a=x$' for 'Fx', viz. let us replace 'Fa' by '$a=a$' and 'Fb' by '$a=b$'. There results the sentence '$a=a \supset a=b$'. Since '$a=a$' is L-true, '$a=b$' follows. Thus $\mathfrak{S}_2$ L-implies $\mathfrak{S}_1$. We conclude from all these observations that $\mathfrak{S}_1$ and $\mathfrak{S}_2$ are L-equivalent.

Because of the L-equivalence between $\mathfrak{S}_1$ and $\mathfrak{S}_2$, we can define the identity sign in the following way:

D17-1. **a.** $(x=y) \equiv (F)(Fx \supset Fy)$.
b. $(x \neq y) \equiv \sim(x=y)$.

The first theorem below expresses the familiar fact that identity is totally reflexive, symmetric and transitive. The second theorem tells us that, given a sentence expressing an identity, one member of this identity can be replaced at any of its occurrences in any sentence by the other member of the identity (in view of the symmetry of identity (T1b) this remark applies indifferently to either the first or the second member of the identity, as the phrasing indicates).

T17-1. The following sentential formulas are L-true:

a. $x=x$.
b. $(x=y) \supset (y=x)$.
c. $(x=y).(y=z) \supset (x=z)$.

T17-2. Suppose '...*a*...' is a sentence containing (one or more occurrences of) '*a*'. Suppose '...*b*...' is a sentence obtained from '...*a*...' by replacing '*a*' by '*b*' at one or more (but not necessarily all) occurrences of '*a*'. Then '...*b*...' is L-implied by '$a=b$' and '...*a*...'.

17b. Examples. Many concepts that naturally fall within the system of family relations specified in **15c** (II) can only be defined there with the help of such auxiliary devices as the identity sign or the use of predicate variables in quantifiers. Consider e.g. the relation Brother, where "a is a brother of b" is written '*Bro*(*a*,*b*)'. It might be thought that '*Bro*(*a*,*b*)' could be explained within the system of **15c** (I) simply by saying: '*Bro*(*a*,*b*)' means the same as "*a* is a son of *b*'s father, and *a* is a son of *b*'s mother". However, this explanation is inadequate, for it is also the case that *a* is a son of *a*'s father and *a* is a son of *a*'s mother—and we do not wish to count *a* as a brother of himself. The definition of '*Bro*' must therefore be so formulated as to exclude this possibility of identity. A definition which does so is the following:

$$Bro(x,y) \equiv (\exists u)(Son(x,u) . Fa(u,y)) . (\exists v)(Son(x,v) . Mo(v,y)) . x \neq y.$$

(A simpler definition of '*Bro*' is put forward in language C, **30c**.)

Exercises. Continuing in the fashion of the previous paragraph, define: **1.** "Sister". — **2.** "Sibling" (without using "Sister"). — **3.** "Cousin". — **4.** Recalling **15c** (I), translate the sentence "2 is the only even prime" into various symbolic forms, viz. the symbolic counterparts of: (a) "2 is an even prime, and every other (number) is not ..." ("other" or "distinct" is symbolized by '$\neq$'); (b) "2 is ..., and there is no other ..."; (c) "if x is identical with 2, x is an even prime; and conversely (i.e. if x is an even prime, then ...)"; (d) the biconditional that results from combining (according to T8-6f(1)) the two conditionals of (c). — Translate: **5.** "Every (natural number) has at most one predecessor" ('*Pred*'); i.e. "If x is a predecessor of z and y is a predecessor of z, then x and y are the same (number)". — **6.** "Every (natural number) precedes one and only one (number)", i.e. "... is a predecessor of at least one ..., and ... of at most one ..." (the second part here being analogous to (5)). — **7.** "For (two) distinct (numbers) x and y it is the case that either x is less than y or y is less than x". (Hint: in many situations of this sort, "two" can be expressed by "not-identical".)

17c. Cardinal numbers. First, with a view to simplifying the verbal explanations that follow, let us introduce several new turns of phrase into the English word-language. (Note that these new phrases are introduced into the word-language, not into the symbolic language.) Instead of "*a* has the property *P*", we shall sometimes say "*a* is a *P*-individual", or briefly "*a* is a *P*"; or again "*a* is an element of the class of those individuals having property *P*", or briefly "*a* is an element of class *P*". Instead of "there are exactly five individuals with property *P*" or "there are exactly 5 *P*-individuals", we shall also say "the property *P* (or: the class *P*) has the *cardinal number* 5", or briefly "*P* has cardinal number 5".

Our ultimate purpose in this section is to explicate the cardinal numbers 0, 1, 2, etc., i.e. to establish precise definitions that comprehend the usual meanings of these number-signs or numerals '0', '1', '2', etc. But, by the remarks just above, having e.g. the cardinal number 5 is a property of certain properties (or classes); hence, this property of having cardinal number 5 is to be symbolized by a predicate of the second level. Let us simply choose the numeral '5' as this predicate. Thus for "*P* has cardinal number 5" we may write '5(*P*)', a formulation clearly indicating that '5' is a second-level predicate with '*P*' as its argument-expression. By analogy, we write '0(*P*)' for "*P* has cardinal number 0" (i.e. "there are no *P*-individuals"); '1(*P*)' for "*P* has cardinal number 1" (i.e. "there is exactly one *P*-individual"); etc.

The precise definitions of predicates '0', '1', '2', etc., appear in D3 below. To simplify the formulation of these definitions, it is convenient to introduce first (in D2) certain auxiliary predicates '1_m', '2_m', etc., which will seldom be used hereafter. By '$1_m(P)$' we mean "there is at least one *P*-individual"; by '$2_m(P)$' we mean "there are at least two *P*-individuals"; etc. This last sentence is not to be construed as meaning simply "there are individuals *x* and *y* such that *x* is *P* and *y* is *P*", which would be true even if there were but one individual (say *a*) having property *P*, since '*a*' could be put in place of both '*x*' and '*y*'; if, therefore, we explain '$2_m(P)$' as "there are individuals *x* and *y* such that *x* is *P* and *y* is *P*", we must add "and *x* is not identical with *y*". This is the reason for the last component of the operand in D2b. — Finally, as a general basis for D3, we agree that "there are *n* *P*-individuals" means the same as "there are at least *n* *P*-individuals, and there are not at least $n+1$ *P*-individuals".

D17-2. **a.** $1_m(F) \equiv (\exists x)Fx$.
b. $2_m(F) \equiv (\exists x)(\exists y)(Fx.Fy.x \neq y)$.
c. $3_m(F) \equiv (\exists x)(\exists y)(\exists z)(Fx.Fy.Fz.x \neq y.x \neq z.y \neq z)$.
Definitions for '4_m', '5_m', etc., are made analogously.

D17-3. **a.** $0(F) \equiv \sim 1_m(F)$.
b. $1(F) \equiv 1_m(F). \sim 2_m(F)$.
c. $2(F) \equiv 2_m(F). \sim 3_m(F)$.
Definitions for '3', '4', etc., are made analogously.

Exercises. Define '*P*' to be such that '*Pb*' means "*b* is a child of *a*" (in this connection, use the predicate '*Par*').—Now, with the help of '*P*', translate the following sentences into our symbolic language: **8.** (a) "*a* has at least 3 children"; (b) "... at most 3 ..." (i.e. "... not at least 4 ..."); (c) "... exactly 3 ...". — **9.** The exercise 8(b) suggests that "at most 2" may be defined by "not at least 3". Define '2_M', where '$2_M(P)$' means "there are at most two *P*-individuals". This last is now to be construed as meaning that if individuals *x*,*y*,*z* have the property *P*, then *x* and *y*, or *x* and *z*, or *y* and *z* must be the same individual. — **10.** Show that '$2_M(P)$' is L-equivalent to '$\sim 3_m(P)$'. Use the theorems in **8** and **14**. — **11.** Show that the following formulas are L-true, using the theorems in **8** and **14**: a) '$3_m(F) \supset 1_m(F)$'; b) '$2(F) \supset 1_m(F)$'; c) '$\sim 1(F) \supset (1_m(F) \supset 2_m(F))$'.

18. FUNCTORS

18a. Functors. Domains of a relation. We begin with an example. Take for the domain of individuals the natural numbers (in so doing, we construe the number signs '1', '2', etc., as individual constants, and not as second-level predicates as in **17c**). Let '*prod*' be such a symbol that '*prod*(a,b)' means "the product of the numbers a and b". The 'a' and 'b' in '*prod*(a,b)' are referred to as the argument-expressions of '*prod*'. Previously we spoke of '*Pa*' as a full sentence of '*P*'; extending this terminology, let us speak here of '*prod*(a,b)' as a *full expression* of '*prod*'. Note that '*prod*' is distinguished from predicates by the fact that a full expression of '*prod*' is not a sentence but a designation for a number, i.e. a zero-level expression in the present context. In this respect, the sign '*prod*' is an instance of a certain kind of sign for which we have a general name: we speak of any sign whose full expressions (involving n arguments) are not sentences as an n-place *functor*.

The full expressions of a functor may (as in the case of '*prod*' above) be expressions of the zero-level, i.e. *individual expressions*—designations for individuals of the domain in question. However, there are also functors whose full expressions are designations of attributes and hence are called *predicate expressions* (of the first or higher level). Functors of this sort appear in the discussion below.

The notions to which we now turn are best introduced by another example. Recall the (two-place) relation Brother. If now a is a brother of b, we say that a is a first-place member of the relation Brother and that b is a second-place member. More generally, any person who bears the relation Brother to someone is a first-place member of the relation, and any person to whom someone bears the relation Brother is a second-place member of the relation. These notions readily extend to any two-place relation R: whatever individual bears the relation R to something is called a *first-place member of* R, and any individual to which something bears the relation R is called a *second-place member of* R.

Now consider an arbitrary two-place relation R. We call the class of all first-place members of R the *first domain of* R and symbolize it (or the corresponding property of being a first-place member of R) by '$mem_1(R)$'. The sentence "a is a first-place member of R" is rendered '$mem_1(R)(a)$'. Notice from the sentence '$mem_1(R)(a)$' that '$mem_1(R)$' is a predicate expression—indeed, a one-place predicate expression of the first level, since it goes over into a sentence when filled by the argument-expression 'a', i.e. by an individual constant. The sign 'mem_1' itself is a functor, since its full expression '$mem_1(R)$' is not a sentence (but a predicate expression).

In analogy with the above, we call the class of all second-place members of R the *second domain of* R and symbolize it (or the corresponding property of being a second-place member of R) by '$mem_2(R)$'. The sentence "a is a second-place member of R" is written '$mem_2(R)(a)$'. As before, the sign 'mem_2' is a functor.

By a *member* of R we mean any individual which is either a first-place member of R or a second-place member of R, or both. The class of all members of R is called the *field of* R, and is designated by '$mem(R)$'. A first-place member of R which is not also a second-place member of R we call an *initial member* of R; and again, a second-place member of R which is not also a first-place member of R we call a *terminal member* (*or final member*) of R. E.g. the relation Predecessor in the domain of natural numbers has for its field the class of natural numbers, has 0 for its (sole) initial member, and has no terminal member.

Now let us introduce the signs 'mem_1', 'mem_2' and 'mem' into our symbolic language by definitions. We shall do so by way of the sentence forms '$mem_1(R)(a)$', '$mem_2(R)(a)$' previously discussed; naturally, however, we must employ variables (say, 'H' and 'x') in place of the constants 'R' and 'a'.

D18-1. $mem_1(H)(x) \equiv (\exists y)Hxy.$

D18-2. $mem_2(H)(x) \equiv (\exists y)Hyx.$

D18-3. $mem(H)(x) \equiv mem_1(H)(x) \vee mem_2(H)(x).$

In the case of an n-place ($n>2$) relation T, we speak of the first domain of T, the second domain of T, ..., the nth domain of T; the union of these n domains is the field of T. It is useful to note that if 'P' is a one-place predicate (i.e. if $n=1$), then '$mem(P)$' and 'P' have the same meaning.

Exercises. Using the functors indicated, translate the following sentences into our symbolic language (in 3–5 and 8, employ the predicate 'Sq'). — **1.** "*a* is a father", i.e. "*a* is a first-place member of ...". — **2.** "Mothers are female". — **3.** "9 is a square (number)". — **4.** "Not every (number) is a square (number)". — **5.** "Every (number) is a square-root" (i.e. "... a second-place member of ..."). — **6.** "Every (number) is a member of the relation Predecessor" (use '*Pred*'). — **7.** If one (number) precedes another, then the product of the two is even". — **8.** "The product of 2 and 18 is a square (number)". — **9.** Translate and give proofs for the following sentences, where R is a two-place relation: a) "If *a* is a member of the first domain of R, then there must be something in the second domain of R"; b) "If there is exactly one member of the second domain of R and there is exactly one member of the first domain of R, then there are at most two members of the field of R"; c) "If *a* is a member of the first domain of R and there are no initial members of R, then *a* is a member of the second domain of R".

18b. Conditions permitting the introduction of functors. Let us admit into our symbolic language the practice of using functors themselves—as well as individual signs and predicates—as argument-expressions of other functors or of predicates. Let us also admit into our symbolic language *functor variables* (e.g. 'f', 'g', etc.), and agree to use them either as free variables or as bound variables (cf. the end of **9a**). Functor variables do not figure prominently in elementary matters; however, functor variables do appear e.g. in the theory of real numbers (a real number can be represented by a functor in the domain of natural numbers; cf. **40d**), while functor variables of higher levels appear in the mathematical theory of functions and in the (symbolic) formulation of certain quite general physical principles (see e.g. **41** and **51**).

It is always possible to supplant an n-place functor by an $(n+1)$-place predicate, but the reverse is not true. Thus e.g. we have the choice of introducing into the language of arithmetic either the two-place functor '*prod*' or the three-place predicate '*Prod*'—the sentence "a is the product of b and c" being rendered '$a=prod(b,c)$' in the first case, and '$Prod(a,b,c)$' in the second. Similarly, we can choose between the one-place functor '*sq*' and the two-place predicate '*Sq*'; the sentence "a is the square of b" (i.e. the sentence "$a=b^2$") is expressed by '$a=sq(b)$' in the first case, and by '$Sq(a,b)$' in the second.

It is possible to supplant an $(n+1)$-place predicate by an n-place functor only when this predicate, say T, satisfies the following conditions: For each sequence $(a_2,a_3,\ldots,a_{n+1})$ of n individuals there is one and only one individual, say a, such that '$T(a_1,a_2,a_3,\ldots,a_{n+1})$' is true. Separating this "one and only one" condition into its two parts, we obtain the two conditions (1), (2) below—where (1) embodies the "at least one" feature, and (2) the "at most one" feature:

(1) $(x_2)(x_3)\ldots(x_n)(x_{n+1})(\exists x_1)T(x_1,x_2,x_3,\ldots,x_{n+1})$;

(2) $(x_1)(y_1)(x_2)(x_3)\ldots(x_n)(x_{n+1})[T(x_1,x_2,\ldots,x_{n+1}).T(y_1,x_2,\ldots,x_{n+1}) \supset x_1=y_1]$.

Otherwise put, condition (1) is that of the existence of a first member; and condition (2) is that of the univalence of T in respect to its first place. (In **19**, this second property will receive the designation 'Un_1'.)

Let us examine conditions (1) and (2) by specifying them to some particular predicates. Can the (two-place) predicate *Pred* (cf. **2c**) be supplanted by a (one-place) functor? The answer is in the negative: for while the predicate '*Pred*' satisfies condition (2), it fails to satisfy condition (1) because 0 has no predecessor in the domain of natural numbers. If now, in spite of this fact, we introduce e.g. '*pred*' as the corresponding functor, we immediately encounter the meaningless expression '*pred*(0)'. Next, consider the relation converse to Predecessor, viz. the relation Successor which we designate by '*Suc*'. For each natural number there is one and only one successor; hence the (two-place) predicate '*Suc*' can be supplanted by a (one-place) functor. We could e.g. introduce '*suc*' as the functor corresponding to '*Suc*', where '$suc(a)$' means "the successor of a", i.e. "$a+1$". Again, consider a relation R which satisfies condition (1), but fails to satisfy condition (2) because, say, each of the sentences 'Rac', 'Rbc' and '$a\neq b$' is true. If, despite this fact, we were to introduce a functor 'k' as surrogate for 'R', then '$k(c)$' would designate indifferently either a or b and so be ambiguous. Such an ambiguity leads to a contradiction: for in place of 'Rac' and 'Rbc' we could write '$a=k(c)$' and '$b=k(c)$' respectively, and hence (by T17-1b,c) infer the sentence '$a=b$' in contradiction to our presupposition '$a\neq b$'.

The considerations above make it evident that to introduce a functor into a language system is a serious step requiring preliminary validation, i.e. requiring a preliminary check to see that conditions (1) and (2) are both satisfied. *If* these two conditions are met, it will generally prove advantageous to supplant the predicate in question by its corresponding functor—especially so because a full expression of the functor can reappear as an argument expression.

Example. By the use of functors, the sentence '$(x)(y)(z)[Suc(y,x) . Prod(z,x,y) \supset Even(z)]$' can be condensed to '$(x)[Even(prod(x,suc(x)))]$'.

19. ISOMORPHISM

The concepts treated in this section are dispensable for many of the simpler applications of symbolic logic, but for many others are of capital importtance. [In the examples of such applications given in Part II, so far as they are formulated in language A, the concepts defined here occur explicitly only in **43a**, **46a**, **51a** and **53a**.]

We say that a two-place relation R is *one-many* (or *single-valued* respecting its first place, or *univalent* respecting its first place) just in case for each second-place member of R there is *exactly one* first-place member of R which bears the relation R to that second-place member. Within our symbolic language the assertion "R is one-many" is rendered '$Un_1(R)$'. Again, we say that R is *many-one* (or *single-valued*, or *univalent*, respecting its second place) provided for each first-place member of R there is *exactly one* second-place member of R to which the first-place member bears the relation R. The assertion "R is many-one" is rendered symbolically by '$Un_2(R)$'. Finally, we say that R is *one-one*, and write '$Un_{1,2}(R)$', whenever R is both one-many and many-one. The formal statement of these definitions follows.

D19-1. $Un_1(H) \equiv (x)(y)(u)(Hxy . Huy \supset x=u)$.
D19-2. $Un_2(H) \equiv (x)(y)(u)(Hxy . Hxu \supset y=u)$.
D19-3. $Un_{1,2}(H) \equiv Un_1(H) . Un_2(H)$.

(Analogous concepts can be defined for relations with three or more places. Thus e.g. we would take '$Un_k(T)$' to mean "the (say, n-place) relation T is univalent (or single-valued) respecting its kth place", which is to say: it is not the case that there are two n-tuples of individuals that satisfy relation T and that differ only at the kth individual.)

Examples. The relation *Fa* (Father) is one-many, and we may correctly write '$Un_1(Fa)$', because each person has exactly one father; however, *Fa* is not many-one and hence not one-one. The relation *Sq* (Square) in the domain of natural numbers is both one-many and many-one (hence one-one, and we may write '$Un_{1,2}(Sq)$') since each number has at most one square-root. Contrariwise, the relation Square in the domain of real numbers is one-many, but is not many-one because a positive number is the square of two different numbers; hence it is not one-one. The relation *Pred* in the domain of natural numbers is one-one because no number has more than one predecessor and no number is the predecessor of more than one number. Similarly, the relation Successor converse to *Pred* is one-one. And finally, in the domain of persons constituting a monogamous society, the relation *Hus* (Husband) is one-one.

Let T_1 and T_2 be three-place relations. Let the two-place relation R be such that R *maps* T_1 onto T_2, i.e. let R be such that the following four conditions are satisfied: (1) R is one-one; (2) the members of T_1 are first-place members of R; (3) the members of T_2 are second-place members of R; and (4) if any three members, say a_1, b_1, c_1, constitute a triple satisfying T_1 (i.e. are such that '$T_1a_1b_1c_1$' is true), then the members, say a_2, b_2, c_2, related to them respectively by R constitute a triple satisfying T_2; and conversely. Now when R maps T_1 onto T_2 (i.e. when R satisfies the four conditions just given), we call R a *correlator* between T_1 and T_2. The definition of this concept depends on the number of places encompassed by T_1 and T_2 (in our illustration: three). In what follows we set up a definition scheme from which can be obtained at will definitions for '$Corr_1$' (correlator for one-place attributes, i.e. for properties or for classes), for '$Corr_2$' (correlator for two-place relations), etc., simply by substituting for 'n' the numerals '1', '2', etc., as desired. In all these instances the correlator itself is a two-place relation.

D19-4. $Corr_n(K,H_1,H_2) \equiv Un_{1,2}(K) \,.\, (x)(mem(H_1)(x) \supset mem_1(K)(x)) \,.\, (x)(mem(H_2)(x) \supset mem_2(K)(x)) \,.\, (x_1)(y_1)(x_2)(y_2)\ldots(x_n)(y_n) [Kx_1y_1 \,.\, Kx_2y_2 \,.\ldots.\, Kx_ny_n \supset (H_1x_1x_2 \ldots x_n \equiv H_2y_1y_2 \ldots y_n)]$.

From D19-4 we obtain the definition of '$Corr_1$' (class correlator) as a special case by setting $n=1$. Recalling (from the end of **18a**) that a one-place predicate 'P' has the same meaning as '$mem(P)$', this definition of '$Corr_1$' comes out as follows:

D19-4$_1$. $Corr_1(K,F_1,F_2) \equiv Un_{1,2}(K) \,.\, (x)(F_1x \supset mem_1(K)(x)) \,.\, (x)(F_2x \supset mem_2(K)(x)) \,.\, (x)(y)[Kxy \supset (F_1x \equiv F_2y)]$.

If there exists a correlator between two n-place attributes T_1 and T_2 ($n=1,2,\ldots$), we say that T_1 and T_2 are (n-place) *isomorphic* to each other, or: T_1 and T_2 have the same (n-place) *structure*. Again, the definition of isomorphism depends on the number n of places; as before, so here we give a definition scheme from which particular definitions can be obtained by substituting for 'n' the numerals '1', '2', etc., as desired.

D19-5. $Is_n(H_1,H_2) \equiv (\exists K)Corr_n(K,H_1,H_2)$.

Up to now the terms "isomorphic" and "structure" have been applied mainly to attributes with two or more places, i.e. to relations. In the case of one-place attributes (properties or classes), isomorphism means the existence of a one-one correspondence between the two classes, viz. that the two classes are equinumerous; and thus the structure of a class is the same as its cardinal number (cf. **34c**).

Example 1. In a group of married couples, let P be the class of men in the group and Q the class of women. The relation Husband establishes a one-one correspondence between P and Q. Hence '$Corr_1(Hus,P,Q)$' holds. From this in turn it follows that P and Q are equinumerous, i.e. '$Is_1(P,Q)$' follows.

Example 2. We have chosen '*Pred*' to designate the relation Predecessor in the whole domain of natural numbers (the class comprising 0,1,2,3, etc.); now let '*Pred'*' be used to designate the relation Predecessor in the restricted domain of natural numbers excluding zero (the class comprising 1,2,3, etc.). The two relations *Pred* and *Pred'* are readily seen to be isomorphic, in view of the following coordination: let 0 (as a member of '*Pred*') be coordinated with 1 (as a member of '*Pred'*'), 1 (as a member of '*Pred*') be coordinated with 2 (as a member of '*Pred'*'), 2 with 3, 3 with 4, etc. Here the correlator is '*Pred*' itself, and so actually coincides with one of the two relations being correlated. We have '$Corr_2(Pred,Pred,Pred')$', and so '$Is_2(Pred,Pred')$'.

[*Note.* The symbol 'Ism_n' appearing in *Carnap-Bachmann* [Extremalaxiome] does not correspond to our 'Is_n' here, but designates the more complicated concept of n-level isomorphism; for this last concept we might perhaps use the symbol, '${}^nIsm$', which has the advantage of saving the subscript position for the place number.]

Exercises. 1. For each of the following two-place relations, decide whether it is one-many, many-one, or neither: a) Sister; b) Youngest Son; c) Identical; d) Having as Father; e) Mother; f) Grandfather. — **2.** Let D be the relation which holds between any natural number x and the natural number $2x$. Is D one-many? Is D many-one? What are the first and second domains of D? What is the field of D? — **3.** Show each of the following by informal reasoning: a) '$Is_2(R_1,R_2) \supset Is_2(R_2,R_1)$'; b) '$Is_2(R_1,R_2) . Is_2(R_2,R_3) \supset Is_2(R_1,R_3)$'; c) '$Is_2(R_1,R_1)$'. — **4.** What properties of the relation Is_2 do 3(a), 3(b), and 3(c) express? (See **16c.**)

Herewith ends our presentation of the simple symbolic language A. So far as they are formulated in this language A, the axiom systems and other illustrative applications given in Part II can now be taken up (see the explanations in **42e**).

Chapter B

The Language B

In Chapter A we developed a simple symbolic language A. In Chapter C we chall construct an extended language C containing not only all the signs of A (except sentential variables), but many additional expressions as well.

In the present chapter, B, we describe a symbolic language B and address ourselves to a number of methodological questions. In particular, we indicate by examples the methods by which syntactical and semantical systems can be constructed. We begin with a brief general elucidation of the character of such systems. Thereafter, as illustrations, we construct both a syntactical system (**21-24**) and a semantical system (**25**) for language B. Lastly, the connections between the two systems are explained (**26**).

Our language B is so chosen that all sentences of C, and therefore of A, can be translated into it. To avoid undue complication in its rules, we omit from language B many modes of expression found in A and especially in C; however, the omitted expressions are inessential and serve merely as abbreviations.

Chapter B is more abstract than our previous chapter, and by this token probably less understandable to the beginner. Furthermore, it is not absolutely necessary for an understanding of what follows, viz. construction of the extended language C (in Chapter C) and application of the symbolic logic (in Part II). Hence it is feasible to omit Chapter B on a first reading of this book.

20. SEMANTICAL AND SYNTACTICAL SYSTEMS

In the investigation of languages, either historical natural ones or artificial ones, the language which is the object of study is called the *object language*. The object languages of this book are the three languages A, B and C comprising letters and artificial symbols. The language we use in speaking *about* the object language is called the *metalanguage*. In this book, the English language, augmented by certain technical signs (including German letters), serves as a metalanguage. The rules for the object language in question—notably the syntactical and semantical rules—are formulated in the metalanguage, as are the theorems which follow from these rules.

Every situation in which a language is employed involves three principal factors: (1) the *speaker*, an organism in a determinate condition within a determinate environment; (2) the linguistic *expressions* used, these being sounds or shapes (e.g. written characters) produced by the speaker (for instance, a sentence consisting of certain words of the French language); and (3) the objects, properties, states of affairs, or the like, which the speaker intends to designate by the expressions he produces—and which we term the *designata* of the expressions (thus e.g. the color red is the designatum of the French word 'rouge'). The entire theory of an object language is called the *semiotic* of that language; this semiotic is formulated in the metalanguage. Within the semiotic of a language, three regions may be distinguished according to which of the three aforementioned factors receive attention. Thus, an investigation which refers explicitly to the speaker of the language—no matter whether other factors are drawn in or not—falls in the region of *pragmatics*. If the investigation ignores the speaker, but concentrates on the expressions of the language and their designata, then the investigation belongs to the province of *semantics*. Finally, an investigation which makes no reference either to the speaker or to the designata of the expressions, but attends strictly to the expressions and their forms (the ways expressions are constructed out of signs in determinate order), is said to be a formal or syntactical investigation and is counted as belonging to the province of (logical) *syntax*.

A pragmatical description of, say, the French language tells how this or that language usage depends on the circumstances of the speaker and his context. Certain modes of expression are used in one period but not another; or they are used when the speaker has certain feelings and images, and evoke from the hearer certain feelings and images; or they are used when the whole situation—comprising speaker, hearer, and environment—satisfies certain conditions. All this is disregarded by the semantics of the French language, which presents (in, say, the form of a dictionary) the relation between French words and compound expressions on the one hand and their designata on the other. Thus, whereas pragmatics includes consideration of historical, sociological and psychological relations within the language community where French is spoken, semantics confines itself simply to giving an interpretation of this language. The semantical description of French contains all the specifications necessary to understand this language and to use it correctly. The syntactical description of the French language, on the other hand, contains still less than the semantical: the syntactical description specifies rules by which it can be decided whether or not a given sequence of words is a sentence of the French language (without it being presupposed that the sentence is understood). Beyond this, as we shall see, syntax may include rules which determine certain logical relations between sentence, e.g. the relation of derivability.

A natural language is given by historical fact, hence its description is based on empirical investigation. In contrast, an artificial language is given by the construction of a system of rules for it. The rules for an object language, as well as theorems based on these rules, are formulated in the metalanguage. A *syntactical system* for an object language L is a theory about L based on syntactical rules for L; and a *semantical system* for L is a theory about L based on semantical rules for L. A language for which syntactical rules are given is sometimes called a *calculus*; it is called an *interpreted* calculus if, in addition thereto, semantical rules are given for it, otherwise an *uninterpreted* (or *formal*) calculus. A language for which semantical rules are given (with or without syntactical rules) is sometimes called an *interpreted language*. In subsequent sections we give examples for both kinds of systems for the object language B. First we construct a syntactical system for B by stating syntactical rules for B. Then semantical rules for B will be given; these constitute the basis of a semantical system for B.

21. RULES OF FORMATION FOR LANGUAGE B

21a. The language B. In sections **21** through **24** we formulate syntactical rules for language B; and in section **25**, semantical rules for B.

The language B is sufficiently comprehensive that all the sentences of language C (a language that will be explained in the next chapter) can be translated into it. Since all the sentences of language A also appear in language C, the sentences of A are likewise all translatable into B. Language B contains each sort of variable found in C, but it does not contain the sentential variables found in A (this sort of variable occurs in A only in open sentential formulas, and not in sentences). However, language B does omit most of those logical constants of A and C that serve mainly to make formulations more concise and do not contribute in an essential way to the scope of these languages. We omit these signs from B so that we can give simpler versions of the syntactical and semantical rules for B.

Language B contains as primitive signs the five connectives of **3**, and the sign of identity for expressions of all types. [The two connectives '$\sim$' and '$\vee$' alone would suffice, since in terms of these two the other three can be defined in accordance with T8-6g(6),j(1),f(1). Again, '$=$' can be dispensed with, in view of D17-1 and the techniques of raising levels (**16b**). However, by taking all five connectives and the identity sign as primitive we can simplify our formulation of the primitive sentences and the rules of inference for B.] Also B contains universal quantifiers with variables of all kinds that occur; the existential quantifier is then definable in B, in accordance with T14-2a(4) and the technique of raising levels. And further, B contains the λ-operator (see **33**). With the exception of this λ-operator, B contains none of the other logical constants (chiefly predicates and functors of higher levels) which were introduced into language A in **17c-19** of the preceding chapter or will appear in language C; these other constants are reducible, by definitions or other rules of transformation laid down for them, to the constants now included in B.

The rules of formation for B governing the construction of expressions of various sorts, particularly sentences, are the same for both the syntactical system and the semantical system for B. Further, these rules agree with the explanations given in Chapters A and C—explanations that are often imprecise and mostly non-formal—of the way the different signs occur in sentences of language A and language C respectively.

In the metalanguage, we use the following German letters (some of which have already been so employed) as designations for signs and expressions of the object languages A, B and C: '$\mathfrak{a}$' for arbitrary signs; '$\mathfrak{v}$' for variables; '$\mathfrak{A}$' for arbitrary expressions; and '$\mathfrak{S}$' for sentential formulas. As designations for a specified sign or a specified expression, we use the appropriate German letter with a numerical subscript. E.g. '$\mathfrak{a}_1$' might serve as a designation for '*R*', '$\mathfrak{a}_2$' for '*a*', '$\mathfrak{a}_3$' for '*c*'; in which case '$\mathfrak{a}_1(\mathfrak{a}_2,\mathfrak{a}_3)$' would designate the sentence '*R*(*a*,*c*)'. A German letter with '*i*' or '*j*' or the like as subscript is used in speaking of expressions in general. Thus e.g. we write "If $\mathfrak{v}_i$ occurs in $\mathfrak{S}_j$, then ..." for "If a certain (unspecified) variable occurs in a certain (unspecified) sentential formula, then ...". Note that '$\mathfrak{v}_i$', '$\mathfrak{S}_j$', etc., are variables of the metalanguage, and that '$\mathfrak{v}_1$', '$\mathfrak{S}_2$', etc., are corresponding constants of the metalanguage.

21b. The system of types. Each sign of language B belongs to one of the following kinds:

1. Connective signs: (a) one-place ('$\sim$'), (b) two-place ('$\vee$', '.', '$\supset$', '$\equiv$').
2. Special signs: '(', ')', ',', '=', 'λ'.
3. Sentential constants.
4. Individual signs: (a) constants; (b) variables.
5. Predicates: (a) constants; (b) variables.
6. Functors: (a) constants; (b) variables.

Signs of the sorts 4b, 5b and 6b are called variables ($\mathfrak{v}$). All other signs are constants. Signs of the sorts 4, 5 and 6 are called *signs of the type system*. From 2 we see there is *only one* kind of bracketing signs; in practice, however, we employ both round and square as well as brackets of different sizes, with the understanding that these differences have no syntactical significance and serve only to facilitate reading.

Each sign of B is either taken as a primitive sign or else introduced by a definition. As *primitive signs* of language B we take the indicated separate signs of sorts 1 and 2, and all the variables. Further, we agree that any constant of sort 3, 4, 5, or 6 can at will be taken as a primitive sign of B. We also agree that other constants of these sorts can be introduced at will by way of definitions; rules governing the form of such definitions will be stated later.

Individual expressions, predicate expressions and functor expressions are classified into levels (or orders), and then further into types, in accordance with the following rules; hence expressions of these kinds are called *expressions of the type system*.

1. Every *individual expression* is said to be of type 0.
2. A compound *n*-place *argument expression* $\mathfrak{A}_{i_1}, \mathfrak{A}_{i_2},..., \mathfrak{A}_{i_n}$ (here $n \geq 2$) with $\mathfrak{A}_{i_1}$ of type t_{i_1}, $\mathfrak{A}_{i_2}$ of type t_{i_2},..., $\mathfrak{A}_{i_n}$ of type t_{i_n}, is said to be of type $t_{i_1}, t_{i_2},..., t_{i_n}$.
3. A *predicate expression* $\mathfrak{A}_i$ which can be completed by a one- or many-place argument expression $\mathfrak{A}_j$ of type t_j is said to be of type (t_j).
4. A *function expression* $\mathfrak{A}_i$ which can be completed by an argument expression $\mathfrak{A}_j$ of type t_j and which upon such completion becomes a full expression $\mathfrak{A}_i(\mathfrak{A}_j)$ of type t_k is said to be of type $(t_j : t_k)$.
5. If the type designation of an expression $\mathfrak{A}_i$ contains at least one numeral '0' surrounded by *n* pairs of brackets and no '0' surrounded by more than *n* such pairs, then $\mathfrak{A}_i$ is said to be an expression of the *n*th level.

The application of these rules can be clarified by some examples.

Examples. By rule (1) the expressions '*a*', '*x*', '*moon*' (recall **2c**) are of type 0; hence by rule (2) the argument expressions '*b,c*' and '*x,y*' are both of type 0,0. By rule (3) the predicate expression '*Sph*' is of type (0), and '*Fa*' is of type (0,0). The argument expression '*a,Sph*' of the sentence '*M*(*a,Sph*)' is of type 0,(0); hence by rule (3) *M* is of type (0,(0)) and by rule (5) belongs to the second level, whereas both '*Sph*' and '*Fa*' belong to the first level (in agreement with our previous non-formal explanation in **16**). In view of D17-3, we see that '0', '1', etc., are predicates of type ((0)) and of the second level. Contrariwise, the predicates '*Trans*' and '*Sym*' introduced in **16c** are of type ((0,0)) because argument expressions that can complete them (e.g. '*Fa*') are of type (0,0). The expression '*prod*(*a,b*)' used in **18a** is an individual expression, hence is of type 0; its argument expression '*a,b*' is of type 0,0; hence the functor '*prod*' is by rules (4) and (5) a functor of type (0,0 : 0) and of level one. The expression '*mem*(*Fa*)' (cf. D18-3) is a predicate expression of type (0), since the argument expression '*x*' can complete it; thus, in view of the fact that '*Fa*' is of type (0,0), we see by rules (4) and (5) that the functor '*mem*' is of type ((0,0) : (0)) and of the second level.

It follows from the rules above that a given predicate expression always takes argument expressions of one and the same type. Two predicate expressions $\mathfrak{A}_i$ and $\mathfrak{A}_i'$ are of the same type if and only if (1) they have the same number of arguments, and (2) argument expressions in corresponding places are of the same type. [E.g. each predicate may be a two-place predicate, so that their full sentences appear as $\mathfrak{A}_i(\mathfrak{A}_j,\mathfrak{A}_k)$ and $\mathfrak{A}_i'(\mathfrak{A}_j',\mathfrak{A}_k')$ respectively; then $\mathfrak{A}_i$ and $\mathfrak{A}_i'$ are of the same type provided $\mathfrak{A}_j$ and $\mathfrak{A}_j'$ are of the same type and similarly for $\mathfrak{A}_k$ and $\mathfrak{A}_k'$. The separate argument expressions $\mathfrak{A}_j$ and $\mathfrak{A}_k$ may be of the same type, or of different types; in the first case, both the predicate expression and the relation it designates are called *homogeneous*, in the second case *inhomogeneous*. The predicate '*M*' appearing in the examples just above is inhomogeneous.]

As will be fully explained in **33**, λ-expressions are either predicate-expressions or functor-expressions. A λ-expression has the form $(\lambda\mathfrak{A}_i)(\mathfrak{A}_j)$, where $\mathfrak{A}_i$ is either a variable or a sequence of n different variables separated by commas; $(\lambda\mathfrak{A}_i)$ is called a *λ-operator*, and $\mathfrak{A}_j$ its *operand*. Taking $\mathfrak{A}_i$ to be of type t_i, two cases arise: (1) $\mathfrak{A}_j$ is a sentential formula, in which case the λ-expression is a predicate expression of type (t_i); and (2) $\mathfrak{A}_j$ is an expression of type t_i, in which case the λ-expression is a functor expression of type $(t_i : t_j)$.

Exercises. **1.** Determine the type and level of each of the following expressions (cf. **2c**): a) '*a*'; b) '*Par*'; c) '$mem_1(Par)$'; d) '*x,b,x*'; e) '*Sm*'; f) '*Refl*' (cf. **16c**); g) '*mem,x,Fa*'; h) 'Un_1' (cf. **19**); i) '$Corr_2$' (cf. **19**, example 2); j) 'Is_2' (cf. **19**, example 2); k) '*suc*' (cf. **18b**).

21c. Russell's antinomy. The distinction between types was introduced by Bertrand Russell in order to avoid the so-called logical antinomies. One such antinomy e.g. is the Russell antinomy centering on the concept of those properties which do not apply to themselves. So long as no distinction is made between predicates of different levels, it will appear meaningful to say of a property F that either it applies to itself or it does not. Thus we might make some such definition as the following: a property is *impredicable* in case it does not apply to itself; symbolically, '$Impr(F) \equiv {\sim}F(F)$'. Substituting for the free variable 'F' of this definitional formula the defined predicate '*Impr*' itself, we obtain '$Impr(Impr) \equiv {\sim}Impr(Impr)$'. But this sentence, like every sentence of the form '$p \equiv {\sim}p$', is L-false. Our definition thus leads to a contradiction; this is the Russell antinomy. If, however, the distinction of types is introduced, then the expression '$F(F)$' is not an admissible sentential formula because a predicate must always be of higher level than its argument expression. I.e. the definition above cannot be set up, and the antinomy vanishes with it.

Concerning the *antinomies*, see: [P.M.] vol. I, 60 ff.; Russell [Introduction] 135 ff.; Ramsey [Foundations]; Fraenkel [Einleitung] §§ 13–15, with an account of the literature; Carnap [Syntax E] § 60a–c. On the *system of types*, see: [P.M.] vol. I, 39 ff., 168 ff.; Russell [Introduction] 131 ff.; Ramsey [Foundations]. Russell originally undertook a further subdivision of the types, which led to the so-called ramified system of types; in connection with this ramified system certain fresh difficulties arose, for whose elimination he required the so-called axiom of reducibility. Ramsey showed that a further subdivision of types is unnecessary, and that the so-called simple system of types (the one presented here) is sufficient; thus the axiom of reducibility becomes superfluous (cf. [P.M.] vol. I², p. xiv; Ramsey [Foundations] 275 ff.).

Many-sorted languages. Sometimes it is useful to subdivide the class of zero-level expressions itself into sorts or types. The usual occasion for this is when there are various kinds of individuals for which the same predicates are not uniformly meaningful. A language with n individual types is said to be n-sorted. Most of the usual symbolic languages are one-sorted. A language with individual expressions which are either designations of objects (e.g. things, points, or the like) or numerical expressions is a two-sorted language; an example of such is the language form employed in **46c** for D19 through D22. When, in a system of geometry, it is desired to view lines and planes as separate individuals and not as classes of points, a useful procedure is to take points, lines and planes as different types of individuals, i.e. to adopt a three-sorted language (as in **47**).

Languages with no type distinctions. In a language of this kind, individuals, classes of individuals, classes of classes of individuals, etc., can each occur as values of the same variable—and thus also as elements of the same class ("inhomogeneous classes"). Such languages have been constructed in analogy to axiom systems of set theory (cf. Fraenkel's axiom system in **43**, and the references there to certain other axiom systems such as those of von Neumann, Bernays, and Gödel). Systems of logic with this form have been developed and thoroughly investigated, especially by Quine ([Logistic], [Types], [Math. Logic]). A language with no type distinctions has among its advantages that of avoiding a multiplicity of arithmetics; this last will be mentioned later (see **29b**). On the other hand, a language of this kind seems unnatural with regard to non-logical sentences. For since in such a language a type-differentiation is also omitted for descriptive signs, formulas turn up that can claim admission into the language as meaningful sentences and that have verbal counterparts running as follows: "The number 5 is blue", "The relation of friendship weighs three pounds", "5% of those prime numbers, whose father is the concept of temperature and whose mother is the number 5, die within a period of 3 years after their birth either of typhoid or of the square root of a democratic state constitution". As to the possibility of using transfinite levels to avoid the cited disadvantage in both language forms, cf. **29b**.

The system of types can be extended by *inclusion of sentences*. Suppose that sentential formulas are assumed to be of type s and level 0. Connectives are then predicates of the first level—a one-place connective having type (s), and a two-place one having type (s,s). *Operator signs* (of language C) also can be included; such a sign $\mathfrak{a}_i$ is said to be of the type $(t_j;t_k;t_m)$ provided $(\mathfrak{a}_i\mathfrak{A}_j)(\mathfrak{A}_k)$ is of type t_m, $\mathfrak{A}_j$ is a variable (or a sequence of variables separated by commas) of type t_j, and $\mathfrak{A}_k$ is of type t_k. Thus e.g. the '$\exists$' in '$(\exists x)(...)$' is of type $(0;s;s)$; the '$\imath$' (of **35**) is of type $(0;s;0)$; and the 'λ' in '$(\lambda x,y)(prod(x,y))$' is of type $(0,0;0;(0,0{:}0))$.

21d. Sentential formulas and sentences in B. An expression of the language B is called a *sentential formula* ($\mathfrak{S}$) provided it has one of the following six forms:

(1) A sentential constant.
(2) $\mathfrak{A}_i(\mathfrak{A}_j)$, where $\mathfrak{A}_j$ is of arbitrary type t_j and $\mathfrak{A}_i$ is of type (t_j) (i.e. a predicate expression).
(3) $\mathfrak{A}_i = \mathfrak{A}_j$, where $\mathfrak{A}_i$ and $\mathfrak{A}_j$ are expressions of the same type.
(4) $\sim(\mathfrak{S}_i)$, where $\mathfrak{S}_i$ is a sentential formula.
(5) $(\mathfrak{S}_i)\mathfrak{a}_k(\mathfrak{S}_j)$, where $\mathfrak{S}_i$ and $\mathfrak{S}_j$ are sentential formulas and $\mathfrak{a}_k$ is one of the signs '$\vee$', '$.$', '$\supset$', and '$\equiv$'.
(6) $(\mathfrak{v}_i)(\mathfrak{S}_j)$, where $\mathfrak{S}_j$ is a sentential formula.

Suppose $\mathfrak{v}_i$ occurs at some particular place in $\mathfrak{A}_j$. We say $\mathfrak{v}_i$ is *bound* at this place in $\mathfrak{A}_j$ provided $\mathfrak{A}_j$ (or a part of $\mathfrak{A}_j$ that includes the position in question) has the form $(\mathfrak{v}_i)(\mathfrak{S}_k)$ or the form $(\lambda\mathfrak{A}_i)(\mathfrak{A}_k)$, where $\mathfrak{A}_i$ is either $\mathfrak{v}_i$ or a sequence of variables separated by commas and containing $\mathfrak{v}_i$, and $\mathfrak{A}_k$ is a sentential formula or an expression of the type system. When this condition is not satisfied, we say $\mathfrak{v}_i$ is *free* in $\mathfrak{A}_j$. The expressions $(\mathfrak{v}_i)$ and $(\lambda\mathfrak{A}_i)$ used above are called *operators*, with $\mathfrak{S}_k$ and $\mathfrak{A}_k$ respectively their operands. If at least one of the variables in $\mathfrak{A}_j$ is free, we say that $\mathfrak{A}_j$ is *open*; otherwise, we say $\mathfrak{A}_j$ is *closed*. A closed sentential formula is called a *sentence*.

Our rules of formation, established for expressions of the type system and for sentential formulas, envisage expressions written out fully with all the requisite parentheses. In practice, of course, we follow previous custom and omit parentheses in accordance with earlier rules [see **3c** and **9a**].

21e. Definitions in B. A *definition* in B is a sentence of the form $\mathfrak{a}_i \equiv \mathfrak{S}_j$, or $\mathfrak{a}_i = \mathfrak{A}_j$, where the *definiendum* $\mathfrak{a}_i$ is the constant to be defined and the *definiens* ($\mathfrak{S}_j$ or $\mathfrak{A}_j$, respectively) is a closed expression containing only primitive signs or signs which were previously defined.

All definitions in the language B can be phrased in this simple way, with the definiendum consisting only of the new sign, because in B the λ-operator can be employed. In other languages the usual practice is to admit open sentential formulas as definitions, the definiendum there containing variables as well as the new constants. [For definitions of this latter sort, it is required that (a) each variable in the definiendum be free, and (b) occur not more than once; and that (c) no variable occur free in the definiens which does not also occur free in the definiendum (cf. [Syntax] §8).] It was in accord with this practice that we introduced into language A e.g. the functor 'mem_1': we utilized in D18-1 the *open* definitional formula '$mem_1(H)(x) \equiv (\exists y)Hxy$'. In contradistinction to this, language B allows us to write instead the definitional *sentence* '$mem_1 = (\lambda H)[(\lambda x)[(\exists y)Hxy]]$'; see **33a**, example 2. From this last definition there may be obtained (as we shall see in **33c**) the sentence '$(x)(H)[mem_1(H)(x) \equiv (\exists y)Hxy]$', whence it appears both forms of the definition lead to the same results. Language C likewise permits the use of the λ-operator in definition. Usually, however, we will adhere to the open formula kind of definition because such definitions are more readily comprehended.

22. RULES OF TRANSFORMATION FOR LANGUAGE B

22a. Primitive sentence schemata. The rules of formation laid down in the preceding section are taken to be part of both the syntactical system and the semantical system for language B. Now let us turn to the rules of transformation which constitute the characteristic feature of the syntactical system for B. They consist of rules specifying primitive sentences and rules of inference. On this basis—the primitive sentences, together with the rules of inference—additional sentences can be proved, and other sentences derived from any given sentences; this will be established in the next section. Our choice of primitive sentences and rules of inference will turn out to square with the interpretation we intend to make of language B. This interpretation was suggested in the earlier non-formal explanations of language A (and will be appropriately extended in **33** to include the λ-operator); it will be presented exactly and systematically in the semantical system. Only after the intended interpretation has been so presented can the question of its agreement with the syntactical system be posed and answered adequately (**26**). Naturally, however, the rules of transformation themselves must not refer in any way to any interpretation. Since in fact we wish here to regard these rules of transformation strictly as syntactical rules, we must take care to phrase them formally without any reference to the intended interpretation.

Each sentence of language B whose form is one of the list P1 through P12 below is called a *primitive sentence* of B. The sign '( )' signifies a sequence of universal quantifiers, one for each of the variables occurring free in the operand; if no variables occur free in the operand, '( )' is understood to vanish.

Connectives:

P1. $(\,)[\mathfrak{S}_i \vee \mathfrak{S}_i \supset \mathfrak{S}_i]$.

P2. $(\,)[\mathfrak{S}_i \supset \mathfrak{S}_i \vee \mathfrak{S}_j]$.

P3. $(\,)[\mathfrak{S}_i \vee \mathfrak{S}_j \supset \mathfrak{S}_j \vee \mathfrak{S}_i]$.

P4. $(\,)[(\mathfrak{S}_i \supset \mathfrak{S}_j) \supset (\mathfrak{S}_k \vee \mathfrak{S}_i \supset \mathfrak{S}_k \vee \mathfrak{S}_j)]$.

Universal quantifiers:

P5. *Specialization.* $(\,)[(\mathfrak{v}_i)(\mathfrak{S}_j) \supset \mathfrak{S}_k)$, where $\mathfrak{S}_k$ is obtained from $\mathfrak{S}_j$ by substituting at each free occurrence of $\mathfrak{v}_i$ in $\mathfrak{S}_j$ an expression $\mathfrak{A}_i$ of the same type; $\mathfrak{A}_i$ must contain no free variable which would become bound at one of the substitution places in $\mathfrak{S}_j$.

P6. *Distribution of the universal quantifier.*
$(\,)[(\mathfrak{v}_i)(\mathfrak{S}_j \supset \mathfrak{S}_k) \supset ((\mathfrak{v}_i)(\mathfrak{S}_j) \supset (\mathfrak{v}_i)(\mathfrak{S}_k))]$.

P7. *Vacuous universal quantifier.*
$(\,)[\mathfrak{S}_k \supset (\mathfrak{v}_i)(\mathfrak{S}_k)]$, where $\mathfrak{v}_i$ has no free occurrence in $\mathfrak{S}_k$.

Identity:

P8. $(\mathfrak{v}_i)(\mathfrak{v}_j)[(\mathfrak{v}_i = \mathfrak{v}_j) \equiv (\mathfrak{v}_k)(\mathfrak{v}_k(\mathfrak{v}_i) \supset \mathfrak{v}_k(\mathfrak{v}_j))]$, where $\mathfrak{v}_k$ is a one-place predicate variable.

Extensionality (this will be explained in **29c**):

P9. $(\mathfrak{v}_i)(\mathfrak{v}_j)[(\mathfrak{v}_{k_1})(\mathfrak{v}_{k_2})\ldots(\mathfrak{v}_{k_n})(\mathfrak{v}_i(\mathfrak{v}_{k_1},\mathfrak{v}_{k_2},\ldots,\mathfrak{v}_{k_n})\mathfrak{a}_m\ \mathfrak{v}_j(\mathfrak{v}_{k_1},\mathfrak{v}_{k_2},\mathfrak{v}_{2k},\ldots,\mathfrak{v}_{k_n})) \supset \mathfrak{v}_i = \mathfrak{v}_j]$; here either (a) $\mathfrak{v}_i$ and $\mathfrak{v}_j$ are n-place predicate variables ($n \geq 1$) and $\mathfrak{a}_m$ is '$\equiv$', or (b) $\mathfrak{v}_i$ and $\mathfrak{v}_j$ are n-place functor variables and $\mathfrak{a}_m$ is '$=$'.

λ-operator (this will be explained in **33**):

P10. $(\ )[(\lambda \mathfrak{v}_{k_1},\mathfrak{v}_{k_2},\ldots,\mathfrak{v}_{k_n})(\mathfrak{A}_i)(\mathfrak{v}_{m_1},\mathfrak{v}_{m_2},\ldots,\mathfrak{v}_{m_n})\ \mathfrak{a}_j\ (\mathfrak{A}_k)]$; here the $\mathfrak{v}_{k_\mathfrak{p}}$ ($p=1$, ..., n; $n \geq 1$) are n different variables of arbitrary types; the $\mathfrak{v}_{m_\mathfrak{p}}$ are n other different variables which do not occur in operators in $\mathfrak{A}_i$; for any p, $\mathfrak{v}_{m_\mathfrak{p}}$ is of the same type as $\mathfrak{v}_{k_\mathfrak{p}}$; either $\mathfrak{A}_i$ is a sentential formula and $\mathfrak{a}_j$ is '$\equiv$', or $\mathfrak{A}_i$ is an expression of the type system and $\mathfrak{a}_j$ is '$=$'; and $\mathfrak{A}_k$ is obtained from $\mathfrak{A}_i$ by substituting $\mathfrak{v}_{m_\mathfrak{p}}$ for $\mathfrak{v}_{k_\mathfrak{p}}$ (for each p, $p=1,\ldots, n$).

Principle of choice:

P11. $(\mathfrak{v}_i)[(\mathfrak{v}_j)[\mathfrak{v}_i(\mathfrak{v}_j) \supset \sim(\mathfrak{v}_l)(\sim\mathfrak{v}_j(\mathfrak{v}_l))] \,.\, (\mathfrak{v}_j)(\mathfrak{v}_k)[\mathfrak{v}_i(\mathfrak{v}_j) \,.\, \mathfrak{v}_i(\mathfrak{v}_k) \,.\, \sim(\mathfrak{v}_l)\sim(\mathfrak{v}_j(\mathfrak{v}_l) \,.\, \mathfrak{v}_k(\mathfrak{v}_l)) \supset (\mathfrak{v}_m)(\mathfrak{v}_j(\mathfrak{v}_m) \equiv \mathfrak{v}_k(\mathfrak{v}_m))] \supset \sim(\mathfrak{v}_k)\sim(\mathfrak{v}_j)[\mathfrak{v}_i(\mathfrak{v}_j) \supset \sim(\mathfrak{v}_m)\sim(\mathfrak{v}_n)(\mathfrak{v}_j(\mathfrak{v}_n) \,.\, \mathfrak{v}_k(\mathfrak{v}_n) \equiv (\mathfrak{v}_n = \mathfrak{v}_m))]]$; here $\mathfrak{v}_l$, $\mathfrak{v}_m$ and $\mathfrak{v}_n$ have the same (arbitrary) type, say t_l; $\mathfrak{v}_j$ and $\mathfrak{v}_k$ are predicate variables of type (t_l); and $\mathfrak{v}_i$ is a predicate variable of type $((t_l))$.

Number of individuals:

P12. See the note that follows, and **37e**.

22b. Explanatory notes on the separate primitive sentences. It should be remarked at the outset that the list above comprises *primitive sentence schemata*, and *not* single primitive sentences. Such schemata describe sentential forms with the help of the metalanguage. All the (infinitely many) sentences of the forms listed are primitive sentences of B. Instead of schemata P1 to P4 we could, had we admitted sentential variables, set up four single sentential formulas ('$p \vee p \supset p$', etc.). On the other hand, schemata P5 to P11 are necessary as they stand; they cannot be replaced by single formulas, because each scheme refers to infinitely many types.

Schemata P1 to P4, together with the two rules of inference (see the next section), describe the *sentential calculus* (or the propositional calculus) which is part of B. With the help of these primitive sentences and rules of inference, every tautology (recall **5a**) of language B can be proved: and further, for each tautological open sentential formula $\mathfrak{S}_i$ of B (thus $\mathfrak{S}_i$ contains no sentential variables), the sentence $(\ )(\mathfrak{S}_i)$ can be proved.

Schema P5 is the primitive schema of *specialization* (or instantiation). From it we see that when the variable in question is an individual variable, there may be substituted for it either an individual constant or another individual variable (examples are: '$(x)(Px) \supset Pa$', '$(y)[(x)(Px) \supset Py]$'). If the variable is a predicate variable, the schema countenances simple substitution for it, but not formula-substitution (cf. **12c**). In particular, for a predicate variable there may be substituted a closed or open predicate expression, e.g. a predicate, another predicate variable, or a λ-predicate-expression. Instead of the earlier formula-substitution, what is permitted here is the simple substitution of a λ-expression (see **33** below). Finally, if the variable is a functor variable, there may be substituted for it a closed or open functor expression, e.g. a functor, another functor variable, or a λ-functor-expression. — **Schema P6** corresponds to our earlier T14-1d(1), but refers to arbitrary types. — **Schema P7** is seldom invoked; it allows e.g. the derivation of '$(x)(Pa)$' from 'Pa'.

The following are examples of primitive sentences conforming to schema P8:

$\mathfrak{S}_1$: '$(x)(y)[x=y \equiv (F)(Fx \supset Fy)]$';

$\mathfrak{S}_2$: '$(F)(G)[F=G \equiv (N)(N(F) \supset N(G))]$';

$\mathfrak{S}_3$: '$(f)(g)[f=g \equiv (N)(N(f) \supset N(g))]$',

where 'f' and 'g' are functor variables. Since B contains the *sign of identity* as a primitive sign, $\mathfrak{S}_1$ would appear in B in lieu of the definition of this sign respecting individual expressions of A (see D17-1a); similarly, $\mathfrak{S}_2$ would appear respecting first-level predicate expressions, and $\mathfrak{S}_3$ respecting first-level functor expressions. Analogous sentences hold for expressions of any other type. Speaking generally, what P8 indicates is that any two individuals (or attributes or functions) of whatever type are identical provided each has all the properties that the other has. E.g. two physical bodies a and b are identical if they have all their properties in common, among these properties being their space-time relations to other bodies.

The following is an example of a primitive sentence conforming to schema P11, the *principle of choice* (or selection). The sentence is formulated at the lowest level permitted by the principle; to facilitate reading it, we write '$(\exists x)$' for '$\sim(x)\sim$'.

$$(N)[(F)[N(F) \supset (\exists x)Fx].(F)(G)[N(F).N(G).(\exists x)(Fx.Gx) \supset (x)(Fx \equiv Gx)] \supset (\exists H)(F)[N(F) \supset (\exists x)(y)(Fy.Hy \equiv y=x)]].$$

In the terminology of classes, this sentence says: If N is such a second-level class that its element classes are non-empty and mutually exclusive, then there exists such a first-level class H that with each element class of N the class H has precisely *one* individual in common. (This class H is sometimes called the "selection class of N".) Schema P11 allows similar sentences to be constructed for expressions of any other type. The principle of choice was enunciated first by Zermelo. Regarding the much-disputed questions about it, cf. [P.M.] I 536 ff.; Russell [Introduction] 117 ff.; Fraenkel [Grundlagen] 80 ff., and [Einleitung] 288´ff. together with full discussion and bibliography; Rosser [Logic] ch. xiv.

Under the heading P 12 one primitive sentence is to be given—a sentence which specifies the *number of individuals* that constitute the domain of language B. If that domain is fixed in advance, this primitive sentence depends on the domain; in any case, of course, the sentence speaks only of the structure of the domain and says nothing about its content. In connection with most axiom systems, what is useful is to establish that the corresponding domain is not finite, i.e. that the domain is at least denumerable—its cardinal number is at least $\aleph_0$ (*axiom of infinity*; cf. **37e**). For some axiom systems, however—e.g. projective or metric (Euclidean or non-Euclidean) geometries in their usual form—a higher cardinal number, viz. that of the continuum, is required for the domain. Since it is desirable to give at least *one* example of a primitive sentence bearing on the number of individuals, we do so below in terms of a domain having the cardinal number 2—since for this cardinal the corresponding primitive sentence can be quite simply formulated with the primitive signs of language B. This illustrative primitive sentence runs as follows:

$$`\sim(x)(y)[x=y \vee \sim(z)(z=x \vee z=y)]';$$

in words: "There are exactly two individuals". (In language A, this sentence is L-equivalent to '$(\exists x)(\exists y)[x \neq y \,.\, (z)(z=x \vee z=y)]$'; cf. **17c**.)

As mentioned earlier, we always make the presupposition (familiar in other systems of logic) that the domain of individuals is not empty. Thus e.g. '$(x)Fx \supset (\exists x)Fx$' is L-true in A (T14-1c), hence so also are the sentential formulas '$(\exists x)(Gx \vee \sim Gx)$' and '$(\exists x)(x=x)$' (which come from the first by substitution for 'Fx' of '$Gx \vee \sim Gx$' and '$x=x$' respectively); these last two formulas may be viewed as formulations of the word-sentence "There is at least one individual". The corresponding sentences '$(G)[\sim(x)\sim(Gx \vee \sim Gx)]$' and '$\sim(x)\sim(x=x)$' are provable in B. That an existential assumption is thus built into the logical foundation of our present system appears unobjectionable (this certainly, so far as we are concerned with the practical application of our system in a scientific theory or an axiom system), for it is hardly ever required to consider empty domains. Should it be desired to free the logical system from such existential assumptions, the rules must be altered in a certain way (cf. [Syntax E] § 38a).

22c. Rules of inference. The rules of inference for B are two in number, as follows:

R1. *Modus ponens.* From $\mathfrak{S}_i$ and $\mathfrak{S}_i \supset \mathfrak{S}_j$, $\mathfrak{S}_j$ is directly derivable.

R2. *Rule for connectives.* $\mathfrak{S}_j$ is directly derivable from $\mathfrak{S}_i$ provided $\mathfrak{S}_j$ is obtained from $\mathfrak{S}_i$ by replacing an expression $\mathfrak{A}_i$ in one place by the expression $\mathfrak{A}_j$, or conversely, where:

a. $\mathfrak{A}_i$ is $\mathfrak{S}_k \supset \mathfrak{S}_m$; $\mathfrak{A}_j$ is $\sim\mathfrak{S}_k \vee \mathfrak{S}_m$.

b. $\mathfrak{A}_i$ is $\mathfrak{S}_k \,.\, \mathfrak{S}_m$; $\mathfrak{A}_j$ is $\sim(\sim\mathfrak{S}_k \vee \sim\mathfrak{S}_m)$.

c. $\mathfrak{A}_i$ is $\mathfrak{S}_k \equiv \mathfrak{S}_m$; $\mathfrak{A}_j$ is $(\mathfrak{S}_k \supset \mathfrak{S}_m) \,.\, (\mathfrak{S}_m \supset \mathfrak{S}_k)$.

Explanations of these rules. Rule R1 conforms with the truth-table technique of language A: $\mathfrak{S}_i$ and $\mathfrak{S}_i \supset \mathfrak{S}_j$ together L-imply $\mathfrak{S}_j$ (cf. T6-14a). Rule R2 refers the *connectives* '$\supset$', '$.$' and '$\equiv$' back to the connectives '$\sim$'

and '$\vee$', again in accordance with the truth-tables for these signs in language A (cf. T8-6j(1), g(6) and f(1)). If the connectives '$\supset$', '.' and '$\equiv$' were eliminated from language B, rule R2 would be dropped.

23. PROOFS AND DERIVATIONS IN LANGUAGE B

23a. Proofs. In setting up a syntactical system for a language L, generally there is in view a certain interpretation of L which motivates the selection of syntactical rules but is not explicitly mentioned in the rules. The primitive sentences of L are so chosen that they are true sentences in the intended interpretation; and the rules of inference for L are so chosen that they lead invariably from true sentences to other true sentences. Thus, all sentences of L which can be "proved", i.e. can be obtained by means of the primitive sentences and the rules of inference, turn out true in the intended interpretation. Of course, the choice of primitive sentences and rules of inference can be made in different ways, even though the totality of provable sentences remains the same. What dictates a particular choice is, usually, some technical requirement, e.g. the requirement that proofs and derivations be simple. Primitive sentences are not required to have any kind of preferred character of a logical or epistemological sort.

By a *proof* in L we understand not a train of thoughts of a particular kind, but a sequence of sentences of L which in a certain sense corresponds to such a train of thoughts. The correctness of a given step from the preceding sentences of such a sequence to some subsequent sentence thereof is not tested on the ground that it is a more or less plausible inference in the train of thought, but rather on the ground that it does or does not conform to the transformation rules for L. Primitive sentences can be utilized freely in a proof, and the same is true of any definition (so far as it conforms to the formation rules established earlier for definitions)—since definitions are simply conventions regarding the use of new signs. The rules of inference for L specify conditions under which a sentence may be derived from one or more sentences. It is in this way that the rules of inference make possible a movement from primitive sentences or definitions to new sentences. Thus we arrive at the following definition: a *proof* in L is a (finite) sequence of sentences of L, each of which is either a primitive sentence or a definition, or else is directly derivable from sentences preceding it in the sequence. The final sentence of a proof in L is said to be *provable* in L. If the negation of a sentence is provable in L, we say the sentence itself is *refutable* in L. A sentence which is either provable or refutable in L we call *decidable* in L; otherwise, *undecidable* in L.

Example of a proof in language B. The successive sentences comprising the proof below are numbered consecutively in the right margin. In the left margin we enter notations that facilitate a final test of the proof by indicating the use of a primitive sentence, or a definition, or a rule of inference respecting certain previous sentences. Strictly speaking, neither the entries in the right margin nor those in the left are to be regarded as part of the proof.

P1	$A \vee A \supset A$	(1)
P4 (with '$A \vee A$' as $\mathfrak{S}_i$, 'A' as $\mathfrak{S}_j$, and '$\sim A$' as $\mathfrak{S}_k$)	$(A \vee A \supset A) \supset [\sim A \vee (A \vee A) \supset \sim A \vee A]$	(2)
(1) (2) R1	$\sim A \vee (A \vee A) \supset \sim A \vee A$	(3)
(3) R2a	$(A \supset (A \vee A)) \supset \sim A \vee A$	(4)
P2	$A \supset A \vee A$	(5)
(5) (4) R1	$\sim A \vee A$	(6)
(6) R2a	$A \supset A$	(7)
P3	$\sim A \vee A \supset A \vee \sim A$	(8)
(6) (8) R1	$A \vee \sim A$	(9)

Inasmuch as we could at will break off the proof with step (6), or step (7), or step (9), each of the sentences '$\sim A \vee A$, '$A \supset A$', and '$A \vee \sim A$' is provable in B.

Exercises. Give a proof in B for each of the following sentences on the basis of the suggestions:

a) $(B \supset C) \supset [(A \supset B) \supset (A \supset C)]$
Use an appropriate sentence of the form P4, and then apply R2.

b) $A \supset \sim \sim A$
The proof should be modeled on that of the example; however, lines (1) and (2) should be appropriately modified so that '$\sim A$' replaces 'A' throughout. R2 should then be used on the resulting line (9).

c) $\sim (A . \sim A)$
Applying R2 to '$A \supset \sim \sim A$' (which has been shown to be provable), obtain '$\sim A \vee \sim \sim A$'. Now by modeling a proof on that of '$A \supset \sim \sim A$', a proof can be obtained for '$(\sim A \vee \sim \sim A) \supset \sim \sim (\sim A \vee \sim \sim A)$'. Applications of R1 and R2 then yield the desired result.

d) $A \supset (B \supset A)$
Use P3 with 'A' as $\mathfrak{S}_i$ and '$\sim B$' as $\mathfrak{S}_j$. Then use P2 with 'A' as $\mathfrak{S}_i$ and '$\sim B$' as $\mathfrak{S}_j$ and, by applying R2 to this result, obtain '$\sim A \vee (A \vee \sim B)$'. Using an appropriate sentence of the form P4, the two results can be used with R1 twice to obtain '$\sim A \vee (\sim B \vee A)$'. The desired sentence now results from two uses of R2.

e) $(A \supset B) \supset (\sim B \supset \sim A)$
From '$B \supset \sim \sim B$' (which is obtainable as in (b)) and an appropriate sentence of the form P4, '$\sim A \vee B \supset \sim A \vee \sim \sim B$' can be obtained. Next secure '$\sim A \vee B \supset \sim \sim B \vee \sim A$' with the help of P3 and a provable sentence of the form of exercise (a). Now apply R2.

f) $\sim \sim A \supset A$
First obtain '$\sim A \supset \sim \sim \sim A$' and '$A \vee \sim A$', then a suitable sentence of the form P4 will yield '$A \vee \sim \sim \sim A$'. Then use P3, and R2.

g) $A . B \supset A$

Through the use of P2 and a sentence of the form of exercise (e), '$A . B \supset \sim\sim A$' can be obtained.

h) $(x)(Px) \supset (x)(Px \vee Qx)$

Use P6 with 'Px' as $\mathfrak{S}_j$, '$Px \vee Qx$' as $\mathfrak{S}_k$, and 'x' as $\mathfrak{v}_i$. R1 can then be applied to this result and the result of an appropriate instance of P2 to yield the desired sentence.

i) $(x)(A \supset Px) \supset (A \supset (x)Px)$

Here it is supposed that 'x' does not occur in 'A'. Use P6 and P7.

j) $(x)(Px \supset Qx) \supset (\sim Px \vee Qx)$

23b. Derivations. Use of the primitive sentences, the definitions, and the rules of inference is not restricted to proofs, i.e. to showing that certain sentences are provable—and hence true in the intended interpretation. It is also legitimate to employ these rules of transformation, when what is wanted is a derivation of certain sentences from certain other sentences (generally not provable). The sentences from which the derivation proceeds are called the *premisses* of the derivation. We define: in a language L, a *derivation* with given premisses is a (finite) sequence of sentences of L, each of which is either a premiss, a primitive sentence, or a definition, or else is directly derivable from sentences preceding it in the sequence. If $\mathfrak{S}_n$ is the last sentence of a derivation in L with premisses $\mathfrak{S}_i, \ldots, \mathfrak{S}_k$, we say $\mathfrak{S}_n$ is *derivable* in L from $\mathfrak{S}_i, \ldots, \mathfrak{S}_k$.

Examples of derivations. Below are four derivations in B. Entries in the two margins have the same role as in the case of proofs, and similarly are not part of the derivation.

1. Premiss:	$A \vee B$	(1)
P3	$A \vee B \supset B \vee A$	(2)
(1) (2) R1	$B \vee A$	(3)

Thus '$B \vee A$' is derivable from '$A \vee B$'. In general, $\mathfrak{S}_j \vee \mathfrak{S}_i$ is derivable from $\mathfrak{S}_i \vee \mathfrak{S}_j$.

2. Premisses:	1. A	(1)
	2. $\sim A$	(2)
P2	$\sim A \supset \sim A \vee B$	(3)
(2) (3) R1	$\sim A \vee B$	(4)
(4) R2a	$A \supset B$	(5)
(1) (5) R1	B	(6)

Thus an arbitrary sentence 'B' is derivable from 'A' and '$\sim A$'. Generally: from $\mathfrak{S}_i$ and $\sim\mathfrak{S}_i$ any sentence is derivable.

3. Premiss:	$(x)Px$	(1)
P5	$(x)Px \supset Pa$	(2)
(1) (2) R1	Pa	(3)

Thus '*Pa*' is derivable from $(x)Px$'. This operation is called *specialization* or instantiation.

4. Premiss:	$(x)Px$	(1)
P7	$(x)Px \supset (y)(x)Px$	(2)
(1) (2) R1	$(y)(x)Px$	(3)
P5	$(y)[(x)(Px) \supset Py]$	(4)
P6	$(y)[(x)(Px) \supset Py] \supset [(y)(x)Px \supset (y)Py]$	(5)
(4) (5) R1	$(y)(x)Px \supset (y)Py$	(6)
(3) (6) R1	$(y)Py$	(7)

Thus '$(y)Py$' is derivable from '$(x)Px$'. Earlier, we called this operation the *revision* of a bound variable (see T12-2a).

Exercises. 1. Show that '$B.A$' is derivable in B from '$A.B$'. First prove '$[(\sim A \vee \sim B) \supset (\sim B \vee \sim A)] \supset [\sim(\sim B \vee \sim A) \supset \sim(\sim A \vee \sim B)]$', modeling your proof on that given in exercise 1e, **23a**. Then use P3 and R2. — **2.** Show that '$\sim A$' is derivable in B from '$A \supset B$' and '$\sim B$'. (See exercise (e), **23a.**) — **3.** Show that 'B' is derivable in B from '$\sim A$' and '$A \vee B$'. First prove '$A \vee B \supset \sim\sim A \vee B$'. — **4.** Show that '$B \vee D$' is derivable in B from '$A \supset B$', '$C \supset D$', and '$A \vee C$'. First derive '$A \vee D$', and use P3. Then derive '$D \vee B$' and use P3 again. — **5.** Show that '$(x)Qx$' is derivable in B from '$(x)(Px \supset Qx)$' and '$(x)Px$'. Use P6. — **6.** Show that '$\sim Pa$' is derivable in B from '$(x)(Px \supset Qx)$' and '$\sim Qa$'. Use P5. — **7.** Show that '$(x)(Px \supset Qx)$' is derivable in B from '$(x)\sim Px$'. Use P2. — **8.** Show that '$(z)Rzz$' is derivable in B from '$(x)(y)Rxy$'. Use P5 twice. — **9.** Show that '$(x)(Qx \vee Pa)$' is derivable in B from '$(x)Px$'. Use P5.

24. THEOREMS ON PROVABILITY AND DERIVABILITY IN LANGUAGE B

24a. General theorems for B.

T24-1. If $\mathfrak{S}_i$ is derivable from provable sentences, then $\mathfrak{S}_i$ itself is also provable.

T24-2. From $\mathfrak{S}_i$ and $\sim\mathfrak{S}_i$ any sentence whatever is derivable. (Recall example 2, **23b**.)

T24-3. If $\sim\mathfrak{S}_i$ is provable, then any sentence whatever is derivable from $\mathfrak{S}_i$. (By T2.)

T24-4. If $\mathfrak{S}_i \supset \mathfrak{S}_j$ is provable, then $\mathfrak{S}_j$ is derivable from $\mathfrak{S}_i$.

T24-5. If $\mathfrak{S}_i \equiv \mathfrak{S}_j$ is provable, then each of $\mathfrak{S}_i$ and $\mathfrak{S}_j$ is derivable from the other.

T24-6. **a.** Every tautology (recall **5a**) is provable.

b. If, on the basis of truth-tables, a sentence is L-implied by one or more other sentences, then the sentence is derivable from these other sentences.

T6a says that for any tautology in B there is a proof in B. But this theorem does not tell us how to construct a proof for an arbitrary given tautology. There is a method for doing this, which, however, cannot be described here. (The method makes use of the so-called conjunctive normal form; cf. Hilbert [Logic].)

More generally, the following is the case: *all theorems regarding language A* (see especially **8**, **13**, **14** and **15a**) *have valid counterparts for language B.* This means: 1. All sentences of language A that have been identified as L-true are provable in B (insofar as they are sentences of B, otherwise their translations into B); 2. If it is known that a certain sentence of A is L-implied by certain other sentences of A, then in B that sentence is derivable from the others. In this connection, special emphasis is to be given the theorem on raising levels (T16-1).

24b. Interchangeability. As was the case for language A (recall T15-3), so here in language B equivalent formulas are mutually interchangeable in a sentential formula. Additionally, in B the same interchange of equivalent formulas can take place in expressions of the type system which contain sentential formulas, e.g. a λ-predicate expression of the form $(\lambda\mathfrak{v}_i)(\mathfrak{S}_j)$. And further, in B expressions of the type system that are linked by an identity sign are mutually interchangeable in a sentential formula (this connects with our earlier definition D17-1 of the identity of individuals, and with the theorem on raising levels), as well as in a larger expression of the type system. Theorem T7 below refers to all four cases.

T24-7. Suppose that $\mathfrak{A}_i$, $\mathfrak{A}_j$, $\mathfrak{A}_i'$, and $\mathfrak{A}_j'$ are expressions of language B, that $\mathfrak{a}_k$ and $\mathfrak{a}_k'$ are signs of B, and that these expressions and signs satisfy the following three conditions. (1) Either: (a) $\mathfrak{A}_i$ and $\mathfrak{A}_j$ are sentential formulas and $\mathfrak{a}_k$ is '≡'; or: (b) $\mathfrak{A}_i$ and $\mathfrak{A}_j$ are expressions of the same type and $\mathfrak{a}_k$ is '='. (Hence, in either case $\mathfrak{A}_i\mathfrak{a}_k\mathfrak{A}_j$ is a sentential formula.) (2) The same condition holds for $\mathfrak{A}_i'$, $\mathfrak{A}_j'$ and $\mathfrak{a}_k'$ ($\mathfrak{a}_k'$ is not necessarily the same sign as $\mathfrak{a}_k$). (3) $\mathfrak{A}_j'$ is obtained from $\mathfrak{A}_i'$ by replacing in $\mathfrak{A}_i'$ an occurrence of $\mathfrak{A}_i$ by $\mathfrak{A}_j$ (without regard to other possible occurrences of $\mathfrak{A}_i$ in $\mathfrak{A}_i'$). Then the following hold in B:

a. $(\,)(\mathfrak{A}_i\mathfrak{a}_k\mathfrak{A}_j) \supset (\,)(\mathfrak{A}_i'\mathfrak{a}_k'\mathfrak{A}_j')$ is provable.
b. $(\,)(\mathfrak{A}_i'\mathfrak{a}_k'\mathfrak{A}_j')$ is derivable from $(\,)(\mathfrak{A}_i\mathfrak{a}_k\mathfrak{A}_j)$.
c. If $(\,)(\mathfrak{A}_i\mathfrak{a}_k\mathfrak{A}_j)$ is provable, then so is $(\,)(\mathfrak{A}_i'\mathfrak{a}_k'\mathfrak{A}_j')$.

Illustrative applications of this theorem appear in the four examples below.

Examples. The following examples are phrased under the supposition that in B definitions have been introduced for '3', 'mem_1', 'mem_2', 'Is_1' and 'str_1' in analogy with D17-3, D18-1 and 2, D19-5 and D34-2.

1. *Interchanging a sentential formula in a sentential formula.*
   a. Given '$A \equiv B$' as a premiss 'A' can be interchanged with 'B' in e.g. '$C.\sim A$'; the result is '$C.\sim B$'. In other words, '$C.\sim A \equiv C.\sim B$' is derivable from '$A \equiv B$'; hence '$C.\sim B$' is derivable from '$A \equiv B$' and '$C.\sim A$' together.
   b. (Recall example 1 in connection with T15-3.) From '$(x)(Rxa \equiv Sbx)$' the formula '$\sim(x)(Px \vee Rxa) \equiv \sim(x)(Px \vee Sbx)$' is derivable.

2. *Interchanging a sentential formula in an expression of the type system.*
   a. Given, as in example 1b above, '$(x)(Rxa \equiv Sbx)$', the formula '$(\lambda x)(Px \vee Rxa) = (\lambda x)(Px \vee Sbx)$' is derivable.
   b. (Recall example 2 for T15-3.) The formula '$(x)(y)[(Px \supset Rxy) \equiv (\sim Rxy \supset \sim Px)]$' is provable. Hence, by T7c, the sentence '$(\lambda y)[(x)(Px \supset Rxy) \vee Qy] = (\lambda y)[(x)(\sim Rxy \supset \sim Px) \vee Qy]$' is also provable.
3. *Interchanging an expression of the type system in a sentential formula.*
   a. '$\sim(x)Rxa \equiv \sim(x)Rxb$' is derivable from '$a=b$'.
   b. Assume $\mathfrak{S}_1$: '$(x)(Qx \equiv P_1x.\sim P_2x)$' is given. Notice that '$(x)[(\lambda y)(P_1y.\sim P_2y)x \equiv P_1x.\sim P_2x]$' is a primitive sentence conforming to schema P10. Then '$(x)[Qx \equiv (\lambda y)(P_1y.\sim P_2y)x]$' is derivable from $\mathfrak{S}_1$, and from this in turn (with the help of P9a) may be derived the sentence $\mathfrak{S}_2$: '$Q=(\lambda y)(P_1y.\sim P_2y)$'. [If '$P_1$' and '$P_2$' are primitive signs of language A, then either $\mathfrak{S}_1$ or its operand may be laid down in A as the definition of 'Q'. The corresponding definition in B would be $\mathfrak{S}_2$.] Hence 'Q' may be interchanged anywhere with the λ-expression; e.g. '$3(Q) \equiv 3((\lambda y)(P_1y.\sim P_2y))$' is derivable from $\mathfrak{S}_2$.
   c. Let $\mathfrak{S}_1$: '$(x)[mem_2(R)x \equiv mem_1(S)x]$' be given. Then by P9a the sentence '$mem_2(R)=mem_1(S)$' is derivable from $\mathfrak{S}_1$. Hence '$Is_1(P,mem_2(R)) \equiv Is_1(P,mem_1(S))$' is also derivable from $\mathfrak{S}_1$.
4. *Interchanging an expression of the type system in another such expression.*
   a. '$(\lambda x)(Rxa) = (\lambda x)(Rxb)$' is derivable from '$a=b$'.
   b. Recalling example 3c above, let '$mem_2(R) = mem_1(S)$' be assumed. From it can be derived '$str_1(mem_2(R)) = str_1(mem_1(S))$', which says: "The cardinal number of the second domain of R is the same as that of the first domain of S."

From T7 we also see the possibility of manipulating definitions in the customary way, viz. the introduction into any context—or the elimination therefrom—of the defined signs; for in B a definition has one of the forms $\mathfrak{a}_i \equiv \mathfrak{S}_j$ or $\mathfrak{a}_i = \mathfrak{A}_j$, where $\mathfrak{a}_i$ is the sign being defined (cf. **21e**).

Exercises. Use T7 (among others) for the following: **1.** Show that 'Pb' is derivable in B from 'Pa' and '$a=b$'. Use P8. — **2.** Show that 'B' is derivable in B from '$B \equiv \sim B \vee A$'. Use P2, T7, P4, P1. — **3.** Show that '$\sim Qa$' is derivable in B from '$\sim Pb$' and '$(y)(Py \equiv Qa \vee Ry)$'.

25. THE SEMANTICAL SYSTEM FOR LANGUAGE B

25a. Value-assignments and evaluations. Now let us establish the rules of the semantical system for language B, rules which systematize the intended interpretation of B.

To begin with, the semantical system B contains the same rules of formation as the syntactical system B (**21**); hence we do not repeat them here. What we understand is that the interpreted language B as described in the semantical system contains the same signs, expressions of the type system, sentential formulas, sentences, and definitions, as the uninterpreted language B described in the syntactical system.

The meaning of individual constants of a language L will depend on the domain of things to which L is applied. These things may be space-time points, events extended in space-time, physical bodies, persons (of any historic epoch), persons now alive, etc. Later (in Part II) we shall give examples of various domains of individuals. In the present chapter we leave the choice of domain open, and phrase the semantical rules for **B** in terms of "individuals" without specifying what they are.

Of the primitive signs of B we count as *descriptive* the sentential constants and the constants (individual constants, predicates, and functors) belonging to the type system. All other primitive signs are *logical.* A defined sign is descriptive if a descriptive sign occurs in its definiens, otherwise logical. [Strictly speaking, this division of the primitive signs into descriptive and logical depends on the kind of domain of individuals chosen. E.g. the division we have made above holds if the domain of individuals is taken to be all space-time points, or all space-time regions, or all processes of the physical world. Other choices of the domain may, under certain circumstances, compel a modification of this division; e.g. if the domain of individuals is taken to be all numbers, and the undefined predicates and functors are interpreted as arithmetical concepts, then all the primitive signs are logical. Concerning this problem of division, which is still not yet fully clarified, cf. [Semantics] § 13, [Meaning] § 21.]

Value-assignments. The rules below agree with those given earlier (**11**) for language A, but have received a broader formulation suitable to B. Like A, our language B is extensional. Hence here, too, it is sufficient to take for the assigned values extensions of appropriate types. This is what the following rules do.

Rules governing value-assignments:

1. Possible values for sentential formulas are the two truth-values: Truth (T) and Falsity (F).

2. Possible values for an expression of type t_i of the type system are the values of type t_i hereinafter specified:

- **a.** A value of type 0 (i.e. a possible value for an individual expression) is any individual of the chosen domain.
- **b.** A value of type $t_{i_1},t_{i_2},...,t_{i_n}$ with $n \geq 2$ (i.e. a possible value for an n-place argument expression) is any n-tuple of values whose pth entry ($p=1,...,n$) is a value of type t_{i_p}.
- **c.** A value of type (t_i) (i.e. a possible value for a predicate expression) is any class of values of type t_i.
- **d.** A value of type $(t_i:t_j)$ (i.e. a possible value for a functor expression) is any function-extension by which with each possible value of type t_i (as argument of the function) there is coordinated exactly one value of type t_j (as value of the function).

Explanation of "function-extension." Suppose f_1 and f_2 are functions of the same type $(t_1 : t_2)$. We say that f_1 has the same function-extension as f_2 whenever f_1 has for each argument the same function-value as f_2. If this condition is satisfied only as a matter of contingent fact but not logically, the function-extension f_1 is still the same as the function-extension f_2 while the function f_1 is *not* the same as the function f_2. In this case, f_1 and f_2 count for a value-assignment as the same value of type $(t_1 : t_2)$.

As *value-bearing signs* in an expression $\mathfrak{A}_i$ of system B we count all descriptive signs and all variables occurring free in $\mathfrak{A}_i$. A *value-assignment* for $\mathfrak{A}_i$ consists in associating with each value-bearing sign of $\mathfrak{A}_i$ one of the possible values of that sign (the associated value having, of course, the same type as the sign). Given a determinate value-assignment $\mathfrak{B}_i$ for the value-bearing signs of $\mathfrak{A}_i$, the *evaluation* of $\mathfrak{A}_i$ at $\mathfrak{B}_i$, i.e. the establishment of the value of $\mathfrak{A}_i$ relative to $\mathfrak{B}_i$, is made in accordance with the following rules. These rules permit the evaluation of any component expression—be it a sentential formula or an expression of the type system—so that, by beginning with the smallest components and proceeding step-wise through successively larger ones, we can arrive finally at the value of expression $\mathfrak{A}_i$ itself. In the following Rules 1 and 2 we write simply "value" for "value at $\mathfrak{B}_i$".

Rules governing evaluation:

1. *Of expressions of the type system.*

a. A compound argument expression $\mathfrak{A}_{i_1},\mathfrak{A}_{i_2},\ldots,\mathfrak{A}_{i_n}$ with $n \geq 2$ has as its value the n-tuple comprising successively the values of $\mathfrak{A}_{i_1}$, of $\mathfrak{A}_{i_2}$,..., and of $\mathfrak{A}_{i_n}$.

b. A predicate expression of the form $(\lambda\mathfrak{v}_i)(\mathfrak{S}_j)$ has as its value the class of those values of $\mathfrak{v}_i$ which satisfy $\mathfrak{S}_j$ (i.e. those values of $\mathfrak{v}_i$ which, together with the values assigned by $\mathfrak{B}_i$, give formula $\mathfrak{S}_j$ the value T).

c. A predicate expression of the form $(\lambda\mathfrak{v}_{i_1},\mathfrak{v}_{i_2},\ldots,\mathfrak{v}_{i_n})(\mathfrak{S}_j)$ with $n \geq 2$ has as its value the class of those n-tuples of values of $\mathfrak{v}_{i_1},\mathfrak{v}_{i_2},\ldots,\mathfrak{v}_{i_n}$ that satisfy $\mathfrak{S}_j$.

d. A functor expression of the form $(\lambda\mathfrak{v}_i)(\mathfrak{A}_j)$ has as its value that function-extension by which with each possible value of $\mathfrak{v}_i$ there is coordinated that value taken on by $\mathfrak{A}_j$ at this assignment to $\mathfrak{v}_i$.

e. A functor expression of the form $(\lambda\mathfrak{v}_{i_1},\mathfrak{v}_{i_2},\ldots,\mathfrak{v}_{i_n})(\mathfrak{A}_j)$ with $n \geq 2$ has as its value that function-extension by which with each n-tuple of possible values of $\mathfrak{v}_{i_1},\mathfrak{v}_{i_2},\ldots,\mathfrak{v}_{i_n}$ there is coordinated that value taken on by $\mathfrak{A}_j$ at this assignment to the variables named.

f. A full expression $\mathfrak{A}_i(\mathfrak{A}_j)$ of the functor expression $\mathfrak{A}_i$ has that value which the function-extension which is the value of $\mathfrak{A}_i$ coordinates with the value of $\mathfrak{A}_j$.

2. *Of sentential formulas.*

a. A sentential formula of the form $\mathfrak{A}_i(\mathfrak{A}_j)$, comprising the predicate expression $\mathfrak{A}_i$ (of arbitrary type) and the (simple or compound) argument expression $\mathfrak{A}_j$, has the value T provided the value of $\mathfrak{A}_j$ belongs to the class which is the value of $\mathfrak{A}_i$; otherwise, the value F.

b. A sentential formula of the form $\mathfrak{A}_i = \mathfrak{A}_j$ has the value T provided $\mathfrak{A}_i$ has the same value as $\mathfrak{A}_j$; otherwise, the value F.

c. $\sim\mathfrak{S}_i$ has the value T provided $\mathfrak{S}_i$ has the value F; otherwise, the value F.

d. $\mathfrak{S}_i \vee \mathfrak{S}_j$ has the value T provided at least one of $\mathfrak{S}_i$ and $\mathfrak{S}_j$ has the value T; otherwise, the value F.

e. $\mathfrak{S}_i . \mathfrak{S}_j$ has the value T provided both $\mathfrak{S}_i$ and $\mathfrak{S}_j$ have the value T; otherwise, the value F.

f. $\mathfrak{S}_i \supset \mathfrak{S}_j$ has the value F provided $\mathfrak{S}_i$ has the value T and $\mathfrak{S}_j$ the value F; otherwise, the value T.

g. $\mathfrak{S}_i \equiv \mathfrak{S}_j$ has the value T provided $\mathfrak{S}_i$ and $\mathfrak{S}_j$ have the same value; otherwise, the value F.

h. $(\mathfrak{v}_i)(\mathfrak{S}_j)$ has the value T provided $\mathfrak{S}_j$ has the value T at every possible value-assignment to the free variable $\mathfrak{v}_i$ in $\mathfrak{S}_j$ (together with the given value-assignment $\mathfrak{B}_i$ to the other value-bearing signs); otherwise, the value F.

In agreement with earlier practice, we say: the value-assignment $\mathfrak{B}_i$ (or the values assigned by $\mathfrak{B}_i$) *satisfies* the formula $\mathfrak{S}_i$ provided $\mathfrak{S}_i$ has the value T at $\mathfrak{B}_i$. The concept of *range* and the *L*- and *F-concepts* are then defined for language B in the same way they were earlier defined for language A (cf. **5b** and **6a**); we shall not repeat these definitions here.

25b. Rules of designation. Whereas L-concepts are among the most important concepts of logic and so occur frequently in the theorems of this book, the concept of *truth* has less importance within logic: it appears here mostly in conditional contexts such as "if $\mathfrak{S}_i$ is true, then $\mathfrak{S}_j$ is true". However, the concept of truth has quite an important role in epistemology and the methodology of science. As a basis for our subsequent definition of truth we lay down rules governing variables and descriptive constants. The first step here is to specify the value domains of variables of all types; this we do by means of the following two rules. These rules are phrased on the assumption that the domain of individuals of language B is the domain of physical things.

Rules governing the values of variables:

1. The values of individual variables are physical things.
2. The values of predicate variables and functor variables of arbitrary type are all possible values of the type in question drawn from the specified domain of individuals (cf. Rule 1) in conformity with our earlier rules (those of 2c, d) governing value-assignments.

[The formulation of these rules is in terms of value-extensions (**10b**), and makes no reference to value-intensions; however, these extensions furnish an adequate basis for our definition of truth.]

Next we turn to the rules of designation governing the descriptive primitive signs of the system. Suppose that, for a certain application, language B contains only the following descriptive primitive signs: three individual constants '*a*', '*b*', '*c*'; two one-place predicates '*P*' and '*Q*' of the first level; and, finally, a single two-place predicate '*R*' of the first level. By way of illustration, let us now lay down rules of designation which correlate with these signs and designata listed in the second column of the table below. These designata are intensions (concepts), not extensions (cf. **10b**); by means of these intensions the corresponding extensions (given in the third column) are determined.

Rules of designation:

Primitive sign	Designatum (Intension)	Extension
'*a*'	(the individual concept) moon	(the thing) moon
'*b*'	(the individual concept) sun	(the thing) sun
'*c*'	(the individual concept) Africa	(the thing) Africa
'*P*'	the property of being spherical	the class of spherical things
'*Q*'	the property of being blue	the class of blue things
'*R*'	the relation greater than	the class of pairs *x*,*y* such that *x* is greater than *y*

This choice of designata conforms with our previous agreement regarding the domain of values of variables. From the designata of the primitive signs there result in an obvious way the designata of closed expressions, viz. on one hand certain concepts (properties, relations, etc.) as designata of expressions of the type system, on the other certain propositions as designata of sentences. (Rules governing the determination of these derived designata are omitted; such rules are not necessary for the definition of truth.) Similarly there results for each defined constant its designatum, viz. the designatum of its definiens.

25c. Truth. Rules (1) and (2) above fix the possible values for each kind of variable in language B. Now we fix a special value-assignment $\mathfrak{B}_1$ to all the descriptive primitive signs of language B: to each of these signs, $\mathfrak{B}_1$ assigns as its value the extension of that sign specified by the rules of designation above. Further, we say: the *extension* of a closed expression $\mathfrak{A}_i$ of language B is the value $\mathfrak{A}_i$ takes on at the value-assignment $\mathfrak{B}_1$ (the evaluation being accomplished in accordance with our previous rules governing evaluation, **25a**).

Example. In view of evaluation rules 2a, 2e and 1b, the value of '$(\lambda x)(Px \, . \, Qx)$' at $\mathfrak{B}_1$ is the class of all those things which are both spherical and blue; thus, this class is the extension of the expression '$(\lambda x)(Px \, . \, Qx)$'.

At this point we can define the concept of truth: a sentence $\mathfrak{S}_i$ is *true* in language B provided its extension is the value T. I.e. a sentence is true provided its evaluation (according to the rules of evaluation) at $\mathfrak{B}_1$ (fixed by the rules of designation) produce the value T. [This definition of 'true' in terms of 'T' does not involve us in a vicious circle; for 'T' and 'F' are here to be construed simply as technical terms whose use is governed by the rules of evaluation—'1' and '0', or any other pair of neutral terms, could just as well have been used in place of 'T' and 'F'.]

The following theorem states truth conditions for sentences of the simplest form; these conditions are general, i.e. they make no reference to particular rules of designation. The theorem is an immediate consequence of rules 2a,b governing evaluation and the definition of the value-assignment $\mathfrak{B}_1$.

T25-1. **a.** A one-place atomic sentence $\mathfrak{a}_i(\mathfrak{a}_j)$ is true if and only if the individual which is the extension of $\mathfrak{a}_j$ belongs to the class which is the extension of $\mathfrak{a}_i$, i.e. if the individual designated by $\mathfrak{a}_j$ has the property designated by $\mathfrak{a}_i$.

b. An n-place atomic sentence $\mathfrak{a}_i(\mathfrak{a}_{j_1},\mathfrak{a}_{j_2},...,\mathfrak{a}_{j_n})$ with $n \geq 2$ is true if and only if the n-tuple comprising those individuals which are successively the extensions of $\mathfrak{a}_{j_1}$, of $\mathfrak{a}_{j_2}$, ..., and of $\mathfrak{a}_{j_n}$ belongs to the class which is the extension of $\mathfrak{a}_i$; i.e. if the relation designated by $\mathfrak{a}_i$ holds between the n individuals designated by $\mathfrak{a}_{j_1}$, by $\mathfrak{a}_{j_2}$, ..., and by $\mathfrak{a}_{j_n}$ respectively.

c. An identity sentence $\mathfrak{a}_i = \mathfrak{a}_j$ involving the individual constants $\mathfrak{a}_i$ and $\mathfrak{a}_j$ is true if and only if each of these two constants has the same individual as its extension.

Suppose we wish to decide by means of our definition of truth whether a given sentence $\mathfrak{S}_i$ of language B is true or false. Evidently we are required to go back to the specific value-assignment $\mathfrak{B}_1$, i.e. in effect to the rules of designation, as well as to the rules of evaluation. Even this does not suffice, however, if $\mathfrak{S}_i$ is a factual sentence—which is to say, neither L-true nor L-false. Here we must also bring to bear factual knowledge about the individuals of the domain in question. E.g. should $\mathfrak{S}_1$ be the atomic sentence '*Pa*', then T1a indicates that $\mathfrak{S}_1$ is true if and only if the moon is spherical. No more than this can be extracted from the semantical rules. What these rules have furnished here—and the same situation will obtain for any other factual sentence—is simply a *truth-condition*, i.e. a necessary and sufficient condition for the truth of the sentence. A final decision as to the truth or falsity of the factual sentence in question (whether the truth-condition given by the semantical rules is in fact satisfied or not) lies outside the province of semantics; it lies in the province of empirical science (more particularly here in astronomy).

26. RELATIONS BETWEEN SYNTACTICAL AND SEMANTICAL SYSTEMS

26a. Interpretation of a language. We have constructed two systems for language B—first a syntactical system, then a semantical system. The semantical system furnishes an *interpretation* of language B, since it contains rules which yield for each sentence $\mathfrak{S}_i$ of B a truth-condition p_i such that $\mathfrak{S}_i$ is true if and only if p_i. Once this truth-condition p_i is obtained, we "understand" $\mathfrak{S}_i$, we know what it "says" about the individuals of the domain in question, what its "meaning" is. $\mathfrak{S}_i$ says that p_i, i.e., $\mathfrak{S}_i$ says the individuals are of such a nature that the truth-condition is satisfied. The meaning of the sentence $\mathfrak{S}_i$ or, in technical terms (see **20**), its designatum is the proposition p_i. What we found to be the case respecting the illustrative system of the previous section, viz. the sentence '*Pa*' is true if and only if the moon is spherical, appears in these earlier terms as: the sentence '*Pa*' designates the proposition that the moon is spherical.

One who constructs a syntactical system usually has in mind from the outset some interpretation of this system. (This interpretation need not itself have a prior representation as a semantical system; and indeed, what prior representation it may have is normally non-systematic.) While this intended interpretation can receive no explicit indication in the syntactical rules—since these rules must be strictly formal—the author's intention respecting interpretation naturally affects his choice of the formation and transformation rules of the syntactical system. E.g. he chooses primitive signs in such a way that certain concepts (perhaps those of some given unsystematized theory) can be expressed. He chooses sentential formulas in such a way that their counterparts in the intended interpretation can appear as meaningful declarative sentences. His choice of primitive sentences must meet the requirement that these primitive sentences come out as true sentences in the interpretation. And his rules of inference must be such that if by one of these rules the sentence $\mathfrak{S}_j$ is directly derivable from a sentence $\mathfrak{S}_i$ (or from $\mathfrak{S}_{i_1}$ and $\mathfrak{S}_{i_2}$, where $\mathfrak{S}_i$ is $\mathfrak{S}_{i_1} . \mathfrak{S}_{i_2}$), then $\mathfrak{S}_i \supset \mathfrak{S}_j$ turns out to be a true sentence under the customary interpretation of '$\supset$'. These last requirements ensure that all provable sentences also come out true.

If in particular the purpose in constructing a syntactical system is to represent formally a part of logic, not a part of empirical science, then the transformation rules must be so chosen that each primitive sentence is *logically* true and $\mathfrak{S}_i$ *logically* implies $\mathfrak{S}_j$ whenever $\mathfrak{S}_j$ is directly derivable from $\mathfrak{S}_i$. A language for which rules of this kind are given is often called a "logical calculus"; e.g. in view of our syntactical rules, language B is a logical calculus of this sort.

Now the interpretation we intend for our language B has been presented systematically in the semantical system. Our syntactical system is so constructed that it mirrors formally certain logical relations holding between the sentences of B, but no factual knowledge expressible in B. Indeed, the following can be established: each primitive sentence of B is L-true in virtue of the semantical rules (admittedly, this is still controversial insofar as it concerns primitive sentences P12 about the number of individuals); and if by a rule of inference of B the sentence $\mathfrak{S}_j$ is directly derivable from $\mathfrak{S}_i$, then the sentence $\mathfrak{S}_i$ L-implies $\mathfrak{S}_j$. From this in turn it follows that every provable sentence is also L-true; and that if $\mathfrak{S}_j$ is derivable from $\mathfrak{S}_i$, then $\mathfrak{S}_i$ L-implies $\mathfrak{S}_j$.

However, the converse does not hold: not all L-true sentences of B are provable. In fact, it is impossible to construct a syntactical system of the usual kind—one with a finite number of primitive sentences or primitive sentential schemes, and with a finite number of rules of inference each of which applies only to finitely many premisses—whose provable sentences are all and only the L-true sentences of B. The general result here is that no syntactical system of the usual kind can encompass the arithmetic of natural numbers (with variables for natural numbers, and recursive definitions of arithmetical functions). However, the converse referred to above does hold under certain limitations: if an L-true sentence of B consists only of primitive signs and contains no variables except possibly individual variables, then this sentence is also provable on the basis of our rules of transformation.

For more detailed consideration of the relations between syntactical and semantical systems see [Semantics] and [Formalization]. The results of the last paragraph above are due to *Kurt Gödel*; see Hilbert and Bernays [Grundlagen] vol. II, and Kleene (Metamathematics].

26b. On the possibility of a formalization of syntax and semantics. In this chapter we have discussed syntactical systems and semantical systems in general, and especially such systems for language B. Our explanations were phrased in a non-formalized metalanguage, viz. English supplemented by some technical symbols. However, it is possible to formalize both syntax and semantics; and this is sometimes desirable for greater precision. We shall now illustrate this possibility by giving some basic definitions and axioms. Since the present topic goes beyond the boundaries of an introductory book, we shall restrict the exposition to some brief indications without detailed explanations. For the purpose of this formalization it would also be possible and useful to employ a symbolic metalanguage; for the sake of simplicity, however, we shall proceed as before and give our formulations in ordinary English except for a few technical symbols. The reader may omit this subsection **26b** since no reference will be made to it later.

The main purpose of the exposition which follows is to show a way of defining more exactly such concepts as *language*, *calculus*, *interpreted language*, and *interpreted calculus*. Earlier in this text we referred informally to the class $\mathfrak{a}$ of the signs of a language L and the class $\mathfrak{S}$ of the sentences of L; but we did not say what a language is. Now in a formalized system we may define the *language* L as the ordered pair $(\mathfrak{a},\mathfrak{S})$. The class $\mathfrak{A}$ of the expressions of L is defined as the class of all finite sequences whose members are elements of the class $\mathfrak{a}$. (An *n*-place sequence can be defined as a many-one relation between the *n* first natural numbers and the members of the sequence.) Then a syntactical axiom is adopted to the following effect: For any class $\mathfrak{a}$ and any class $\mathfrak{S}$, if $(\mathfrak{a},\mathfrak{S})$

is a language, then every element of $\mathfrak{S}$ is a finite sequence of elements of $\mathfrak{a}$, and every element of $\mathfrak{a}$ occurs as a member of some element of $\mathfrak{S}$.

A *calculus*, i.e., a language with syntactical rules of deduction, can then be defined as an ordered triple $(\mathfrak{a},\mathfrak{S},Dd)$, where Dd is the relation of direct derivability. This relation is here understood in a comprehensive sense such that the primitive sentences of the calculus are taken as directly derivable from the null class of sentences (cf. [Semantics] §26.). Direct derivability is a relation between a sentence and a finite (possibly empty) class of sentences; in this connection, therefore, axioms are laid down to the effect that every first-place member of Dd is an element of the class $\mathfrak{S}$, and that every second-place member is a finite subclass of $\mathfrak{S}$.

An *interpreted language*, i.e. a language for which a sufficient system of semantical rules is given, can be defined as an ordered triple $(\mathfrak{a},\mathfrak{S},D)$. Here an axiom would say that the first domain of the relation D is identical with the class $\mathfrak{S}$. If, as is usually done, an extensional metalanguage is used for semantics, then D is the relation of value assignment **(25a)** for the sentences of the language. E.g. "$D(\mathfrak{S}_1$, the moon is spherical)" means as much as "The sentence $\mathfrak{S}_1$ is true if and only if the moon is spherical". An axiom is stated to the effect that the relation D is many-one in a certain sense; more exactly, that for any p and q and any element $\mathfrak{S}_i$ of the class $\mathfrak{S}$: if $D(\mathfrak{S}_i,p)$ and $D(\mathfrak{S}_i,q)$, then p if and only if q. If on the other hand an intensional metalanguage, containing a modal operator, e.g. "it is necessary that", is used for semantics, then D is taken as the relation of *designation* (i.e. the relation between an expression and its intension, see **25b**) for sentences. E.g. "$D(\mathfrak{S}_1$, the moon is spherical)" means here as much as "The sentence $\mathfrak{S}_1$ designates the proposition that the moon is spherical". The axiom last mentioned is now replaced by the following: For any p and q and any element $\mathfrak{S}_i$ of the class $\mathfrak{S}$, if $D(\mathfrak{S}_i,p)$ and $D(\mathfrak{S}_i,q)$, then the propositions p and q are identical (i.e. it is logically necessary that p if and only if q). In either of these two metalanguages (extensional or intensional), truth with respect to any given interpreted language $(\mathfrak{a},\mathfrak{S},D)$ can be defined as follows: A sentence $\mathfrak{S}_i$ is *true* if and only if: for some p, $D(\mathfrak{S}_i,p)$ and p. (Cf. [Semantics] D12-1.)

[An alternative method applicable in either of the two metalanguages takes the relation D in a more comprehensive sense, as applying not only to sentences but to a more comprehensive class $\mathfrak{D}$ of so-called designators. (E.g. in language B the relation D may also apply to all closed expressions of the type system (see **21b**).) By this method an interpreted language is an ordered quadruple $(\mathfrak{a},\mathfrak{S},\mathfrak{D},D)$. Here axioms are laid down to the effect that every element of $\mathfrak{D}$ is a finite sequence of elements of the class $\mathfrak{a}$; that the class of the first-place members of D is the class $\mathfrak{D}$; and that $\mathfrak{S}$ is a subclass of $\mathfrak{D}$. A third and still more explicit method demands for the specification of an interpreted language the indication also of the class $\mathfrak{d}$ of descriptive signs of the language **(5b)**. In this method, an interpreted language is a quintuple $(\mathfrak{a},\mathfrak{d},\mathfrak{S},\mathfrak{D},D)$. Then an axiom is added which says that $\mathfrak{d}$ is a subclass of $\mathfrak{a}$. This most explicit form is convenient as a basis for definitions of the concepts of models, of value-assignments, of the range of a sentence, of L-truth, and other L-concepts (see **11**).]

Finally, an *interpreted calculus* is a language for which both syntactical rules of deduction and semantical rules of interpretation are given. An interpreted calculus can therefore be defined as an ordered quadruple $(\mathfrak{a},\mathfrak{S},Dd,D)$. Here axioms are stated like those for a calculus and others like those for an interpreted language. Sometimes we wish to require that the relation Dd be truth-preserving, i.e., that any sentence which is directly derivable from true sentences is itself true. We can formulate an axiom to this effect in the following way, without use of the term "true". For any $\mathfrak{S}_{i_1},\ldots,\mathfrak{S}_{i_n}$, $\mathfrak{S}_j$, $p_1,\ldots,p_n$, q, if $Dd(\mathfrak{S}_j,\{\mathfrak{S}_{i_1},\ldots,\mathfrak{S}_{i_n}\})$, $D(\mathfrak{S}_{i_1},p_1)$ and ... and $D(\mathfrak{S}_{i_n},p_n)$ and $D(\mathfrak{S}_j,q)$ and p_1 and ... and p_n, then q. [For the concept of an interpreted calculus, there are alternative, more explicit definitions which include $\mathfrak{D}$, or both $\mathfrak{d}$ and $\mathfrak{D}$, in analogy to the alternative definitions of the concept of an interpreted language given above.]

Incidentally, it is possible to give a definition of a calculus in a simpler form, using instead of the triple $(\mathfrak{a}, \mathfrak{S}, Dd)$ simply the relation Dd. The class $\mathfrak{S}$ can be defined as the class containing all first-place members of Dd and all elements of second-place members of Dd and nothing else. The class $\mathfrak{a}$ can then be defined as the class of all members of the sequences which are elements of the class $\mathfrak{S}$. Similarly, $\mathfrak{a}$ and $\mathfrak{S}$ may be omitted in the definitions of an interpreted language and of an interpreted calculus. However, the earlier formulations of the definitions, the ones which refer explicitly to the classes $\mathfrak{a}$ and $\mathfrak{S}$, seem to be easier to understand and to work with.

Chapter C

The extended language C

27. THE LANGUAGE C

The language A described in Chapter A contains sufficiently many forms of expression to allow the formulation of most axiom systems and scientific theories. To carry out such a formulation, arbitrary constants of suitable types are chosen as primitive signs for some of the concepts of the theory or system in question, in such a way that constants for the other concepts can be introduced by definition. Later, in Part II, we will give examples of the formulation of axiom systems in language A.

The present chapter, C, describes an extended language C. This language C contains all the forms of expression of language A except sentential variables. [Such variables appeared in A because they facilitate the statement of tautological formulas; they occur in A not in sentences, but only in open sentential formulas. Sentential variables are seldom useful in the formulation of scientific theories.] Thus, all the *sentences* of A are also sentences of C. Language C contains in addition a number of other forms of expression which often permit briefer and more perspicuous formulations of axioms and scientific sentences than can be obtained in A. All the illustrative sentences (axioms and the like) of Part II are formulated in language C. Most of these sentences are also formulated in language A, so the two formulations can readily be compared in abbreviation and simplicity. Some sentences will only be formulated in C because their formulation in A is too cumbersome.

Language B, dealt with in the preceding chapter, contains all the forms of expression found in A except sentential variables and the constants defined in **17c**, **18a**, and **19**. Since these latter constants can always be eliminated from any context (by means of their definitions), it is clear that every sentence of A is translatable into a sentence of B. With the exception of 'λ', the new constants that appear in language C can similarly be eliminated from any sentence by means of definitions or analogous rules given for them; since B contains the λ-operator to begin with, all sentences of C are likewise translatable into B.

In Chapter B we laid down rules of formation for expressions in language B; these rules specified what forms of expression (sentences, sentential formulas, and expressions of the type system) were to be admitted into B. We do not explicitly lay down corresponding rules of formation for language

C. Instead, we simply assume that all forms of expression admitted into B are also admitted into C, as well as forms resulting from the introduction of new signs in C. Again, we laid down syntactical rules of transformation for expressions in B, and by means of these rules defined the concepts of provability and derivability in B. If in the present chapter we say that a certain sentence of language C is provable (or derivable from certain other sentences of C), we mean that the translation of this sentence into B is provable in B (or derivable in B from the corresponding premisses). Finally, we laid down semantical rules for B, on the basis of which we defined L-concepts, F-concepts, truth, and other semantical concepts. When any such concept is applied in the present chapter to a sentential formula of C, we mean again that the concept in question applies in B to the translation of the cited formula into B.

As we did for language A, so here for language C we frequently state theorems about the L-truth of certain sentential formulas. Here, as before, if an open sentential formula $\mathfrak{S}_i$ is L-true, so is each formula obtained from $\mathfrak{S}_i$ by prefixing arbitrary universal quantifiers—and in particular the sentence (sometimes called the "closure" of $\mathfrak{S}_i$) obtained by prefixing a universal quantifier for each variable occurring free in $\mathfrak{S}_i$. Further, each formula is L-true which is obtained from this $\mathfrak{S}_i$ by arbitrary substitutions for its free variables—in particular, each sentence so obtained by substituting closed expressions for the free variables of $\mathfrak{S}_i$. Is is to be noted that every sentence of language C specified in this chapter as L-true has a translation into language B which is L-true in virtue of the semantical rules for B and provable in virtue of the syntactical rules for B.

In giving illustrative sentences of language C we often omit brackets, just as we did in language A. (These brackets, it need hardly be said, cannot be omitted from any such sentence when its formulation is to be complete in the strict sense of the rules.) This omission of brackets is governed, in the first instance, by the conventions given in **3c** and **9a**; additional conventions will be specified later.

28. COMPOUND PREDICATE EXPRESSIONS

28a. Predicate expressions. We now introduce compound predicate expressions. These expressions are formed from predicates or predicate expressions with the help of connectives heretofore used only for combining sentential formulas. The new compounds are as follows: '$(P \vee Q)a$' is counted as an abbreviation of the sentence '$Pa \vee Qa$', and '$P \vee Q$' treated as a predicate expression of the same type as 'P', viz. a one-place predicate expression of the first level and of type (0); '$(P\,.\,Q)a$' is counted as an abbreviation of '$Pa\,.\,Qa$', and '$P\,.\,Q$' as a predicate expression; '$(P \supset Q)a$' as an abbreviation of '$Pa \supset Qa$', with '$P \supset Q$' a predicate expression; '$(P \equiv Q)a$' as an abbreviation of '$Pa \equiv Qa$' and '$P \equiv Q$' a predicate expression; and '$(\sim P)a$' as a reformulation of '$\sim(Pa)$' or '$\sim Pa$', with '$\sim P$' a predicate expression. We agree to employ such abbreviations only when the two predicates written separately have the same argument-expressions.

Sometimes the technique of abbreviation just illustrated can be applied more than once in the same sentence, in which case it leads to still other compound predicate expressions. E.g. '$P_1a \vee P_2a \supset \sim P_3a$' abbreviates first into '$(P_1 \vee P_2)a \supset (\sim P_3)a$', and then into '$((P_1 \vee P_2) \supset (\sim P_3))a$'; this last, by the conventions of **3c** regarding omission of parentheses, comes to '$(P_1 \vee P_2 \supset \sim P_3)a$', whence we have the more elaborate predicate expression '$P_1 \vee P_2 \supset \sim P_3$'. The compounding of predicate expressions is also possible when the predicates are of any other type; e.g. '$(R_1 \supset R_2)(a,b)$' is the abbreviation for '$R_1(a,b) \supset R_2(a,b)$', as is '$(M_1 \vee M_2)(P)$' for '$M_1(P) \vee M_2(P)$'.

To legitimatize this use of connectives in building up compound predicate expressions we introduce the following definitions for predicate expressions of the simplest type. Analogous definitions are understood to hold for each other type of predicate variable.

D28-1. **a.** $(\sim F)x \equiv \sim(Fx)$.
b. $(F \vee G)x \equiv Fx \vee Gx$.
c. $(F.G)x \equiv Fx.Gx$.
d. $(F \supset G)x \equiv (Fx \supset Gx)$.
e. $(F \equiv G)x \equiv (Fx \equiv Gx)$.

Compound predicate expressions can appear as argument-expressions of higher-level predicates. In language A we could only use the cardinal predicates '0', '1', etc., on predicates, not on compound predicate expressions. Thus e.g. to translate into A the sentence "There are 5 (individuals) which are P_1 and P_2", we first had to define a predicate 'Q' by '$Qx \equiv P_1x.P_2x$' before giving the formulation: '$5(Q)$'. In language C we can avoid the introduction of any new predicate, and simply use the predicate expression '$P_1.P_2$' to write: '$5(P_1.P_2)$'.

28b. Universality. A property of individuals is called *universal* provided every individual has this property. Correspondingly, in the terminology of classes: a class of individuals is universal provided every individual belongs to this class. Generally, a class of any type is said to be *universal* if each entity of that type belongs to this class. Our symbol for universality is 'U'; and "the class (or property) P is universal" is rendered '$U(P)$'. Since '$U(P)$' is clearly synonymous with '$(x)Px$', the following definition is natural:

D28-2. $U(F) \equiv (x)Fx$.

Analogous definitions are understood to hold for predicates of any other type, be they one-place or many-place. E.g. '$(x)(y)Rxy$' can be abbreviated '$U(R)$'. In general, given an n-place predicate expression $\mathfrak{A}_j$ (with $n \geq 1$) of arbitrary type and arbitrary composition, we take $U(\mathfrak{A}_j)$ to be the abbreviation for $(\mathfrak{v}_{i_1})(\mathfrak{v}_{i_2})...(\mathfrak{v}_{i_n})[\mathfrak{A}_j(\mathfrak{v}_{i_1},\mathfrak{v}_{i_2},...,\mathfrak{v}_{i_n})]$. We make frequent use of 'U' in such abbreviated formulations, especially in connection with compound predicate expressions. E.g. the sentence '$(x)(Px.\sim Qx)$' is first written '$(x)(P.\sim Q)x$' and then '$U(P.\sim Q)$'.

If '$(x)(Px \supset Qx)$', i.e. '$U(P\supset Q)$', holds, then we say that P is *contained* (or: *included*) in Q; in the terminology of classes, P is a *subclass* of Q. Our notation for this is '$P\subset Q$'. Similarly, when '$(x)(y)(Rxy \supset Sxy)$' or '$U(R\supset S)$' holds, we say that R is included in S, or R is a *subrelation* of S, and write '$R\subset S$'. The definition:

D28-3. $(F\subset G) \equiv U(F\supset G)$.

Analogous definitions are understood to hold for predicate expressions of any other type. Generally: if $\mathfrak{A}_i$ and $\mathfrak{A}_j$ are n-place predicate expressions of the same type, then $\mathfrak{A}_i\subset\mathfrak{A}_j$ is taken as an abbreviation for $U(\mathfrak{A}_i\supset\mathfrak{A}_j)$, which in turn is an abbreviation for $(\mathfrak{v}_{k_1})...(\mathfrak{v}_{k_n})[\mathfrak{A}_i(\mathfrak{v}_{k_1},...\mathfrak{v}_{k_n}) \supset \mathfrak{A}_j(\mathfrak{v}_{k_1},...,\mathfrak{v}_{k_n})]$.

Suppose $U(\mathfrak{A}_i)$ is a sentential formula, where $\mathfrak{A}_i$ is any (open or closed) predicate expression. If $U(\mathfrak{A}_i)$ stands alone, i.e. is not a part of a larger formula, then we consider it legitimate to suppress 'U' and write simply $\mathfrak{A}_i$. Thus e.g. we write '$P\vee Q$' instead of the sentence '$U(P\vee Q)$' and '$F.\sim G$' instead of the sentential formula '$U(F.\sim G)$'. If a list of L-true sentential formulas given in a theorem includes a predicate expression $\mathfrak{A}_i$, what is indicated thereby is that $U(\mathfrak{A}_i)$ is an L-true sentential formula. When $U(\mathfrak{A}_i)$ is a component of a larger formula, the 'U' must not be suppressed, because otherwise the difference between the following two cases would be obliterated: (1) '$\sim U(P)$', an abbreviation for '$\sim(x)(Px)$', which says "not every individual is P"; and (2) '$U(\sim P)$', an abbreviation for '$(x)(\sim Px)$', which says "no individual is P". We may suppress the 'U' in case (2) and write simply '$\sim P$'; we may not suppress the 'U' in case (1). One last remark. Since 'U' is applicable to predicate expressions of arbitrary type, we take our notational suppression of 'U' to be, too. E.g. taking 'M' to be a one-place predicate of the second level, we can abbreviate '$(F)M(F)$' to '$U(M)$' and this in turn—provided it stands alone—to 'M'.

A class (or property) is said to be *empty* or *null* provided no entity of the appropriate type belongs to it; and otherwise, *non-empty* or *non-null*. Our symbol for non-emptiness is '$\exists$'; and "the class P is not empty" is rendered '$\exists(P)$'. Since '$\exists(P)$' is clearly synonymous with '$(\exists x)Px$', and '$\exists(R)$' with '$(\exists x)(\exists y)Rxy$', the following definition is natural:

D28-4. $\exists(F) \equiv (\exists x)Fx$.

Analogous definitions are understood to hold for predicates of any other type. Generally: if as above $\mathfrak{A}_j$ is an arbitrary predicate expression, we take $\exists(\mathfrak{A}_j)$ to be the abbreviation for $(\exists\mathfrak{v}_{i_1})...(\exists\mathfrak{v}_{i_n})[\mathfrak{A}_j(\mathfrak{v}_{i_1},...,\mathfrak{v}_{i_n})]$. In contradistinction to 'U', the symbol '$\exists$' may not be suppressed in *any* case.

Again, the use of '$\exists$' for abbreviated formulations has special advantage with compound predicate expressions. E.g. the sentence '$(\exists x)(Px.Qx)$' is first transformed into '$(\exists x)(P.Q)x$', and then into '$\exists(P.Q)$'; similarly, '$(\exists x)(\exists y)(Rxy \vee Sxy)$' can be abbreviated '$\exists(R \vee S)$'.

The formulas displayed in the following theorem represent simple applications of D1, D2, D3 and D4 to the formulas given in T14-1,2. Analogous results obtain, of course, for predicate variables of any other type.

T28-1. The following sentential formulas are L-true:

a. $\sim U(F) \equiv \exists(\sim F)$.
b. $\sim\exists(F) \equiv U(\sim F)$.
c. $U(F) \equiv \sim\exists(\sim F)$.
d. $\exists(F) \equiv \sim U(\sim F)$.
e. $U(F \supset G) \equiv \sim\exists(F.\sim G)$.
f. $U(F.G) \equiv U(F).U(G)$.
g. $\exists(F \vee G) \equiv \exists(F) \vee \exists(G)$.
h. $F \subset G \supset [U(F) \supset U(G)]$.
i. $F \subset G \supset [\exists(F) \supset \exists(G)]$.
j. $U(F \equiv G) \equiv (F \subset G).(G \subset F)$.
k. $U(F \equiv G) \supset [U(F) \equiv U(G)]$.
l. $U(F \equiv G) \supset [\exists(F) \equiv \exists(G)]$.
m. $\exists(F.G) \supset \exists(F).\exists(G)$.
n. $U(F) \vee U(G) \supset U(F \vee G)$.
o. $(F \subset G).\exists(F.H) \supset \exists(G.H)$.
p. $U(F) \supset \exists(F)$.
q. $(F \subset G).(G \subset H) \supset (F \subset H)$.

28c. Class terminology. In the word language we sometimes speak of properties, sometimes of the "corresponding" classes. The difference, however, is only in our mode of speech; hence it is unnecessary to include in our symbolic object language, parallel with predicates, other special expressions for their corresponding classes. Any predicate expression of language C may be used both as an expression for a property and as an expression for the corresponding class. In translating e.g. the sentence '*Pa*' into the word language, we may at our pleasure use either the terminology of properties (thus reading '*Pa*' as "*a* has property *P*") or the terminology of classes (reading '*Pa*' as "*a* belongs to class *P*" or "*a* is an element of class *P*"). It is because these two word-language versions have the same meaning that we can dispense with two different symbolic paraphrases of them in C. Word-language versions of predicate expressions are often more compact and perspicuous when phrased in the terminology of classes. Thus e.g. the sentence '$(P \vee Q)a$' is customarily translated "*a* belongs to the *union* of classes *P* and *Q*", the sentence '$(P.Q)a$' as "*a* belongs to the *intersection* of classes P and Q", and the sentence '$(\sim P)a$' as "a belongs to the *complement* of class P".

Suppose we pick from language A an arbitrary tautologous sentential formula with sentential variables, e.g. '$\sim(p \vee q) \equiv \sim p . \sim q$'. By substitution we can obtain from this tautology another tautological sentential formula with predicate variables, e.g. '$\sim(Fx \vee Gx) \equiv \sim Fx . \sim Gx$'. Prefixing to this last the universal quantifier '(x)' (in view of T13-1e) and using our 'U' abbreviation, we obtain the L-true formula '$U[\sim(F \vee G) \equiv \sim F . \sim G]$' and so finally '$\sim(F \vee G) \equiv \sim F . \sim G$'. In this way there can be associated with each tautology containing sentential variables a precisely analogous L-true formula containing predicate variables of any type. Thus we obtain formulas of the so-called *calculus of classes* of earlier systems—with the difference, however, that here we obtain them in a direct simple way from the predicates themselves, without the use of any special class expressions. Examples of such formulas are given in the following theorem; each of these formulas is secured from a tautology of language A in the way indicated above, the tautologies employed being those of T8-1,2,6. (Note: when '$\supset$' occurs as a principal sentential connective in a tautology, it is transformed into '$\subset$' in accordance with D3.) We remark again that formulas analogous to those below also hold for predicate variables of any other type.

+T28-2. The following sentential formulas of language C are L-true:

a. $F \vee \sim F$.
b. $\sim(F . \sim F)$.
c. $F \subset F \vee G$.
d. $F . G \subset F$.
e. $F . \sim F \subset G$.
f. $(F \vee G) . \sim F \subset G$.
g. $F \vee G \equiv G \vee F$.
h. $F . G \equiv G . F$.
i. $\sim(F \vee G) \equiv \sim F . \sim G$.
j. $\sim(F . G) \equiv \sim F \vee \sim G$.
k. $F . (G \vee H) \equiv (F . G) \vee (F . H)$.
l. $F \vee (G . H) \equiv (F \vee G) . (F \vee H)$.
m. $F \equiv (F \vee G) . (F \vee \sim G)$.
n. $F \equiv (F . G) \vee (F . \sim G)$.
o. $F \equiv F \vee (F . G)$.
p. $F \equiv F . (F \vee G)$.
q. $F \vee G \equiv F \vee (G . \sim F)$.
r. $F . G \equiv F . (G \vee \sim F)$.

28d. Exercises. Translate the following sentences, using compound predicate expressions and omitting 'U' wherever possible. — **1.** "Every book is blue". — **2.** "Not every book is blue". — **3.** "No book is blue" (i.e. "every book is not-blue"). — **4.** "There is a blue book". — **5.** "There is a not-blue book". — **6.** "There are (exactly) 5 blue books".

— **7.** "Fathers are male" (use 'mem_1'; cf. **18a**). — **8.** "There are even (numbers) and odd (numbers)". — **9.** "There are no (numbers) which are both even and odd". — **10.** "Every (natural number) belongs to the first-domain of the predecessor relation". — **11.** "Not every (natural number) belongs to the second-domain of the predecessor relation" (e.g., 0 does not). — **12.** "2 is an even prime (number)".

29. IDENTITY. EXTENSIONALITY

29a. Identity. In language C we use the identity sign '=' not only (as in language A; cf. **17**) between individual expressions, but also (as in language B) between predicate expressions and between functor expressions. What identity means here is what it meant in language A, viz. agreement in all properties. E.g. the sentence '$P=Q$' says that every property of properties possessed by property P is also possessed by property Q, and conversely; thus '$P=Q$' is synonymous with the sentence '$(N)[N(P) \equiv N(Q)]$'. If, therefore, '$P=Q$' and some other sentence '..P..P..' about P both hold, then the corresponding sentence '..Q..Q..' about Q also holds. Similarly, if 'k_1' and 'k_2' are functors, the sentence '$k_1=k_2$' says that the function k_1 has all the properties that function k_2 has, and conversely; so that again '$k_1=k_2$' is synonymous with '$(N)[N(k_1) \equiv N(k_2)]$'. And further, given '$k_1=k_2$' and another sentence '..k_1..k_1..' about k_1, then the sentence '..k_2..k_2..' about k_2 also holds. Correspondingly, the theorem regarding interchangeability on the basis of identity (it is T24-7(b)) holds in C.

The identity principle P8 of language B (see **22a,b**) is in accord with what has just been said. With its help e.g. '$Pa \supset Pb$' is derivable from '$a=b$' on the one hand, and (by substituting '$\sim P$') '$\sim Pa \supset \sim Pb$' on the other; from this last by transposition (cf. T8-6i(1)) comes '$Pb \supset Pa$', which together with '$Pa \supset Pb$' leads us to '$Pa \equiv Pb$'. Thus we see that it is adequate to phrase P8 with the conditional sign.

The following theorem tells us that identity is (totally) reflexive, symmetric and transitive.

+T29-1. Suppose $\mathfrak{A}_i$, $\mathfrak{A}_j$ and $\mathfrak{A}_k$ are expressions of the type system; then a sentential formula having one of the following forms is L-true:

a. $\mathfrak{A}_i = \mathfrak{A}_i$.
b. $\mathfrak{A}_i = \mathfrak{A}_j \supset \mathfrak{A}_j = \mathfrak{A}_i$.
c. $(\mathfrak{A}_i = \mathfrak{A}_j \,.\, \mathfrak{A}_j = \mathfrak{A}_k) \supset \mathfrak{A}_i = \mathfrak{A}_k$.

As earlier (see D17-1b), so here we write '$\neq$' for "non-identical"; in the present context, of course, '$\neq$' can stand between two expressions of any one type. Non-identity is frequently used when the word "two" appears in a verbal text. E.g. "For any two points, there are ..." is rendered '$(x)(y)[Pt(x).Pt(y).(x \neq y) \supset (\exists z)(...)]$'.

Instances of the use of the identity sign between predicate expressions may be found in T29-3, T30-1, and D30-2; and of similar usage respecting functor expressions in **33c**.

Sometimes we find it convenient to use '*I*' as a conventional predicate designating identity, and similarly '*J*' for non-identity—a practice that has proved advantageous in connection with other two-place predicates. Moreover, we can use '$J_3(a,b,c)$' as a compact way of saying that *a*, *b* and *c* are three different individuals; 'J_4' can have a corresponding role respecting four arguments, etc.

D29-1. $Ixy \equiv x=y$.

D29-2. **a.** $Jxy \equiv x \neq y$.

b. $J_3xyz \equiv (x \neq y \,.\, x \neq z \,.\, y \neq z)$.

c. $J_4xyzu \equiv (x \neq y \,.\, x \neq z \,.\, x \neq u \,.\, y \neq z \,.\, y \neq u \,.\, z \neq u)$.

29b. Regarding the types of logical constants. In view of D29-1, the predicate '*I*' should only occur with individual arguments, as e.g. in '*Iab*'. However, since '=' is also used between expressions of higher levels, we agree to extend the scope of '*I*' correspondingly and so to admit sentences like '$I(P,Q)$', '$I(R,S)$', '$I(M_1,M_2)$', etc. These arguments being of different types in different contexts, the same is true of '*I*'. [Thus in the sentence '*Iab*', '*I*' is of type (0,0) and the first level; in '$I(P,Q)$', '*I*' is of type ((0),(0)) and the second level; in '$I(R,S)$', it is of type ((0,0),(0,0)) and the second level; and in '$I(M_1,M_2)$', it is of type (((0)),((0))) and the third level.] Strictly speaking, the type rules we laid down earlier do not permit such an extension: to write sentences like '*Iab*', '$I(P,Q)$', etc., in strict adherence to our formation rules, we would have to use in place of a single sign '*I*' a series of different signs—one for each of the types in which it is used. (These different signs might e.g. be made up by adding the particular type designation to '*I*' as a subscript: '$I_{(0,0)}$', '$I_{((0),(0))}$', etc.) However, entirely parallel theorems would hold for these different signs; hence in practice it is convenient to suppress distinctive notations (e.g. the type-designation subscripts) and simply use '*I*'. The type of '*I*' in any given sentence is to be gathered from the context.

What we have just noted about the predicate '*I*' applies with equal force to the cardinal numbers '0', '1', '2', etc. introduced in **17c**. Theoretically, there are cardinal numbers of the second level, of the third level, etc.; and we ought properly to give them distinctive notations—e.g. '$3_{((0))}(P)$', '$3_{(((0)))}(M)$', etc. For each such kind of cardinal number however, the familiar theorems of arithmetic hold in the same way. Hence we suppress the notational distinctions, write simply '3(*P*)', '3(*M*)', etc., and leave it to the context (viz. the type of the argument-expression) to determine precisely the type of '3'. The same applies to arithmetical theorems: we give them only once, without regard to type distinctions; e.g. we write simply '*sum*(2,3)=5'—i.e. "2+3=5" (see **37b**). Strictly, the expression '*sum*(2,3)=5' is not a sentence of language C; it represents rather the infinite class of sentences obtainable from it by subjoining to each of the signs '2', '3' and '5' the same type index [of the form $((t_i))$] and simultaneously to the functor sign '*sum*' another suitable index [viz. $(((t_i)),((t_i)):((t_i)))$].

For each logical constant written without a type index (more properly, for each family of related logical constants differing only in type) there is a simplest type which we will call the *basic type* of this constant. E.g. the basic type of '*I*' is (0,0), that of '2' is ((0)), and that of '*sum*' is (((0)),((0)) : ((0))). [The types specified for the various logical constants given in the Examples of **21b** are their basic types.]

Theoretically, therefore, language C (as well as the usual systems with a distinction of types) has an infinite multiplicity of arithmetics: one for the cardinal numbers of the second level (referring to classes of the first level), another for the cardinal numbers of the third level (referring to classes of the second level), and so on. The question arises, Can this multiplicity of arithmetics be avoided without giving up the distinction of types? Our device of suppressing the type-index subscripts has the practical merit of reducing the multiplicity of arithmetics to a *single* system of arithmetical formulas, but naturally the theoretical multiplicity remains. One possible way of avoiding this multiplicity consists in adding *transfinite levels*. Using the transfinite ordinal numbers of set theory, we designate the lowest level that is higher than all the finite levels as level ω, the next beyond this as level $\omega+1$, etc. There is then put forward a rule of formation specifying that a predicate of any transfinite level can take argument-expressions of any lower level whatever. As before, descriptive predicates continue to be assigned finite levels (e.g. '*P*' the first level, '*M*' the second, etc.); but now, logical constants are assigned transfinite levels. As for variables, they can either be assigned fixed levels and types (as we have done in our languages) or left unassigned. If the signs '0', '1', etc., for cardinal numbers are assigned level ω, then '3(*P*)', '3(*M*)', etc., turn out to be proper sentences of the language (and not merely ambiguous abbreviations, as is the case with us). Similarly, '*sum*(2,3)$=$5' is a sentence of the language, and '*sum*' is a functor-expression of level $\omega+1$. In this fashion we arrive at *one* arithmetic applicable to descriptive classes of any finite level. Up to the present, the use of transfinite levels has not been studied very extensively; only brief references to it have been made by Hilbert and Gödel (see [Syntax] §53) and by Tarski ([Wahrheitsbegriff] 136 f, [Metamathematics] 270 ff.). The first attempt at a system of this kind: Frank G. Bruner, *Mathematical logic with transfinite types*, privately printed, Chicago, 1943 (see the review in *Jour. Symb. Logic*, 9, 1944, p. 72).

29c. Extensionality. The sentence '$(x)(Px\equiv Qx)$', more briefly '$P\equiv Q$' in view of **28b**, asserts that properties *P* and *Q* both attach to the same individuals, i.e. *P* and *Q* have the *same extension*. This can be the case (i.e. '$P\equiv Q$' can be true) even when '*P*' and '*Q*' have different meanings. If, however, '$P\equiv Q$' is L-true, then '*P*' and '*Q*' have the same meaning. In contrast to '$P\equiv Q$', which says that *P* and *Q* agree in the individuals they apply to, the sentence '$P=Q$' says *P* and *Q* agree in the properties (of second level) that apply to *them*. Suppose a second-level property, say *M*, is such that *M* applies to every property having the same extension as *P* as soon as it applies to *P*; then we say that *M* is *extensional*, i.e. depends only on the extension or scope. E.g. the cardinal number 5 is an extensional property of the second level, since from '5(*P*)' and '$P\equiv Q$' there follows '5(*Q*)'. Indeed, it can be shown that all second-level properties definable by expressions of language C (either expressions introduced so far, or still to be introduced) are extensional; and the same holds for properties of higher levels. Thus in language C it is the case that from '$P\equiv Q$' follows '$P=Q$'. Since further '$P\equiv Q$' follows from '$P=Q$', it is the case that the two sentences '$P\equiv Q$' and '$P=Q$' are synonymous in language C. Further, '$(x)(Px\equiv Qx)$' and '$P=Q$' are technically L-equivalent in language B, for we have there admitted only extensions as possible values for value-assignment (see **25a**). Moreover, in B these two sentences are derivable from each other with the help of the primitive-sentence schema P9 (cf. **22a**). The same holds for predicate expressions of all other types, as well as for functor expressions of any type. Our object languages are therefore all *extensional languages*.

+T29-2. The sentences $(\mathfrak{v}_{k_1})(\mathfrak{v}_{k_2})...(\mathfrak{v}_{k_n})(\mathfrak{A}_i(\mathfrak{v}_{k_1},\mathfrak{v}_{k_2},...,\mathfrak{v}_{k_n})\ \mathfrak{a}_m\ \mathfrak{A}_j(\mathfrak{v}_{k_1},\mathfrak{v}_{k_2},...,\mathfrak{v}_{k_n}))$ and $\mathfrak{A}_i=\mathfrak{A}_j$ are L-equivalent, where $n\geq 1$ and either (a) $\mathfrak{A}_i$ and $\mathfrak{A}_j$ are *n*-place predicate expressions of the same type and $\mathfrak{a}_m$ is '$\equiv$', or (b) $\mathfrak{A}_i$ and $\mathfrak{A}_j$ are *n*-place functor expressions of the same type and $\mathfrak{a}_m$ is '$=$'.

T29-3. The following sentential formulas are L-true:

a. $U(F\equiv G) \equiv (F=G)$. (From T2.)
b. $U(F).U(G) \supset (F=G)$. (From (a).)
c. $\sim\exists(F).\sim\exists(G) \supset (F=G)$. (From (a).)
d. $(F\subset G).(G\subset F) \equiv (F=G)$. (From (a), T28-1j.)

Any two classes which contain each other are identical.

Non-extensional predicates (whose argument-expressions are sentences, predicate expressions or functor expressions) occur in certain logical systems, e.g. the logic of modalities. If it is desired to introduce such predicates into our object language, then the primitive schema P9 must be omitted from the syntactical system, and intensions (rather than extensions) taken as possible values in the semantical system. As we have previously remarked (in **10b**), non-extensional languages are substantially more complicated than extensional ones. On the other hand, it appears that everything which to date has been expressed in terms of non-extensional predicates can be expressed (in a different way, to be sure) without such predicates, i.e. in an extensional language. I am inclined to think that this is the case not only for the non-extensional predicates known so far, but in general (a conjecture known as the "thesis of extensionality"). On this point, cf. [Syntax] §§65–67; [Meaning] §11 and §32 (Method V).

30. RELATIVE PRODUCT. POWERS OF RELATIONS

30a. Relative Product. This section and the next treat the main concepts in the logic of (two-place) relations and introduce symbols for them. By the *relative product* of the relation *R* by the relation *S* is meant that relation which exists between *x* and *y* if and only if there is a *u* such that *x* bears the relation *R* to *u* and *u* bears the relation *S* to *y*. For "the relative product of *R* by *S*" we use the symbol '$R|S$'. Thus:

D30-1. $(H|K)xy\equiv(\exists u)(Hxu.Kuy)$.

We see that '$(R|S)ab$' means: "*a* is an *R* of an *S* of *b*" (e.g. "... is a son of a brother of ...", "... is greater than half of ...").

The stroke '|' has the same logical character as a functor. It differs from what we have called functors only in the unessential detail that it stands *between* its two argument-expressions rather than *before* them. The same remark applies to certain connective signs we will shortly introduce, and which for the sake of definiteness we exhibit here (in each case, between two letters): '*R*"*P*' (D32-6a), '*k*"*P*' (D32-6b), '*R in P*' (D32-7), and '*R*'*b*' (D35-2).

With a view toward *economy in the use of brackets*, let us agree now that all the new signs mentioned in the previous paragraph are *more cohesive* than the signs '$\vee$', '.', '$\supset$', '$\equiv$', '$\subset$', and '=' between predicate expressions (the last sign also between individual expressions). If, therefore, $\mathfrak{A}_i$ is a full expression of the stroke or of any of the other new connectives, the brackets around $\mathfrak{A}_i$ may be omitted whenever $\mathfrak{A}_i$ enters as a component of one of the familiar connectives last mentioned. [E.g. parentheses may be omitted from the following expressions: '$(R|S) \vee (R \textit{ in } P)$', '$(R``P) \supset (k``Q)$', '$(R`b) = a$'; on the other hand, they may not be omitted from: '$(R_1 \vee R_2)|(S_1 . S_2)$', '$(R \vee S)``(P . Q)$'.] Further, as T1a will tell us, the relative product is associative, i.e. '$(R|S)|T$' is synonymous with '$R|(S|T)$'; hence we may omit parentheses from both expressions and write simply '$R|S|T$'. It should be remarked that the relative product is in general *not* commutative: '$R|S$' is generally not synonymous with '$S|R$'; e.g. "*a* is a friend of a teacher of *b*" is different from "*a* is a teacher of a friend of *b*."

Part (a) of the following theorem is the associative law for the relative product; parts (b) and (c) are distributive laws for the relative product respecting disjunction; parts (d) and (e) are the same respecting conjunction (observe that these parts merely claim inclusion, and not identity).

T30-1. Sentential formulas of the forms (a) through (f) below are L-true:

+**a.** $(H_1|H_2)|H_3 = H_1(H_2|H_3)$.
b. $H|(K_1 \vee K_2) = H|K_1 \vee H|K_2$.
c. $(K_1 \vee K_2)|H = K_1|H \vee K_2|H$.
d. $H|(K_1 . K_2) \subset H|K_1 . H|K_2$.
e. $(K_1 . K_2)|H \subset K_1|H . K_2|H$.
f. $\exists(H|K) \equiv \exists(mem_2(H) . mem_1(K))$.

Exercises. 1. Give informal proofs of the following: a) T1a; b) T1c. — **2.** Give counter-examples to the following: a) '$H|(K_1 . K_2) = H|K_1 . H|K_2$'; b) '$H \subset H|H$'.

30b. Powers of relations. We write 'R^2' as an abbreviation for '$R|R$', 'R^3' for '$R^2|R$', etc., and call these relations the (second, third, etc.) *powers of* R. Of these, the second power is used quite frequently (e.g. "... is a friend of a friend of ...", "... is the father of the father of ..."). Continuing the analogy with arithmetical exponents, let us agree that 'R^1' stands for the relation R itself (consequently the notation 'R^1' is scarcely ever seen), and that 'R^0' stands for the relation of identity between the members of R. Continuing the analogy into negative exponents, we take 'R^{-1}' as a designation for the *converse* (or inverse) of the relation R. I.e., R^{-1} is the relation comprising all the pairs of R, but with their members in the reverse order: '$R^{-1}ab$' is true just in case 'Rba' is. E.g. the relation Parent (i.e. "... is a parent of ...") is the converse of the relation Child (i.e. "... is a child of ..."), and conversely. The converse of Square is Square-root. We can (if we wish) continue on with other negative powers, writing 'R^{-2}' in place of

'$R^{-1}|R^{-1}$', or '$(R^{-1})^2$', or '$(R^2)^{-1}$'—all three of which turn out to be synonymous. Similarly for 'R^{-3}', etc. Our formal definitions follow:

D30-2. **a.** $H^0xy \equiv (x=y).mem(H)(x)$.
b. $H^1=H$.
c. $H^2=H|H$.
d. $H^3=H^2|H$.
Etc.

D30-3. $H^{-1}xy \equiv Hyx$.

The theorem below states properties of the converse relation; in particular, part (a) tells us that the converse of the converse of R is R itself.

T30-2. The following sentential formulas are L-true:

+**a.** $(H^{-1})^{-1}=H$.
b. $(H|K)^{-1}=K^{-1}|H^{-1}$.
c. $(H \vee K)^{-1}=H^{-1} \vee K^{-1}$.
d. $(H.K)^{-1}=H^{-1}.K^{-1}$.

The symbols here defined—and in general the constants of language C defined in this chapter—are intended to be applicable also to expressions of appropriate types on higher levels (cf. **29b**). E.g. '|' can be used between two-place homogeneous predicate expressions of arbitrary (equal) types; '*Sym*' (see D31-1a) likewise can take as an argument-expression a two-place homogeneous predicate expression of any type. And further, the theorems stated have corresponding versions appropriate to other types: which is to say, the technique of raising levels (T16-1) applies to these theorems.

Exercises. **1.** Give informal proofs of the following: a) T2a; b) T2b. — **2.** Decide whether the following are L-true (if so, give an informal proof; if not, give a counter-example): a) '$H|H^{-1}=H$'; b) '$H|H^{-1}=H^{-1}|H$'; c) '$H|H^{-1}=H^{-1}$'; d) '$mem_1(H)(x) \supset (H|H^{-1})(x,x)$'; e) '$U(mem_2(H) \supset mem_2(H^2))$'; f) '$U(mem_2(H)) \supset U(mem_2(H^2))$'.

30c. Supplementary remarks. If our language has a variable 'n' for natural numbers (0,1,2,...), as e.g. the language form specified in **40a** does, then the infinitely many definitions put forward in D30-2 can be contracted into a single recursive definition running as follows:

D30-2*. **a.** $H^0xy \equiv (x=y).mem(H)(x)$,
b. $H^{n+1}=H^n|H$.

Powers with negative exponents are defined thusly:

D30-3*. $H^{-n}xy \equiv H^nyx$.

If our language supplies variables 'm' and 'n' for integers (positive and negative whole numbers, and zero), then we have the following:

1. For any relation R: if m and n are non-negative, then '$R^m|R^n=R^{m+n}$' and '$(R^m)^n=R^{m.n}$' are L-true.
2. For any integers m and n: if R is a one-one relation, then '$R^m|R^n \subset R^{m+n}$' and '$(R^m)^n=R^{m.n}$' are L-true.

Examples for (1) are: '$R^0|R=R$'; '$R^3|R^2=R^5$'; '$(R^3)^2=R^6$'. In connection with (2), note that only inclusion—not identity—is claimed in the first formula; examples for (2) are: '$R^5|R^{-3}\subset R^2$'; '$(R^{-2})^2=R^{-4}$'. The contents of (1) and (2) are the practical basis of our definitions for powers of relations when the exponents are non-positive integers. Each particular instance (involving definite numbers as exponents) of theorems (1) and (2) is L-true in language C; e.g. '$Un_{1,2}(H)\supset(H^5|H^{-3}\subset H^2)$' is L-true in C. On the other hand, (1) and (2) themselves cannot appear in C because numerical variables do not occur as exponents in C.

Examples. Using our notations for relative product and for powers of relations, we can give more concise definitions of certain family relations (recall the discussion in **15c**(II) and **17b**).

("Child")	1. $Ch=Par^{-1}$.
("Brother")	2. $Bro=(Son|Fa).(Son|Mo).J$.
("Grandparent")	3. $GrPar=Par^2$.
("Grandfather")	4. $GrFa=Fa|Par$.
("Grandchild")	5. $GrCh=Ch^2$.
("Grandson")	6. $GrSon=Son|Ch$.
("Wife")	7. $Wif=Hus^{-1}$.
("Spouse")	8. $Sp=Hus\vee Wif$.
("Brother-in-law")	9. $BroL=Bro|Wif\vee Hus|Sis\vee Bro|Hus$.
("Half-brother")	10. $HaBro=Son|Par.\sim Bro.J$.
("Father-in-law")	11. $FaL=Fa|Sp$.
("Uncle")	12. $Unc=(Bro\vee BroL)|Par$.

Exercises. **1.** In the system of family relations just described, define the following relational concepts: a. "Sister"; b. "Grandmother"; c. "Grand-daughter"; d. "Sister-in-law"; e. "Half-sister"; f. "Mother-in-law"; g. "Son-in-law"; h. "Daughter-in-law"; i. "Aunt"; j. "Nephew"; k. "Niece". — Translate the following sentences: — **2.** "*a* is (the) father of a friend of *b*". — **3.** "Sometimes (i.e. there is ...) a friend of a friend (of someone) is the latter's friend (too)" (a. using variables; b. without variables, in accordance with **28**). — **4.** "If a (number) is smaller than the predecessor of another (number), then it is (also) smaller than the other (number)"; (a) with variables, (b) without variables. — **5.** "If a (number) is the predecessor of the predecessor of an even (number), it is (also) even".

31. VARIOUS KINDS OF RELATIONS

31a. Representations of relations. Both an *n*-place predicate of the first level ($n>1$) and the *n*-place relation this predicate designates have for their extension the class of those ordered *n*-tuples of individuals for which the predicate holds. If the extension of a predicate (or its corresponding relation) is finite, we may specify this extension by way of a *list* that comprises the *n*-tuples of the extension. E.g. when it is finite, the extension of a two-place relation may be specified by a list of all the pairs belonging to it. However, the list is just one device for specifying the extension of a finite two-place relation. Two other devices are frequently used to advantage because of their intuitive appeal, viz. the arrow diagram and the matrix.

The *arrow diagram* of a relation R represents the R-members by points and the R-pairs by arrows (see Fig. 1). Thus, if (a,b) is an R-pair (i.e. if it is the case that *Rab*), then an arrow is displayed leading from point *a* to point *b*. If it is the case that both *Rab* and *Rba*, a double arrow is displayed between points *a* and *b*. If it is the case that *Raa*, i.e. if (*a*,*a*) is an *R*-pair, we display an arrow that starts at point *a* and loops back into point *a*.

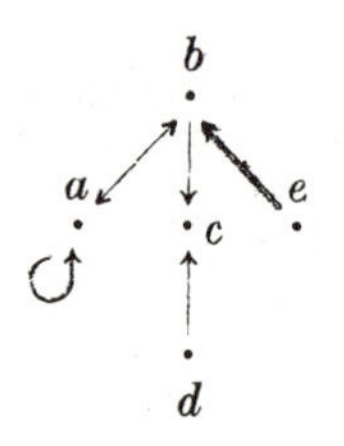

FIG. 1. Arrow diagram of the relation *R*

The *matrix* of a relation *R* having *n* members is constructed as follows: The *n* members are fixed in some (arbitrary) sequence. A square array of *n* rows and *n* columns is put down, and both the *i*th row and the *i*th column ($i=1,\ldots,n$) are coded with the *i*th member of the sequence into which the *n* members were initially ordered (see Fig. 2). If, now, it is the case that *Rab*, we enter the figure '1' at the intersection of the row coded '*a*' with the column coded '*b*'; and we enter the figure '0' there if it is not the case that *Rab*. Places in the square with '1' are called *occupied*, the others *unoccupied*. That diagonal of the square running from upper left to lower right is called the *main diagonal*; it consists of places corresponding to pairs with identical members, i.e. (*a*,*a*), (*b*,*b*), (*c*,*c*), (*d*,*d*), etc. Two places which are symmetrically located with respect to the main diagonal (e.g. the place corresponding to (*b*,*d*) and the place corresponding to (*d*,*b*)) are said to be *converse* to each other.

	a	b	c	d	e
a	1	1	0	0	0
b	1	0	1	0	0
c	0	0	0	0	0
d	0	0	1	0	0
e	0	1	0	0	0

FIG. 2. Matrix of the relation *R*

Example. Suppose a relation *R* is given by the following *list*: "(*a*,*a*), (*a*,*b*), (*b*,*a*), (*b*,*c*), (*d*,*c*), (*e*,*b*)". The arrow diagram of this relation *R* is shown in Fig. 1. (It is obvious, of course, that in an arrow diagram no importance is to be attached to the arrangement of the points: all that counts is the pattern of connections shown by the arrows. Any transformation of Fig. 1 which preserves this pattern yields again an arrow diagram of *R*.)—A matrix of this relation *R* is shown in Fig. 2. Any permutation of the rows of this matrix, together with the same permutation of its columns, produces another matrix of *R*.

31b. Symmetry, transitivity, reflexivity. A relation *R* is called *symmetric* if for each *R*-pair the relation *R* also holds in the inverse direction, i.e. '$(x)(y)(Rxy \supset Ryx)$' or, more concisely, '$R \subset R^{-1}$'. E.g. if *a* is parallel to *b*, then *b* is necessarily parallel to *a*; thus the relation Parallel is a symmetric relation. Examples of other symmetric relations are Similar, Contemporary, Sibling. We say *R* is *non-symmetric* if the condition just stated fails, i.e. if '$\sim(R \subset R^{-1})$' holds; in other words, if there is at least one pair for which *R* holds in a single direction only, i.e. if '$\exists(R . \sim R^{-1})$' holds. And in particular, *R* is said to be *asymmetric* if there is no pair for which *R* holds in both directions, i.e. if *R* and the converse of *R* exclude each other: '$R \subset \sim R^{-1}$'. Examples of asymmetric relations: Father, Less. The relation Brother is an example of a relation which is neither symmetric nor asymmetric. It is to be noted that these three kinds of relations provide a three-part classification of all (homogeneous two-place) relations, as indicated by Fig. 3.

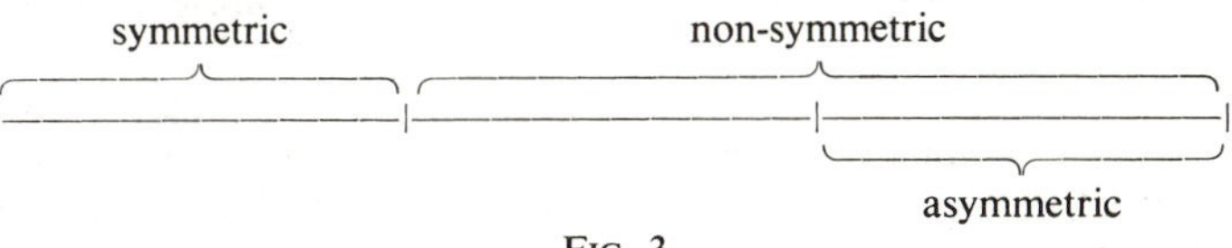

FIG. 3

The arrow diagram of a symmetric relation displays only double arrows (looped arrows count as double arrows), while that of an asymmetric relation contains no double arrows. The matrix of a symmetric relation possesses symmetry respecting the main diagonal, i.e. the place converse to an occupied place is also occupied; the matrix of an asymmetric relation is such that each place converse to an occupied place is unoccupied.

Another three-way classification of all (homogeneous two-place) relations is furnished by the concepts which follow. We say that a relation *R* is *transitive* if the following condition holds: '$(x)(y)(z)(Rxy . Ryz \supset Rxz)$', or in brief '$R^2 \subset R$'. E.g. when *a* is parallel to *b* and *b* is parallel to *c* it is necessarily the case that *a* is parallel to *c*, whence we see that the relation Parallel is transitive. Examples of other transitive relations are Equal, Less, Less-or-Equal, Ancestor. We say *R* is *non-transitive* if the condition just stated fails. And finally, we say *R* is *intransitive* if R^2 and *R* exclude each other, i.e. if the condition '$R^2 \subset \sim R$' holds. Examples of an intransitive relation: Father, Successor (in the sequence of natural numbers). The relations Brother and Friend are neither transitive nor intransitive. The arrow diagram of a transitive relation has the following characteristic property: if a chain of two arrows leads from a to c (i.e. if an arrow leads from a to some other point, and a second arrow leads from that point to c), then the diagram will always have an arrow that leads directly from a to c.

A third three-way classification results from the following definitions. We say a relation R is *reflexive* provided each R-member bears the relation R to itself, i.e. provided '$(x)(mem(R)x \supset Rxx)$', or briefly '$R^0 \subset R$', holds. Examples: Contemporary, Equally-long, Smaller-or-Equal. When the condition just specified is not satisfied, R is called *non-reflexive*. If no R-member bears the relation R to itself, i.e. if R and the identity relation are mutually exclusive, R is called *irreflexive*: '$R^0 \subset \sim R$' or '$R \subset J$'. Examples: Father, Brother, Smaller. The following relations are neither reflexive nor irreflexive: ... votes for ...; ... is murderer of ... . If each individual bears the relation R to itself, i.e. if '$(x)(Rxx)$' or '$I \subset R$', R is said to be *totally reflexive*; clearly, a relation R is totally reflexive if and only if R is reflexive and every individual is an R-member. In the arrow diagram of a reflexive relation, every member-point of the relation has a looped arrow. The same holds for a totally reflexive relation; in this case, moreover, the diagram comprehends all individuals. The diagram of an irreflexive relation exhibits no looped arrows. The matrix of a reflexive relation shows all the main diagonal places occupied; every such place is unoccupied if the relation is irreflexive.

We say R is *connected* provided between any two different R-members either R or R^{-1} holds. Example: the relation Smaller (among natural numbers) is a connected relation since if a and b are different, then either a is smaller than b or b is smaller than a. The arrow diagram of a connected relation shows between any two points an arrow in at least one direction; and the matrix of such a relation shows at least one of every two converse places as occupied.

31c. Theorems about relations. With the help of the following definitions we introduce into language C symbols for the concepts explained above, e.g. '*Sym*' for "symmetric", etc. Since we have to do here with properties of (homogeneous two-place) relations, our signs '*Sym*', etc., —when applied to relations of the first level—are one-place predicates of the second level of type ((0,0)).

D31-1. **a.** $Sym(H) \equiv (H \subset H^{-1})$.
b. $As(H) \equiv (H \subset \sim H^{-1})$.

D31-2. **a.** $Trans(H) \equiv (H^2 \subset H)$.
b. $Intr(H) \equiv (H^2 \subset \sim H)$.

D31-3. **a.** $Refl(H) \equiv (H^0 \subset H)$.
b. $Irr(H) \equiv (H \subset J)$.
c. $Reflex(H) \equiv (I \subset H)$ (totally reflexive).

D31-4. $Connex(H) \equiv (x)(y)[mem(H)x . mem(H)y . (x \neq y) \supset Hxy \vee Hyx]$.

On the basis of these definitions, many results can be obtained. We give some of them here.

T31-1. The following sentential formulas are L-true:

+**a.** $Refl(H) \equiv (x)[mem(H)x \supset Hxx]$.

+**b.** $Reflex(H) \equiv (x)Hxx$.

c. $Reflex(H) \equiv Refl(H) . U(mem(H))$.

+**d.** $Trans . Sym \subset Refl$.
Every relation that is transitive and symmetric is also reflexive.

e. $As(H) \equiv Irr(H^2)$.

+**f.** $As \subset Irr$.
Asymmetric relations are irreflexive.

+**g.** $Trans . As = Trans . Irr$.
Those transitive relations that are asymmetric are also irreflexive, and conversely.

+**h.** $Sym(H) \equiv Sym(H^{-1})$.
A relation is symmetric if and only if its converse is symmetric. (A similar statement can be made for each of the other concepts introduced in D1 through D8.)

i. $As(H) . (K \subset H) \supset As(K)$.
Every subrelation of an asymmetric relation is itself asymmetric. (Analogous assertions hold for '*Intr*' and '*Irr*', and also for the following to be defined below: '*Antis*', 'Un_1', 'Un_2', '$Un_{1,2}$'; but the same is not true of the other relational properties defined in **31**.)

j. $Irr(H^2) \supset Irr(H)$,
$Irr(H^3) \supset Irr(H)$, etc.
If the second power (or any other positive power) of a relation is irreflexive, then the relation itself is irreflexive.

k. $Trans(H) . Irr(H) \supset Irr(H^2)$,
$Trans(H) . Irr(H) \supset Irr(H^3)$, etc.
If a relation is transitive and irreflexive, then every positive power of it is also irreflexive.

In connection with these results we give below proofs of (d), (e), (f), (g), (j), and (k).

Proof of (d). Suppose that (1) *R* is transitive, (2) *R* is symmetric, and (3) *a* belongs to the field of *R*. We must show that *Raa*. In view of (3), there is an individual, say *b*, such that *Rab* or *Rba*. From (2), therefore, it must be the case that *Rab* and *Rba*. Hence by (1), *Raa*.

Proof of (e). **1.** Suppose there are individuals, say a and b, such that each bears the relation R to the other, viz. Rab and Rba. Thus R is not asymmetric; and further, since R^2aa is the case, R^2 is not irreflexive. — **2.** Contrariwise, suppose there is no such pair of individuals. Then R is asymmetric; and also R^2 must be irreflexive, since otherwise there would be an individual (say, a) such that R^2aa—which is to say, there would be an individual (say, b) such that Rab and Rba.

Proof of (f). Suppose R is not irreflexive. Then there is an individual, say a, such that Raa. Thus also $R^{-1}aa$; and hence R is not asymmetric.

Proof of (g). **1.** Suppose R is transitive and asymmetric. Then, by (f), R is irreflexive. — **2.** Suppose R is transitive and irreflexive; and suppose there were two individuals, say a and b, such that each bears the relation R to the other, viz. Rab and Rba. In this case, since R is transitive it would follow that Raa, contrary to the fact that R is irreflexive. Thus there can be no such pair, and R must be asymmetric.

Proof of (j). Suppose R is not irreflexive. Then there is an individual, say a, such that Raa. In this case also R^2aa, R^3aa, etc. Hence R^2, R^3, etc., are not irreflexive.

Proof of (k). Suppose R is transitive and irreflexive, and suppose for some n ($n \geq 2$) that R^n is not irreflexive. Then there is an individual, say a_1, such that $R^n a_1 a_1$. Hence there must be individuals $a_2, a_3, \ldots, a_{n-1}$ such that Ra_1a_2, $Ra_2a_3, \ldots, Ra_{n-2}a_{n-1}$, Ra_{n-1}, a_1. Thus, since R is transitive, a_1 must bear the relation R to a_3, to $a_4, \ldots$, to a_{n-1} and to a_1. This last is impossible because R is irreflexive. Consequently R^n must be irreflexive.

31d. Linear order: series and simple order. We shall now explicate the concept of *linear order* as exemplified by the natural order (i.e., the order of ascending magnitude) of the natural numbers 0, 1, 2, etc., or by the natural order of the integers, or of the rational numbers, or of the real numbers. In each case of this kind, there is a class and a relation ordering the elements of the class. If we wish to specify the ordering, we need not indicate both the class and the relation. To indicate the class alone is obviously not sufficient, because the elements of a given class can be ordered in different ways by different relations. But the specification of the relation is sufficient, since the class is uniquely determined as the field of relation. [The usual term "ordered set" used in set theory for orders of a certain kind is misleading. Actually sets (or classes, as they are called in logic) are not to be classified into ordered and unordered; instead, the relations may be classified into those representing a linear order and those that do not.]

The linear, ascending order of the natural numbers 0, 1, 2, etc., may be represented either by the relation Smaller (for natural numbers) or by the relation Smaller-or-Equal (for natural numbers). The former is irreflexive, the latter reflexive. In terminology now to be introduced, the former relation will be called a "series," the latter a "simple order." In most cases it does not matter which of the two concepts is used. We shall introduce both because each of them has certain distinctive advantages and some authors therefore prefer one to the other.

The concept of series is the older one. It was introduced in [P.M.] and was previously used more frequently. The concept of a simple order may appear to the beginner as less simple than that of a series. But it has the advantage of being slightly more general. It can be used in the degenerate case of a linear order of exactly one member, say a, by way of the relation $\{(a,a)\}$ (see **32e** below); but a series, being irreflexive, cannot have exactly one member. We shall see that the concept of simple order is a more suitable basis for the definition of ordinal number (see **38a** below). Therefore it is frequently preferred recently.

The two concepts just mentioned will be defined as follows. A relation R is said to be a serial relation or, for short, a *series*—in symbols '$Ser(R)$'—provided R is irreflexive, transitive, and connected. A relation R is called *antisymmetric* ('$Antis(R)$') if, for any two distinct members, R and its converse cannot both hold; in other words, if x and y must be identical whenever both R and its converse hold between x and y. A relation R is said to be a partially ordering relation or, for short, a *partial order* ('$POrd(R)$') provided R is reflexive, transitive, and anti-symmetric. A relation R is said to be a simply ordering relation or a *simple order* ('$SOrd(R)$') provided R is a partial order and is connected.

D31-5. $Ser = Irr.Trans.Connex.$

D31-6. $Antis(H) \equiv (H.H^{-1} \subset I).$

D31-7. $POrd = Refl.Trans.Antis.$

D31-8. $SOrd = POrd.Connex.$

Concerning these concepts we have the following theorems.

T31-2. The following sentential formulas are L-true.

+**a.** $Ser = As.Trans.Connex.$ (From T1g.)

b. $Ser(H) \equiv Connex(H).Irr(H^2).Irr(H^3).$

c. $Ser(H) \equiv Connex(H).Irr(H^6).$

d. $Ser(H) \supset \sim 1(mem(H)).$

e. $Antis(H) \supset (H^2.I = H.I).$

f. $POrd(H) \supset (H^2 = H).$

g. $SOrd = Refl.Trans.Antis.Connex.$

h. $SOrd(H) \equiv SOrd(H^{-1}).$

i. $SOrd(H) \supset \big(mem_1(H) = mem_2(H)\big).$

j. $SOrd(H).mem(H)x.mem(H)y.Jxy \supset (Hxy \equiv \sim Hyx).$

k. $SOrd(H) . (K \subset H) . Refl(K) . Connex(K) \supset SOrd(K).$

l. $SOrd(H).(x)(y)\big[Kxy \equiv Fx.Fy.Hxy\big] \supset SOrd(K).$

Proof of (b). **1.** Suppose R is a series. Then, by D5, R is connected and transitive and irreflexive. In view of T1k, R^2 and R^3 are also irreflexive. Hence if R is a series, then R is connected and R^2 is irreflexive and R^3 is irreflexive. — **2.** Conversely, suppose R is connected and R^2 is irreflexive and R^3 is irreflexive. Then R is irreflexive (by T1j) and asymmetric (by T1e). It remains to show that R is transitive, i.e. that if Rab and Rbc, then Rac. Suppose Rab and Rbc. Since R is asymmetric, a and c must be different. Since R is connected, either Rac or Rca. Now Rca cannot hold, for other-wise R^3aa would hold (it having been assumed that Rab and Rbc) in contradiction to the irreflexiveness of R^3. Thus Rac must hold and R must be transitive. Hence, finally, R is a series.

Proof of (c). **1.** Suppose R is a series. Then R is connected, transitive and irreflexive.

Hence, by T1k, R^6 is also irreflexive. — **2.** Conversely, suppose R is connected and R^6 is irreflexive. Now R^6 is the same as $(R^2)^3$ and $(R^3)^2$. Since $(R^2)^3$ is irreflexive, so also is R^2 (by T1j). Since $(R^3)^2$ is irreflexive, so also is R^3 (again by T1j). Thus, by (b), R is a series.

Proof of (e). **1.** Suppose $(H.I)xy$. Then, since $Ixy, x=y$. Hence Hxx, but then it follows that H^2xx. Therefore $(H^2.I)xy$. It is now established that $(H.I)\subset(H^2.I)$. — **2.** Suppose $(H^2.I)xy$. Then, since $Ixy, x=y$; also, since H^2xx, there is some z such that $Hxz.Hzx$. Supposing $z\neq x$, then H is not antisymmetric; therefore $z=x$. Since $z=x$ and Hxz, it is established that Hxx. Since Ixx is trivial, it follows that $(H.I)xy$ (recall that $x=y$). It is now established that $(H^2.I)\subset(H.I)$. — **3.** Thus, since $(H.I)\subset(H^2.I)$ and $(H^2.I)\subset(H.I)$, $(H^2.I)=(H.I)$ (by T29-3d).

Proof of (k). Since *Refl*(K) and *Connex*(K), it remains only to be shown that *Antis*(K) and *Trans*(K) to establish that *SOrd*(K) (by T2g). — 1. In view of the fact that *Antis*(H) (by T2g) and $K\subset H$, *Antis*(K) must hold (by T1i). — 2. To establish that *Trans*(K) assume that Kxy and Kyz. Since *Trans*(H) (by T2g) and Hxy and Hyz, Hxz must hold. Further, if $z\neq x$ then $\sim Hzx$ must hold (by T2j). Thus $\sim Kzx$ must hold. Now since *Connex*(K), either Kxz or Kzx; hence Kxz. If, on the other hand, $z=x$, then Kxz must hold since *Refl*(K).

Proof of (l). Assume that the two conditions of the theorem are fulfilled. — **1.** Then $K\subset H$. Hence *Antis*(K) (by T1i since *Antis*(H)). — **2.** If Kxy, then since *Refl*(H) and Hxy, Hxx and Hyy must hold. But if Kxy, both Fx and Fy. Hence $Fx.Fx.Hxx$ and $Fy.Fy.Hyy$. Thus Kxx and Kyy. Hence *Refl*(K). — **3.** Suppose Kxy and Kyz. Then since Hxy and Hyz and *Trans*(H), Hxz holds. Also, since Kxy and Kyz, Fx and Fz. Hence Kxz, since $Fx.Fz.Hxz$. Therefore *Trans*(K). — **4.** Suppose that *mem*(K)x and *mem*(K)y and $x\neq y$. Then Fx and Fy must both hold. Since $K\subset H$, *mem*(H)x and *mem*(H)y. Also since *Connex*(H), either Hxy or Hyx. Hence either Kxy or Kyx. Therefore *Connex*(K). — **5.** From the results established in (1)–(4) and T2g it follows that *SOrd*(K).

If H is a simple order and the number of its members is not exactly 1, then there is exactly one corresponding series K having the same members as H, viz. $H.J$. (If H is a simple order with exactly one member, then $H.J$ is the empty relation, which is a degenerate case of a series.) If K is a series, then there is exactly one corresponding simple order H having the same members as K, viz. $K\vee K^0$ (this is the relation formed from K by adding all identity pairs of members of K). E.g. let H be the relation Smaller-or-Equal in one of the domains of numbers mentioned at the beginning of **31d**, and let K be the relation Smaller in the same domain; then $K=H.J$, and $H=K\vee K^0$. Parts a, b, and c of T3 below state the results just explained.

T31-3. The sentential formula '*SOrd*(H) . ~ 1(*mem*(H)) . $K=(H.J)$' L-implies each of the following formulas (a) through (e); and '*Ser*(K) . $H=(K\vee K^0)$' does likewise.

a. *Ser*(K).

b. *SOrd*(H).

c. *mem*(H)$=$*mem*(K).

d. $(H|K)\subset K$.

e. $(K|H)\subset K$.

Exercises. 1. Can '*Ser*(R)' and '$mem_1(R)=mem_2(R)$' both be true? Can '*Ser*(R)' and '$mem_1(R)\neq mem_2(R)$' both be true? — 2. Give informal proofs of the following:

a) T2d (recall that $Ser\subset Irr$);
b) T2f (show $H\subset H^2$, since $Refl(H)$; show $H^2\subset H$, since $Trans(H)$);
c) T2g;
d) T2h (show for each of *Refl*, *Trans*, *Antis*, and *Connex* that, if H has that property, then H^{-1} also has it);
e) T2i (recall that $SOrd\subset Refl$);
f) T2j (show $Hxy\supset\sim Hyx$, since $Antis(H)$; show $\sim Hyx\supset Hxy$, since $Connex(H)$).

31e. One-oneness. It is worth noting that the predicates 'Un_1', 'Un_2' and $Un_{1,2}$' defined in language A (see D19-1,2,3) can be defined in language C without employing individual variables, as follows:

D31-9. $Un_1(H) \equiv (H|H^{-1} \subset I)$.
D31-10. $Un_2(H) \equiv (H^{-1}|H \subset I)$.
D31-11. $Un_{1,2}=Un_1 . Un_2$.

In *arrow-diagram* terms: A one-many relation shows at most one arrow proceeding *into* each point; a many-one relation shows at most one arrow proceeding *out* of each point; and a one-one relation shows no two different arrows sharing either the same initial point or the same terminal point. In *matrix* terms: A one-many relation shows at most one occupied place in each column; a many-one relation shows at most one occupied place in each row; and a one-one relation shows at most one occupied place in each row and each column.

+T31-4. The following sentential formulas are L-true:
a. $Un_1(H) \equiv Un_2(H^{-1})$.
b. $Un_2(H) \equiv Un_1(H^{-1})$.
c. $Un_{1,2}(H) \equiv Un_{1,2}(H^{-1})$.

32. ADDITIONAL LOGICAL PREDICATES, FUNCTORS AND CONNECTIVES

32a. The null class and the universal class. Let us so define the one-place predicate of the first level 'Λ_1' that each full sentence thereof, e.g. '$\Lambda_1(a)$', is L-false. For the definition of '$\Lambda_1(x)$' we can employ an arbitrary L-false sentential formula with the free variable x, e.g. '$x\neq x$'. Also, we say that 'Λ' denotes the *null class* or the empty class. (Note that, in accordance with P9 (**22a**), each type has only one null class; see T1b below.) Similarly, let us so define the two-place predicate 'Λ_2' that each full sentence thereof is L-false; Λ_2 is called the *null* (two-place) *relation*. And similarly, 'Λ_3' can be defined for the null (three-place) relation; etc.

Again, let us so define the one-place predicate 'V_1' that every full sentence thereof is L-true. For the definition of '$V_1(x)$' we can employ an arbitrary L-true sentential formula, e.g. '$x=x$'. We say that 'V_1' denotes the *universal class*. Similarly, we so define 'V_2' that each full sentence thereof is L-true; V_2 is called the (two-place) *universal relation*. And similarly we define V_3 as the three-place universal relation, etc.

D32-1. **a.** $\Lambda_1(x) \equiv x \neq x$.
b. $\Lambda_2(x,y) \equiv (x \neq x) \,.\, (y \neq y)$.
Similarly for 'Λ_3', etc.

D32-2. **a.** $V_1(x) \equiv (x=x)$.
b. $V_2(x,y) \equiv (x=x).(y=y)$.
Similarly for 'V_3', etc.

T32-1. The following sentential formulas are L-true (as are analogs phrased with higher indices, e.g. 'Λ_2', etc.):

a. $\sim\exists(\Lambda_1)$.
+b. $\sim\exists(F) \equiv (F=\Lambda_1)$. (From (a), T29-3c.)
c. $U(V_1)$.
d. $U(F) \equiv (F=V_1)$. (From (c), T29-3b.)
+e. $\Lambda_1 \subset F$.
The empty class is contained in every class.
+f. $F \subset V_1$.
Every class is contained in the universal class.
g. $\Lambda_1 = \sim V_1$.
h. Λ_2 belongs to the following classes: *Sym*, *Trans*, *Refl* (but not to *Reflex*), *Irr*, *Connex*, *Ser*, *Antis*, *POrd*, *SOrd*, Un_1, Un_2, $Un_{1,2}$.

Exercises. **1.** Give informal proofs of the following: a) '$\sim Reflex(\Lambda_2)$'; b) '$Refl(\Lambda_2)$'; c) '$SOrd(\Lambda_2)$'; d) '$Un_2(\Lambda_2)$'.

32b. Union class and intersection class. If M is a class of classes, we designate the class of all individuals that belong to at least one of the element classes of M as the *union class* or class-sum of M; this union class is symbolized by '$sm_1(M)$', where 'sm_1' is a functor. Again, if M is a class of two-place relations, we designate that relation which holds for a pair provided at least one of the element relations of M holds for this pair as the *union relation* of M, and denote it by '$sm_2(M)$'. Similarly, the functor 'sm_3' is defined for a class of three-place relations, etc.

If M is a class of classes, we call the class of those individuals that belong to every element class of M the *intersection class* or class-product of M; symbol: '$pr_1(M)$'. The functor 'pr_2' is similarly defined for a class of two-place relations, the functor 'pr_3' for a class of three-place relations, etc.

D32-3. **a.** $sm_1(N)x \equiv (\exists F)(N(F).Fx)$.
b. $sm_2(N)xy \equiv (\exists H)(N(H).Hxy)$.
Similarly for 'sm_3', etc.

D32-4. **a.** $pr_1(N)x \equiv (F)(N(F) \supset Fx)$.
b. $pr_2(N)xy \equiv (H)(N(H) \supset Hxy)$.
Similarly for 'pr_3', etc.

The class of all the subclasses of a given class Q we denote by '$sub_1(Q)$'; the class of all the subrelations of a given (two-place) relation S by '$sub_2(S)$'; etc.

D32-5. **a.** $sub_1(F)(G) \equiv (G \subset F)$.
b. $sub_2(H)(K) \equiv (K \subset H)$.
Similarly for 'sub_3', etc.

32c. Connections between relations and classes. The class of all those individuals that bear the relation R to at least one element of Q we designate by 'R"Q'. 'R"Q' is a one-place predicate expression; a full sentence of this predicate, say '(R"Q)a' would be read "a bears the relation R to an element of Q". The expression 'R"Q' itself can be read: "the R's of the Q's".

If a, b, c, ... are elements of Q and 'k' is a functor, we designate by 'k"Q' the class comprising the individuals $k(a)$, $k(b)$, $k(c)$, etc. (Instead of the constant 'k' the definition will contain a functor variable 'f'.)

D32-6. **a.** $(H``F)x \equiv (\exists y)(Fy.Hxy)$.
b. $(f``F)x \equiv (\exists y)(Fy\,.\,(x=fy))$.

Examples. **1.** '*Fa*"*stud*' designates the property Father-of-a-Student; expressed in the plural fashion: "the fathers of the students". — **2.** '*sq*"*Prime*' reads "the (class of the) squares of the prime numbers".

Given a relation R and a class P, we sometimes consider that subrelation of R obtained by confining the field of R to P, i.e. the relation which holds between x and y provided R holds between x and y, and both x and y belong to P. We designate this new relation by 'R *in* P'. (There are analogous notions for relations of higher degree.)

D32-7. $(H \; in \; F)xy \equiv Hxy.Fx.Fy$.

Examples. **1.** If Q is the class of Englishmen, then '*Fa* in Q' denotes the relation of fatherhood among Englishmen. — **2.** '*Sm in Prime*' denotes the relation Smaller among prime numbers.

The class of all *initial members of* R, i.e. all first-place members of R that are not also second-place members of R, we designate by '*init*(R)'. The corresponding class of terminal members of R requires no new functor; it can simply be designated by '*init*(R^{-1})', since the terminal members of R are just the initial members of the converse of R.

D32-8. $init(H)x \equiv mem_1(H)x \,.\, \sim mem_2(H)x.$

32d. Theorems.

T32-2. The following sentential formulas are L-true (as are analogous formulas phrased with higher indices: 'sm_2', etc.):

a. $N(F) \supset (F \subset sm_1(N))$.

b. $N(F) \supset (pr_1(N) \subset F)$.

c. $Ser(H) \supset Ser(H\ in\ F)$,

Every relation which results from confining the field of a series is again a series. (Analogous theorems hold for '*Sym*', '*As*', '*Trans*', '*Intr*', '*Refl*', '*Irr*', '*Connex*', '*Antis*', '*POrd*', '*SOrd*', 'Un_1', 'Un_2', '$Un_{1,2}$' (see **31**).)

T32-3. In each of the following sentence-pairs, the two given sentences are L-equivalent (the arbitrary constants '*P*', '*M*' and '*R*' may be replaced by arbitrary predicate expressions of the same type):

a. '$sm_1(M) \subset P$' and '$(F)[M(F) \supset (F \subset P)]$'.

b. '$P \subset pr_1(M)$' and '$(F)[M(F) \supset (P \subset F)]$'.

c. '$\sim \exists(init(R))$' and '$mem_1(R) \subset mem_2(R)$'.

Exercises. **1.** Give informal proofs of the following: a) T2a; b) T2b; c) T2c; d) '$Sym(H) \supset Sym(H\ in\ F)$'; e) '$As(H) \supset As(H\ in\ F)$'; f) '$Trans(H) \supset Trans(H\ in\ F)$'; g) $Refl(H) \supset Refl(H\ in\ F)$'; h) '$Connex(H) \supset Connex(H\ in\ F)$'. — **2.** Give a relation *R* and a class *P* such that $Reflex(R)$ but $\sim Reflex(R\ in\ P)$. — **3.** Give informal proofs of the following: a) T3a; b) T3b; c) T3c; d) '$pr_1(N) \subset sm_1(N)$'.

32e. Enumeration classes. The property of being the individual *a*, i.e. the class whose only member is *a*, is called the *unit class of a* and designated by '$\{a\}$'. The property of being either *a* or *b*, i.e. the class whose only members are *a* and *b*, is designated by '$\{a,b\}$'; '$\{a,b,c\}$' is defined similarly, etc. Writing '(a,b)' for the ordered pair comprising *a* and *b* with *a* first and *b* second, we can also use the enumeration notation '{...}' for a relation taken as a class of ordered pairs. Thus the two-place relation whose only pair is (a,b) is denoted by '$\{(a,b)\}$'; '$\{(a,b),(c,d)\}$' designates the relation whose only pairs are (a,b) and (c,d); similarly for '$\{(a_1,a_2),(b_1,b_2),(c_1,c_2)\}$', etc. (Note that the class $\{a,b\}$ is the same as the class $\{b,a\}$, whereas the two relations $\{(a,b)\}$ and $\{(b,a)\}$ are different (provided *a* and *b* are not identical).) Continuing this process, similar definitions can be made for three-place relations (e.g. '$\{(a,b,c)\}$'), for four-place relations, etc.

D32-9. **a.** $\{x\}(u) \equiv (u = x)$.

b. $\{x,y\} = \{x\} \vee \{y\}$.

c. $\{x,y,z\} = \{x\} \vee \{y\} \vee \{z\}$.

In a corresponding way, classes with four or more elements can be defined by enumeration.

Following out the pattern of our introductory remarks, we give next:

D32-10. **a.** $\{(x,y)\}(u.v) \equiv (u=x).(v=y)$.
b. $\{(x,y),(z,w)\} \equiv \{(x,y)\} \vee \{(z,w)\}$.

In a similar way, two-place relations comprising three or more pairs can be defined by enumeration.

Continuing, we have

D32-11. $\{(x,y,z)\}(u,v,w) \equiv (u=x).(v=y).(z=w)$.

In a similar way, three-place relations comprising two or more triples can be defined. And generally, n-place relations comprising a finite number m of given n-tuples can be defined by enumeration of these n-tuples.

T32-4. The following sentential formulas are L-true:

a. $\{x\}x$.
b. $\{x,y\}u \equiv (u=x) \vee (u=y)$.
c. $\{x,y,z\}u \equiv (u=x) \vee (u=y) \vee (u=z)$.
d. $\{(x,y),(z,w)\}uv \equiv ((u=x) \,.\, (v=y)) \vee ((u=z) \,.\, (v=w))$.
e. $Fx \equiv (\{x\} \subset F)$.
f $Hxy \equiv (\{(x,y)\} \subset H)$.

At this point it is possible to read the axiom systems given in language C in **44a** and **46a** of Part Two (Application of symbolic logic) of this book.

Exercises. **1.** Give informal proofs of the following: a) T4a; b) T4b; c) T4d; d) T4f; e) '$\{x\} \subset \{x,y\}$'.

33. The λ-operator

33a. The λ-operator. Let 'M' be a one-place predicate of the second level, i.e. designating a property of properties of individuals. Thus e.g. '$M(P)$' might be rendered "the first-level property P has the second-level property M". (For a concrete example, we might think of a cardinal number, e.g. 5; then '$5(P)$' says that P has cardinal number 5. Here, of course, 5 is regarded as a property of properties.) If we wish to assert that the property predicated of a by the sentence '$Pa \vee Qa$' has the property M, we can do so with the help of the symbolism introduced earlier: for since '$Pa \vee Qa$' can also be written '$(P \vee Q)a$', the proposition named above can be formulated '$M(P \vee Q)$'. [Example: we would read '$5(P \vee Q)$' as "the disjunction of properties P and Q (or: the union of classes P and Q) has cardinal number 5".]

What we have just done in connection with '$Pa \vee Qa$' cannot be extended to more elaborate sentential compounds such as '$Pa \vee (y)Rya$'. The reason is that the symbolism available up to this point furnishes no predicate expressions for the properties predicated of an individual by most compound sentences about the individuals; e.g. we have no predicate expression for the property predicated of individual a by the sentential compound '$Pa \vee (y)Rya$'. The operator sign 'λ' now to be introduced will have the particular role of forming a predicate expression for any property ascribed to an individual by any sentence in language C. Thus it will appear in what follows that the property predicated of individual a by the sentence '$Pa \vee (y)Rya$' is to be designated by the predicate expression '$(\lambda x)(Px \vee (y)Ryx)$'.

An expression of the form '$(\lambda x)(...x...)$' is called a *λ-expression*. In the λ-expression '$(\lambda x)(...x...)$' the portion written '(λx)' is an operator which we call the *λ-operator*; and the portion written '$...x...$' is the *operand* of the λ-operator. Note therefore that the 'x' is bound in '$(\lambda x)(...x...)$'. If '$...x...$' is a sentential formula, then '$(\lambda x)(...x...)$' corresponds, say, to the verbal expression "the property of x such that $...x...$" or the verbal expression "the class of those x such that $...x...$"; and the full expression '$[(\lambda x)(...x...)]a$' is a sentence asserting the individual a has the property $(\lambda x)(...x...)$.

The use of a λ-expression, e.g. '$(\lambda x)(Px \vee (y)Ryx)$', would be superfluous if its purpose were merely to ascribe the property it designated to some individual, say b. For this can be done simply by the sentence '$Pb \vee (y)Ryb$', and the more complicated formulation '$[(\lambda x)(Px \vee (y)Ryx)]b$' can be dispensed with; both formulations say the same thing. Therefore our syntactical system B contains a primitive sentence schema (it is P10 in **22a**) that enables either one of the two sentences just named to be derived from the other; which is to say, we can find in B a sentence in the old symbolism (viz. '$Pb \vee (y)Ryb$') that is synonymous with the full sentence of the λ-expression (viz. '$[(\lambda x)(Px \vee (y)Ryx)]b$'. However, the old symbolism provides *no* expression that is synonymous with the λ-expression itself. Hence the new λ-expression is very useful if we wish to ascribe to the property designated by this λ-expression some property of the second level, for in this case the λ-expression can serve as the argument-expression of the second-level predicate expression.

The particular illustration λ-expression that appears above is a one-place predicate expression. In a similar way, λ-operators with several variables can be used to construct many-place predicate expression. E.g. a λ-expression of the form '$(\lambda xy)(...x...y...)$' whose operand '$...x...y...$' is a sentential formula with free variables 'x' and 'y' is to be recognized as a two-place predicate expression designating that relation which subsists between two individuals x and y just in case they satisfy the condition formulated in the operand. The formulation of λ-predicate expressions with more than two argument-places and of arbitrary type is carried out in an analogous fashion. A variable must not occur in a λ-operator more than once.

While of great importance theoretically, λ-expressions are relatively seldom used in language C. The reason is that in language C other forms of expression (notably, functors) are often available for the construction of predicate expressions. E.g. the property predicated of a by '$Pa \vee (\exists y)Rya$' can be designated '$P \vee mem_2(R)$', hence in this case the less concise λ-expression '$(\lambda x)(Px \vee (\exists y)Ryx)$' can be dispensed with. Again, it often happens that a discussion involves repeated reference to a certain property in a particular connection; in this event it may pay to introduce (by definition) a simple predicate for the property. Thus, reverting to our last example, we can introduce Q, say, by the definition '$Qx \equiv Px \vee (\exists y)Ryx$' and thereafter render as '$M(Q)$' the proposition contemplated about this property. As a general rule, λ-expressions are of use only when there is no advantage either in defining predicates for the properties under consideration, or in defining functors which permit the designation of these properties by compound predicate expressions.

λ-functor expressions. Up to now we have dealt only with λ-expressions which are predicate expressions, i.e. λ-expressions whose operands are sentential formulas. We also wish to admit λ-expressions whose operands are expressions of arbitrary type in the type system. Here, as before, the full expression '$[(\lambda x)(...x...)]a$' is synonymous with '$...a...$', i.e. with what results from substituting 'a' for 'x' in the operand. But whereas formerly this full expression was a sentence, now the full expression is an expression of the type system. For this reason the λ-expressions now under consideration are not predicate expressions, but functor expressions. (It should be noted that the primitive sentence schema P10 of **22a** still serves for the transformation of our present λ-expressions.)

Examples. 1. In accordance with the above, '$[(\lambda x)(prod(3,x))]a$' is synonymous with '$prod(3,a)$' and hence means "the triple of a"; thus '$(\lambda x)(prod(3,x))$' is a functor expression to be read "the triple of" or "the function whose value at x is $3x$". From this example we see that any λ-functor expression '$(\lambda x)(...x...)$' can be read "the function whose value at x is $...x...$". — **2.** The one-place predicate expression '$(\lambda x)[(\exists y)Rxy]$' is read "the class of those x such that there is something y to which x bears the relation R"; hence (in view of D18-1, and the fact that '$[(\lambda x)[(\exists y)Rxy]]a$' means the same as '$(\exists y)Ray$', i.e. '$mem_1(R)a$') it is clear that '$(\lambda x)[(\exists y)Rxy]$' is synonymous with '$mem_1(R)$'. Now suppose we let the λ-expression of this example be the operand of another λ-expression, viz. '$(\lambda H)[(\lambda x)[(\exists y)Hxy]]$'. This new λ-expression is a functor expression; it is read "the function whose value at H is the class of those x which bear the relation H to something" or "the function whose value at H is the class of first members of H"; and hence it is synonymous with 'mem_1'. This last can also be seen as follows: '$[(\lambda H)[(\lambda x)[(\exists y)Hxy]]](R)$' is synonymous with '$(\lambda x)[(\exists y)Rxy]$', which in turn is synonymous with '$mem_1(R)$'; thus '$(\lambda H)[(\lambda x)[(\exists y)Hxy]]$' is synonymous with '$mem_1$'.

According to an earlier rule (**9a**, (4)), those *brackets can be omitted* which immediately enclose an expression consisting of an operator and the operand belonging thereto. This rule permits us to omit e.g. all the square brackets from the illustrative expressions given above; thus '$(\lambda H)(\lambda x)(\exists y)(Hxy)(R)$' can be written in place of '$[(\lambda H)[(\lambda x)[(\exists y)(Hxy)]]](R)$'. (It should be observed that Rule 5 of **9a** does not apply to λ-expressions.)

33b. Rule for the λ-operator. What is said below is a consequence of our explanations of the meaning of λ-expressions. Suppose that immediately after a λ-expression whose λ-operator contains n variables there follows an argument-expression; then the whole complex is a full expression provided this argument-expression is n-place and the member in the kth place thereof ($k = 1,...,n$) is of the same type as the kth variable in the λ-operator. (The argument-expression referred to above is called *the argument-expression belonging to the λ-expression*, or *the argument-expression belonging to the λ-operator*; the λ-expression itself can, of course, be either a predicate expression or a functor expression.) If a λ-expression and its argument-expression together have this character, i.e. if the whole complex is a full expression, then the λ-operator can be eliminated with the help of the λ-rule given below. [So far as the syntactical system B is concerned, this λ-rule follows from the primitive schema P10 of **22a**. So far as the semantical system B is concerned, the λ-rule always produces from a given expression a second that is L-interchangeable with the first; this follows from the fact that sentences of the form P10 are L-true on the basis of the evaluation rules given in **25a**.]

The λ-rule. A full expression of the form

$$[(\lambda \mathfrak{v}_{k_1},\mathfrak{v}_{k_2},...,\mathfrak{v}_{k_n})(\mathfrak{A}_i)](\mathfrak{A}_{m_1},\mathfrak{A}_{m_2},...,\mathfrak{A}_{m_n}),$$

where $\mathfrak{A}_i$ is the operand of the λ-operator, may be transformed into the expression $\mathfrak{A}_k$ which is obtained from $\mathfrak{A}_i$ by substituting in the latter $\mathfrak{A}_{m_1}$ for $\mathfrak{v}_{k_1}$, $\mathfrak{A}_{m_2}$ for $\mathfrak{v}_{k_2}$, ..., and $\mathfrak{A}_{m_n}$ for $\mathfrak{v}_{k_n}$.

The transformation referred to in this λ-rule can be effected whether the displayed λ-expression is an independent sentence or a part of another sentence. In view of the rule, a λ-operator can always be eliminated if there is an argument-expression belonging to it. If an expression consists of a single operand preceded by several λ-operators and followed by several argument-expressions (each of these last is bracketed by itself; their number does not exceed the number of λ-operators), the first argument-expression belongs to the first λ-operator and can be eliminated with it; the second argument-expression belongs to the second λ-operator, and can be eliminated with it; and so on.

Example. By two applications of the λ-rule (the second application involving two variables), the expression '$(\lambda x_1)(\lambda F_2,x_3)(\lambda H_4)(...x_1...F_2...x_3...H_4...)(a_1)(P_2,a_3)$' can be transformed into '$(\lambda H_4)(...a_1...P_2...a_3...H_4...)$'.

Remarks. The use of λ-expressions requires careful attention to *brackets*. According to our earlier stipulation (see the end of **33a**), it is permissible to write '$(\lambda x)(Px)a$' instead of '$[(\lambda x)(Px)](a)$'. On the other hand, brackets enclosing the operand of a λ-operator (e.g. those around 'Px' in the expression just given) are generally not to be omitted; they may be omitted only if some other rule permits. Thus '$(\lambda x)(...x...)(a)$' is to be regarded as an abbreviation for '$[(\lambda x)(...x...)](a)$', but not for '$(\lambda x)[(...x...)(a)]$'. In other words: a predicate expression or functor expression which stands between a λ-operator and an argument-expression belongs to the λ-operator.

Again, the difference between '$(\lambda x,y)$' and '$(\lambda x)(\lambda y)$' should be noticed. Suppose '$...x...y...$' is a sentential formula. Then '$(\lambda x,y)(...x...y...)$' is a predicate expression; and '$(\lambda x,y)(...x...y...)(a,b)$' can be transformed by the λ-rule into '$...a...b...$'. On the other hand, in view of our agreement about omission of brackets, '$(\lambda x)(\lambda y)(...x...y...)$' is an abbreviation for '$(\lambda x)[(\lambda y)(...x...y...)]$' and hence is a functor expression; a full expression of it is e.g. '$(\lambda x)(\lambda y)(...x...y...)(a)$', which by the λ-rule may be transformed into '$(\lambda y)(...a...y...)$' and so recognized as a predicate expression. Using this predicate expression, let us form the full sentence '$(\lambda x)(\lambda y)(...x...y...)(a)(b)$'. This sentence is an abbreviation for '$[(\lambda x)[(\lambda y)(...x...y...)]](a)(b)$', which by two successive applications of the λ-rule transforms first into '$(\lambda y)(...a...y...)(b)$' and then into '$...a...b...$'.

The λ-predicate expressions are entirely analogous to the class expressions of [P.M.]. Here, however, they are genuine predicate expressions, and are used exactly like predicates. Thus e.g. '$(\lambda x)(Px)$' and 'P' are interchangeable in any context whatever. Concerning the line of development which led to this identification of predicate expressions and class expressions, see [Syntax] §37, §38. This development was initiated by Russell (see [P.M.], Introduction to vol. I, 2nd ed., and Chap. VI).—Church was the first to use the λ-operator for functor expressions; he has given the λ-operator a central role in his system ("The calculi of lambda-conversion", *Annals of Math. Studies*, No. 6, Princeton, 1941).

With the background provided by the present section **33b**, we can state the following theorem.

T33-1. The following sentential formulas are L-true:

+**a.** $(\lambda x)(Fx)=F$.

b. $(\lambda x)(Fx)(y) \equiv Fy$.

c. $(\lambda x,y)(Hxy)=H$.

d. $(\lambda x,y)(Hxy)(u,v) \equiv Huv$.

Exercises. **1.** Give an informal proof of T1a based on T29–3a. — **2.** Give an informal proof of the following: '$(z)(w)[(\lambda x)(\lambda y)(Rxy)(z)(w)\equiv(\lambda x,y)(Rxy)(z,w)]$'. — **3.** Decide whether the following is a sentence, restore all parentheses and specify the type of the expressions on each side of the '=': '$(\lambda x)(\lambda y)(Rxy)=(\lambda x,y)(Rxy)$'.

33c. Definitions with the help of λ-expressions. Suppose $\mathfrak{a}_i$ is a predicate or functor of arbitrary type, and suppose that a definition of $\mathfrak{a}_i$ can be formulated in language C. Then there is always a λ-expression $\mathfrak{A}_j$ which comprises only previous signs and which is synonymous with $\mathfrak{a}_i$. Hence, if desired, $\mathfrak{a}_i=\mathfrak{A}_j$ can serve as a definition of $\mathfrak{a}_i$. Such a definition is an explicit definition in the strict sense, viz. its definiendum consists precisely in the sign being defined. If $\mathfrak{a}_i$ is an n-place predicate, a definition of it in the present manner would appear in the form $\mathfrak{a}_i=(\lambda\mathfrak{v}_{i_1},...,\mathfrak{v}_{i_n})(\mathfrak{S}_k)$, in contrast to the usual form $\mathfrak{a}_i(\mathfrak{v}_{i_1},...,\mathfrak{v}_{i_n}) \equiv \mathfrak{S}_k$. Similarly, when $\mathfrak{a}_i$ is a functor its definition can now have the form $\mathfrak{a}_i=(\lambda\mathfrak{A}_{j_1})(\lambda\mathfrak{A}_{j_2})...(\lambda\mathfrak{A}_{j_n})(\mathfrak{S}_k)$ rather than the form $\mathfrak{a}_i(\mathfrak{A}_{j_1})(\mathfrak{A}_{j_2})...(\mathfrak{A}_{j_n}) \equiv \mathfrak{S}_k$ (here the $\mathfrak{A}_{j_1}$, $\mathfrak{A}_{j_2}$,...,$\mathfrak{A}_{j_n}$ are argument expressions consisting of variables). Note that slight notational revisions might be required in some definitions to bring them into this form. Definitions in which the predicate or functor does not precede its arguments must be revised to that form, e.g. D28-3: '$(F \subset G) \equiv U(F \supset G)$' must first be revised to D28-3*: '$\subset(F,G) \equiv U(F \supset G)$', whereupon D28-3* can be replaced by: '$\subset \; = (\lambda F,G)(U(F \supset G))$'; D32-7: '$(H \; in \; F)xy \equiv Hxy.Fx.Fy$' must first be revised to D32-7*: '$in(H,F)xy \equiv Hxy.Fx.Fy$', whereupon D32-7* can be replaced by: '$in = (\lambda H,F)(\lambda x,y)(Hxy.Fx.Fy)$'. (See exercise 2.)

This λ-style of definition can be used in defining any descriptive predicate or functor whatever, once an adequate stock of primitive descriptive signs is available. The same remark applies to all the *logical* predicates and functors previously defined in language A (in **17**, **18**, **19**), and to those additional ones of this chapter which are defined in language C. A few examples will illustrate this possibility. In place of D17-2b, we could use: '$2_m = (\lambda F)(\exists x)(\exists y)(Fx.Fy.(x \neq y))$'; for D18-1: '$mem_1 = (\lambda H)(\lambda x)(\exists y)(Hxy)$'; for D19-1: '$Un_1 = (\lambda H)(x)(y)(u)(...)$'; for D19-4: '$Corr_n = (\lambda K,H_1,H_2)(...)$'; for D29-1: '$I = (\lambda x,y)(x=y)$'; for D31-1: '$Sym = (\lambda H)(H \subset H^{-1})$'; for D32-1a: '$\Lambda_1 = (\lambda x)(x \neq x)$'; for D32-3a: '$sm_1 = (\lambda N)(\lambda x)(\exists F)(N(F).Fx)$'; for D32-5a: '$sub_1 = (\lambda F)(\lambda G)(G \subset F)$'; for D32-8: '$init = (\lambda H)(\lambda x)(...)$'; for D34-2: '$str_n = (\lambda H_1)(\lambda H_2)(Is_n(H_2,H_1))$'; for D36-1: '$Her = (\lambda F,H)[(x)(y)(Fx.Hxy \supset Fy)]$'; and for D37-3: '$sum = (\lambda N_1,N_2)(\lambda F)(\exists G_1)(\exists G_2)(...)$'.

It should be observed, finally, that definitions phrased in the λ-style have the same consequences as the more usual open definitional formulas. Suppose e.g. that the sentence $\mathfrak{S}_1$: '$mem_1 = (\lambda H)(\lambda x)[(\exists y)(Hxy)]$' is taken as defining '$mem_1$' in the syntactical system B. On the basis of $\mathfrak{S}_1$ we can, with the help of the interchangeability theorem (T24-7), replace the second occurrence of 'mem_1' in the provable sentence '$(H)(x)[mem_1(H)(x) \equiv mem_1(H)(x)]$' by the λ-expression given in $\mathfrak{S}_1$. From the resulting sentence '$(H)(x)[mem_1(H)(x) \equiv (\lambda H)(\lambda x)[(\exists y)(Hxy)](H)(x)]$' we obtain '$(H)(x)[mem_1(H)(x) \equiv (\exists y)(Hxy)]$' by two successive applications of the λ-rule and the trivial substitution of 'H' for 'H' and 'x' for 'x'. The sentence standing within the square brackets of this last result is to be recognized as the open definitional formula given for 'mem_1' in language A (see D18-1). Hence we necessarily obtain from the definitional sentence $\mathfrak{S}_1$ in B the same results as we do from the open definitional formula D18-1 in A.

Exercises. **1.** Replace the following with λ-style definitions: a) D29-2c; b) D31-3a; c) D31-4; d) D32-2a. — **2.** For each of the following, decide whether it can be replaced by λ-style definitions (recall that in a λ-style definition the definiendum consists precisely of the sign being defined); if it cannot be so replaced, give a notational revision that might be made in the definiendum which would allow the replacement, and give the replacement for the revised definition: a) D28-2; b) D30-1; c) D32-4a; d) D32-6a; e) D32-9a.

33d. The *R*'s of *b*. The property of bearing the relation R to b, i.e. the class comprising the R's of b, can be designated by the predicate expression

'$(\lambda x)(Rxb)$' formed with the help of the λ-operator. Let us introduce for this predicate expression the shorter form '$R(-,b)$'. Similarly, let us write '$R(a,-)$' as short for '$(\lambda y)(Ray)$', the class of those individuals to which a bears the relation R. E.g. '$Gr(-,3)$' denotes the class of all numbers greater than 3, while '$Gr(3,-)$' denotes the class of all numbers smaller than 3.

Our use of the dash '–' will for the most part be confined to the two sorts of cases just described; see e.g. T2 below. For the sake of theoretical completeness, however, we wish to specify here a general rule governing the use of the dash.

The dash is to occur only in an argument-expression belonging to a predicate expression; an argument expression may contain several dashes. Suppose $\mathfrak{A}_j$ is an n-place argument-expression and $\mathfrak{A}_i$ is an n-place predicate expression, and suppose that p of the argument-places of $\mathfrak{A}_j$ (where $1 \leq p \leq n-1$) are filled by dashes; then $\mathfrak{A}_i(\mathfrak{A}_j)$ is taken to be synonymous with—and hence, in any context, interchangeable with—the λ-expression '$(\lambda \mathfrak{v}_{k_1}, \mathfrak{v}_{k_2}, \ldots, \mathfrak{v}_{k_p})[\mathfrak{A}_i(\mathfrak{A}_j')]$', where $\mathfrak{A}_j'$ is obtained from $\mathfrak{A}_j$ by replacing each successive dash in $\mathfrak{A}_j$ by the corresponding variable in the λ-operator (viz., the first dash is replaced by $\mathfrak{v}_{k_1}$;...; the last, or pth, dash is replaced by $\mathfrak{v}_{k_p}$).

Of course, the λ-expression given above can be a predicate expression only when $\mathfrak{A}_i(\mathfrak{A}_j')$ is a sentential formula, i.e. when the variables that fill the argument-places in question are of the types appropriate to $\mathfrak{A}_i$. Beyond this, the variables in the λ-operator can be arbitrary, provided only that they do not already occur in $\mathfrak{A}_i(\mathfrak{A}_j)$.

The remarks above are illustrated by the following examples concerning the use of two dashes in a three-place argument-expression: '$T(-,-,c)$' is synonmous with '$(\lambda x,y)(Txyc)$'; '$T(-,c,-)$' is synonymous with '$(\lambda x,y)(Txcy)$'; '$T(c,-,-)$' is synonymous with '$(\lambda x,y)(Tcxy)$'; on the other hand, '$(\lambda x,y)(Tyxc)$' cannot be transformed into a full expression of 'T' with dashes.

We are now able to state:

T33-2. The following sentential formulas are L-true:

+**a.** $H(-,y) = (\lambda x)(Hxy)$.

b. $(H(-,y))(x) \equiv Hxy$.

+**c.** $H(x,-) = (\lambda y)(Hxy)$.

d. $(H(x,-))(y) \equiv Hxy$.

Exercises. For each of the following sentences, give (a) a translation in terms of 'λ'; and (b) a translation in terms of '–'. — **1.** "There are four primes which are greater than 10 and less than 20" (use the form '4(...)', with 'Gr'). — **2.** "a is the mother of five children" (use '5(...)'). — **3.** "a has as many brothers as b" (recall D19-5, and use 'Is_1'). — **4.** "The primes greater than 2 are odd". — **5.** "The squares greater than 100 have property P".

At this point it is possible to read three more systems in language C given in Part Two (Application of symbolic logic), viz. that of **47** and of **51a**.

34. EQUIVALENCE CLASSES, STRUCTURES, CARDINAL NUMBERS

34a. Equivalence relations and equivalence classes. If a relation R is symmetric and transitive, it is said to be an *equivalence relation*. Note that by T31-1d, an equivalence relation is always reflexive. (One instance of this kind of relation is the logical relation called "material equivalence" and symbolized '$\equiv$'; another instance is the semantic relation of L-equivalence.) We shall not introduce any special symbol for the concept of an equivalence relation.

If R is an equivalence relation, the field of R may be divided into mutually exclusive classes that satisfy the two following conditions: (1) R holds for each pair of individuals in any one of these classes; and (2) if an individual in one of these classes bears the relation R to another individual, then this second individual belongs to the same class as the first individual. This general fact may be extablished as follows. First, consider (1). Let a be an arbitrary R-member, and let P be the class of all individuals *to which a* bears the relation R (according to **33d**, this class is also designated by '$R(a,-)$'). Suppose, now, that b and c belong to P, i.e. that Rab and Rac; then it must be the case that Rba and Rcb (since R is symmetric), that Rbc and Rcb (since R is transitive), and further that each of Raa, Rbb, and Rcc holds (since R is reflexive); hence, in view of all these results, R holds for every pair in P and condition (1) is satisfied. Next, consider (2). Let a and P be as above, and suppose that b belongs to P, i.e. that Rab holds; if, now, Rbd also holds, then so must Rad (since R is transitive); thus d too belongs to P, and condition (2) is satisfied.

That the class P satisfies conditions (1) and (2) above can be formulated symbolically as follows: '$(x)(y)(Px.Py \supset Rxy).(x)(y)(Px.Rxy \supset Py)$'. A still more concise formulation thereof is: '$(x)(y)(Px \supset (Py \equiv Rxy))$'. Classes that satisfy these conditions we call *equivalence classes* with respect to R:

D34-1. $equ(H) = (\lambda F)[(x)(y)(Fx \supset (Fy \equiv Hxy))]$.

Note that '*equ*' is a functor; that '$equ(R)$' denotes the class of all equivalence classes with respect to R; and that the sentence '$equ(R)(P)$' reads "P is an equivalence class with respect to R". Our definition D1 is quite general in that it specifies the functor '*equ*' with respect to any (two-place, homogeneous) relation. However, the usual practice is to apply this concept only to equivalence relations. It should also be observed that, by D1, the empty class is an equivalence class (cf. T1d below); little use is made of this in practice (compare, however, our remarks in connection with T37-5 below about null cardinals). The discussion which now follows concerns non-empty equivalence classes.

Suppose R is a relation which expresses likeness (or equality, or agreement) in some particular respect, e.g. color. Then obviously R is an equivalence relation; the equivalence classes with respect to R are the maximal classes of individuals having the same color; and each equivalence class corresponds to a particular color. This approach presupposes the separate colors as primitive concepts. If, however, the relation Having-the-Same-Color is taken as a primitive concept, then the several colors can be defined as the equivalence classes of that relation. Our verbal explanation of '*equ*' is phrased in terms of classes only because that phrasing is the customary one; a phrasing in terms of properties is equally possible. E.g. we could use the term "equivalence property": each of two individuals has a certain one of the equivalence properties with respect to an equivalence relation R if and only if each bears the relation R to the other. In the color illustration just given, the separate colors are the equivalence properties relative to color-likeness, i.e. each separate color is characterized by the fact that two individuals have the same color if and only if they are alike in color.

Suppose R is an arbitrary equivalence relation. It is of interest to consider the equivalence classes with respect to R without regard to any prior interpretation of R as likeness in any particular respect. Here the case is that the equivalence classes with respect to R represent certain properties which permit a *subsequent* interpretation of R as a relation of agreement in one of these properties. E.g. let R be the relation of parallelism between the lines of a fixed plane. Then R is an equivalence relation. Now define the equivalence classes with respect to R, i.e. the maximal classes of lines parallel to one another. These classes represent properties of lines which might be called "directions"; these properties are characterized by the fact that two lines have the same direction if and only if they are parallel. Thus it appears that parallelism is identical with sameness of direction. What is to be noted here, however, is that we did *not* begin with the concept of direction and define parallelism in terms of it as sameness of direction; rather, we began with the concept of parallelism and proceeded to a definition of directions as equivalence classes with respect to parallelism. Such a definition of a family of properties by way of the equivalence classes of an equivalence relation is often called *definition by abstraction* (see Russell [Principles] 166; Frege [Grundlagen] 73ff.; H. Scholz and H. Schweitzer, *Die sogenannten Definitionen durch Abstraktion*, Forschungen zur Logistik, No. 3, 1935).

The discussion that has just been concluded allows us to state the following theorems.

T34-1. The following sentential formulas are L-true:

a. $Trans(H).Sym(H) \supset (x)(y)[Hxy \equiv (\exists F)(equ(H)(F).Fx.Fy)]$.
A given equivalence relation holds between two individuals if and only if these individuals belong to the same equivalence class.

b. $Trans(H).Sym(H) \supset (x)(equ(H)(H(-,x)))$.
If R is an equivalence relation, then $R(-,a)$ is an equivalence class. [Note two things here: $R(-,a)$ and $R(a,-)$ are the same; and in view of (d) below it is not necessary to require that a be a member of R.]

c. $Trans(H).Sym(H).equ(H)(F).equ(H)(G).(F \neq G) \supset 0(F.G)$.
Two different equivalence classes with respect to an equivalence relation have no individual in common.

d. $equ(H)(\Lambda_1)$.
The empty class is an equivalence class with respect to any relation.

Exercises. — **1.** Give informal proofs of the following: a) T1a; b) T1c; c) '$Trans(H).Sym(H) \supset sm_1(equ(H)) = mem(H)$'.

34b. Structures. Earlier, in D19-5, we defined the concept of isomorphism; our symbolism was 'Is_n', where for 'n' one of the numerals '1', '2', etc., must be put. From that discussion it is seen that two n-place relations are isomorphic provided there is a 2-place relation which serves as a correlator between the two. If R is a correlator between S_1 and S_2, then the converse of R is a correlator between S_2 and S_1; hence isomorphism is a symmetric relation. Again, if R_1 is a correlator between S_1 and S_2 and R_2 is a correlator between S_2 and S_3, then $R_1|R_2$ is a correlator between S_1 and S_3; hence isomorphism is a transitive relation. In view of these results, isomorphism is an equivalence relation; moreover, it is totally reflexive since identity is a correlator between S_1 and S_1.

+T34-2. The following sentences are L-true:

a. $Sym(Is_n)$.

b. $Trans(Is_n)$.

c. $Reflex(Is_n)$.

If two relations are isomorphic, we say they have the *same structure*. Hence the various relational structures can be represented as the equivalence classes (or equivalence properties) with respect to isomorphism. Following our previous considerations, the structure of a relation is thus the class of relations isomorphic with it (or: the property of being isomorphic with it). Employing a functor 'str_n', we agree to write '$str_n(T)$' for "the structure of the (n-place) relation T":

D34-2. $str_n(H) = Is_n(-,H)$.

That M is a structure of n-place relations or—as we shall also say—an n-place structure, is symbolized by '$Str_n(M)$'; 'Str_n' is a predicate of the third level.

D34-3. $Str_n = equ(Is_n)$.

Definitions D2 and D3 are actually definitional schemes, just as D19-5 was. By supplanting 'n' with such numerals as '1', '2', etc., we obtain from D2 explicit definitions of the functors 'str_1', 'str_2', etc.; and from D3 explicit definitions of the predicates 'Str_1', 'Str_2', etc. The same remark applies to the formulas given in the theorems below: numerals '1', '2', etc., are to be inserted for 'n'.

T34-3. The following sentential formulas are L-true:

a. $str_n(H)(K) \equiv Is_n(K,H)$.
A relation has the structure of another relation if and only if the first relation is isomorphic with the second.

b. $Str_n(str_n(H))$.
For each n-place relation H it is the case that $str_n(H)$, i.e. the structure of H, is an element of the class Str_n, or an n-place structure.

+**c.** $Str_n(N) \equiv (H)(K)[N(H) \supset (N(K) \equiv Is_n(H,K))]$.
(This result follows from D1.)

d. $Str_n(N) \supset (H)(K)[N(H).N(K) \supset Is_n(H,K)]$.
(This result follows from (c).)

e. $Str_n(N) \supset (H)(K)[N(H).Is_n(H,K) \supset N(K)]$.
(This result also follows from (c). Note that (d) and (e) correspond to conditions (1) and (2) in **34a**.)

f. $Str_n(\Lambda_1)$.
The empty class of n-place relations is an n-place structure. (From T1d.)

Exercises. **1.** Domain of individuals: the straight lines of a given plane. Using '*Par*' for "Parallel", define a predicate '*Dir*' (analogous to 'Str_1') where '*Dir*(F)' means "F is a direction". Also define a functor '*dir*' (analogous to 'str_1') where '*dir*(x)' means "the direction of x" (see remark in **34a**). — **2.** Domain of individuals: the points of a given plane. Using '*Eqda*' for "Equidistant from the point a", give an informal proof to show that *Eqda* is an equivalence relation. Define a predicate '*Cira*' where '*Cira*(F)' means "F is an equivalence class with respect to *Eqda*". Also define a functor '*cira*' where '*cira*(x)' means "the equivalence class of x with respect to *Eqda*". Using the language of geometry, what other readings could be given to '*Cira*(F)' and '*cira*(x)'?

34c. Cardinal numbers. As we have already mentioned in **19**, one-place isomorphism of classes (or properties) means that these classes are equinumerous. Hence the one-place structures are the cardinal numbers.

Suppose e.g. there are exactly three individuals having property P; as we learned in **17c,** this fact can be expressed by the sentence '$3(P)$'. It follows from the definition of '3' in D17-3 that a property Q has the second-level property 3 if and only if Q is isomorphic with P. Thus by T3a we have '$3 = str_1(P)$', and hence '$Str_1(3)$'; which is to say, 3 is the cardinal number of P and so 3 is a cardinal number. Similar remarks hold for every other second-level predicate defined in accordance with D17-3. Consequently the results stated in the theorem below are valid.

T34-4. Suppose 'M' is any one of the second-level predicates '0', '1', '2', etc., defined according to D17-3. Then the following sentential formulas are L-true:

a. $M(F).M(G) \supset Is_1(F,G)$.
b. $M(F).Is_1(F,G) \supset M(G)$.
c. $M(F) \supset (M(G) \equiv Is_1(F,G))$. (From (a),(b).)
d. $equ(Is_1)(M)$. (From (c) and D1.)
+**e.** $Str_1(M)$. (From (d) and D3.)
f. $M(F) \supset [M = Is_1(-,F)]$. (From (c).)
g. $M(F) \supset [M = str_1(F)]$. (From (f) and D2.)

Earlier in this book (in **17c**) we called the second-level properties 0,1,2, etc., cardinal numbers. But only here, after defining the general concept of cardinal number ('Str_1'), have we been able to show that 0,1,2, etc., actually are cardinal numbers (T4e).

The empty class, and only the empty class, has the cardinal number 0 (see below T5b,c,d). Therefore 0 itself is not empty (cf. T5e). The contrast between T5e below and T32-1a thus signalizes an important difference between the (first-level) empty class Λ_1 and the (second-level) non-empty class 0; this difference is particularly to be noted since in set theory unfortunately the empty class is often designated by '0'.

T34-5. The following sentential formulas are L-true:

+**a.** $0(F) \equiv \sim\exists(F)$.
+**b.** $0(\Lambda_1)$.
c. $0(F) \equiv (F = \Lambda_1)$.
d. $0 = \{\Lambda_1\}$. (From (c).)
e. $\exists(0)$. (From (b).)
f. $1(F) \equiv (\exists x)(y)(Fy \equiv (y = x))$.
g. $1(F) \equiv (\exists x)(F = \{x\})$.
h. $2(F) \equiv (\exists x)(\exists y)[Jxy.(F = \{x,y\})]$.
i. $3(F) \equiv (\exists x)(\exists y)(\exists z)[J_3xyz.(F = \{x,y,z\})]$.
j. $1\{x\}$.
k. $2\{x,y\} \equiv Jxy$.
l. $3\{x,y,z\} \equiv J_3xyz$.

Later we shall encounter examples of two-place structures: '*Prog*' (D37-1), 'η_{00}' etc., '$ContSer_{00}$' etc., and '$ContOrd_{00}$' etc. (**38**).

Frege was the first to indicate clearly that cardinal numbers are to be attributed to classes (or properties) rather than individuals. He constructed definitions for the separate cardinals, and for the general concept of cardinal number, with which our definitions (in **17c**, and D3 for 'Str_1') in essence agree (Frege [Grundlagen] 79 ff., [Grundgesetze] vol. I, 57). In 1901, independently of Frege, Russell constructed similar definitions and used them in establishing the foundations of arithmetic. Both Frege and Russell considered it necessary to use different forms of expression for classes and for properties, and both defined the cardinal numbers as classes of classes. According to this view, the cardinal number 3 e.g. is the class of all triples of individuals. Such a conception understandably provoked some adverse criticism, especially since classes were usually considered as totalities; and admittedly the totality of triples of, say, all physical things in the world is a vague and extravagant affair. (Criticisms of this kind may be found e.g. in Hausdorff [Grundzüge] 46 and J. König [Logik] 226, note; for further discussion, see Fraenkel [Einleitung] 57 ff.) If, however, a class expression is regarded as an expression which facilitates the making of statements about that which is common to the elements of the class, all semblance of paradox vanishes from the Frege-Russell definitions (cf. Carnap [Aufbau] 54 f.). And if we go on, as we did above, to introduce cardinal numbers as properties of properties, thus e.g. '3' as a predicate designating the property of being a triple, the earlier objections are entirely vacated and so are the criticisms which Wittgenstein and Waismann have leveled against the Frege-Russell definitions (cf. Waismann [Math. Thought] §9B).

Exercises. **1.** Give informal proofs of the following, substituting '2' for 'M' in a) through d): a) T4a; b) T4b; c) T4e; d) T4g; e) T5b; f) T5c; g) T5f; h) T5k.

34d. Structural properties. If R is a symmetric relation, it is easy to show that every relation having the same structure as R is also symmetric. A symbolic phrasing of this statement runs as follows: '$(H_1)(H_2)[Sym(H_1).Is_2(H_1,H_2) \supset Sym(H_2)]$'. (Later, in **36a**, we will say instead: "Symmetry is an hereditary property with respect to isomorphism"; and write: '$Her(Sym,Is_2)$'.) Since the property of being symmetric thus depends only on the structure of the relation, let us call it a *structural property* and write '$Struct_2(Sym)$'. In general we say a property of n-place relations is an (n-place) *structural property* provided it depends simply on the structure, i.e. provided it is preserved under isomorphism.

D34-4. $Struct_n(N) \equiv (H_1)(H_2)[N(H_1).Is_n(H_1,H_2) \supset N(H_2)]$.

T34-6. The following sentences are L-true:

+**a.** $Struct_2(Sym)$.
The same holds for the other predicates defined in **31**: '*As*', '*Trans*', '*Intr*', '*Refl*', '*Irr*', '*Reflex*', '*Connex*', '*Ser*', '*Antis*', '*POrd*', '*SOrd*', 'Un_1', 'Un_2', '$Un_{1,2}$'.

b. $Str_n \subset Struct_n$. (From T34-3e.)

c. $Struct_n(M) \supset Struct_n(\sim M)$.

d. $Struct_n(M).Struct_n(N) \supset Struct_n(M \vee N)$.

e. $Struct_n(M).Struct_n(N) \supset Struct_n(M.N)$.

We learn from T6b that structures are also structural properties. Indeed, they are the strongest structural properties, in the following sense: Suppose $\mathfrak{S}_i$ is a sentence that attributes a definite structure to a given *n*-place relation, and suppose $\mathfrak{S}_j$ is a sentence that attributes to the same relation some arbitrary structural property; then $\mathfrak{S}_i$ L-implies either $\mathfrak{S}_j$ or $\sim\mathfrak{S}_j$. Which is to say, when a relation is assigned a structure, the relation is fully specified so far as its structural properties are concerned. It is to be noted, however, that most structural properties—including those named in T6a—are not structures since they do not satisfy T34-3d.

Exercises. 1. Give informal proofs of T6c, T6d, and T6e. — **2.** Give informal proofs of the following parts of T6a (on the basis of other parts of T6a and T6e): a) '$Struct_2(Ser)$'; b) '$Struct_2(POrd)$'; c) '$Struct_2(SOrd)$'. — **3.** Which of the following expressions designate structural properties?: a) '$(\lambda H)[(\exists x)Hxa]$'; b) '$(\lambda H)[Sym(H).(mem(H) \subset Even)]$' (domain of individuals: natural numbers); c) '$(\lambda H)[(\exists x)(\exists y)Hxy]$'.

35. INDIVIDUAL DESCRIPTIONS

35a. Descriptions. The expressions elucidated in this section are treated chiefly because they occur frequently in the system of [P.M.] and in certain other systems. In our language **C**, however, expressions of this kind will seldom be used.

Our task is the explication of phrases such as "the son of Charles Smith", "the book on my desk", etc. Now the sentence "the book on my desk is black" says two things: (1) that there is exactly one book on my desk, and (2) that it is black. If '*P*' designates the property of being a book on my desk and '*Q*' the property of being black, we symbolize "the book on my desk" by '$(\imath x)(Px)$' and the whole sentence by '$Q[(\imath x)(Px)]$'. The square brackets here may be omitted, in view of rule (4) in **9a**; however, the brackets about '*Px*' must not be omitted.

Next, observe that from the sentence $\mathfrak{S}_1$: '$(x)[Px \equiv (x=a)]$' there follows, on the one hand, '$(x)[(x=a) \supset Px]$' and so '$(a=a) \supset Pa$' and thus '*Pa*'; while on the other, we have '$(x)[Px \supset (x=a)]$' and so '$(x)[(x\neq a) \supset \sim Px]$'. Thus $\mathfrak{S}_1$ says "*a* has property *P* and no other individual does", i.e. "*a* is the only individual having property *P*". Consequently, component (1) in the paragraph above—the part of our original sentence often called the *uniqueness condition*—can be formulated as '$(\exists y)(x)[Px \equiv (x=y)]$'; indeed, it can be written still more concisely as '$1(P)$' (which, in view of T34-5f, is L-equivalent to the formulation just given). Our entire original sentence may therefore also be written: '$(\exists y)[(x)(Px \equiv (x=y)).Qy]$'. The relation between this formulation and the previous one, '$Q[(\imath x)(Px)]$', is exploited in D1a below.

An expression of the form '$(\imath x)(...x...)$' denotes an individual, the denoting being not in the fashion of a proper name (e.g. '*a*', '*b*' or the like) but with the help of a property which attaches to this individual only. Such an expression is called a *description* (or: an individual description). The symbol '$(\imath x)$' is an operator; it is called the $\imath$-operator (read: "iota-operator"; the '$\imath$' is an inverted Greek iota). Because '$(\imath x)$' is an operator, 'x' is bound at each of its occurrences in the description '$(\imath x)(...x...)$'. To avoid complicating unduly the rules governing the use of descriptions, we shall restrict the role of a description to that of an argument-expression for a predicate expression (but not for a functor expression), and to that of a member of an identity formula.

On the basis of the explanations given to date, we construct the following three formal schemata:

D35-1. **a.** $\mathfrak{A}_k \equiv (\exists \mathfrak{v}_j)[(\mathfrak{v}_i)(\mathfrak{A}_i \equiv (\mathfrak{v}_i = \mathfrak{v}_j)).\mathfrak{A}_j]$.

Here $\mathfrak{v}_i$ and $\mathfrak{v}_j$ are two different individual variables; $\mathfrak{A}_i$ is an arbitrary sentential formula in which $\mathfrak{v}_j$ has no free occurrences; $\mathfrak{A}_k$ is a full expression of a predicate expression, and is such that one of its argument-places is occupied by the description $(\imath \mathfrak{v}_i)(\mathfrak{A}_i)$; and finally, $\mathfrak{A}_j$ is like $\mathfrak{A}_k$ except that the former has $\mathfrak{v}_j$ in the place where the latter has the description just cited.

b. $[(\imath \mathfrak{v}_i)(\mathfrak{A}_i) = \mathfrak{A}_j] \equiv (\mathfrak{v}_i)[\mathfrak{A}_i \equiv (\mathfrak{v}_i = \mathfrak{A}_j)]$;

Here $\mathfrak{v}_i$ is an individual variable; $\mathfrak{A}_i$ is a sentential formula; and $\mathfrak{A}_j$ is an individual expression (but not a description) in which $\mathfrak{v}_i$ has no free occurrences.

c. $[(\imath \mathfrak{v}_i)(\mathfrak{A}_i) = (\imath \mathfrak{v}_j)(\mathfrak{A}_j)] \equiv (\exists \mathfrak{v}_m)[(\mathfrak{v}_i)(\mathfrak{A}_i \equiv (\mathfrak{v}_i = \mathfrak{v}_m)) . (\mathfrak{v}_j)(\mathfrak{A}_i \equiv (\mathfrak{v}_j = \mathfrak{v}_m))]$.

Here $\mathfrak{v}_i$, $\mathfrak{v}_j$ and $\mathfrak{v}_m$ are individual variables, with $\mathfrak{v}_m$ different from $\mathfrak{v}_i$ and $\mathfrak{v}_j$; and $\mathfrak{A}_i$ and $\mathfrak{A}_j$ are both sentential formulas having no free occurrences of $\mathfrak{v}_m$.

These three formulas do not have the form used elsewhere for definitions in language C. Nevertheless, they serve the same purpose as the typical definition, viz. to eliminate descriptions from arbitrary contexts of the kind indicated in D1. Thus formulas like D1a are useful in any case involving the occurrence of a description as an argument-expression for a predicate expression. E.g. with the help of the formula '$Q(\imath x)(Px) \equiv (\exists y)[(x)(Px \equiv (x=y)).Qy]$' we can replace '$Q(\imath x)(Px)$' by '$(\exists y)[(x)(Px \equiv (x=y)).Qy]$' in any context, whether '$Q(\imath x)(Px)$' appears there as an independent sentence or as a component sentence. On the other hand, formulas like D1b are useful in cases that involve an identity formula having precisely one of its members in the form of a description. E.g. '$(\imath x)(Px)=a$' can be replaced by '$(x)[Px \equiv (x=a)]$'; of course, if '$a=(\imath x)(Px)$' is the given formula, we first revise it into '$(\imath x)(Px)=a$' and then apply D1b. Lastly, formulas like D1c suit cases involving an identity formula each of whose members is a description. Thus D1c enables us to transform '$(\imath x)(Px)=(\imath y)(Qy)$' into '$(\exists z)[(x)(Px \equiv (x=z)).(y)(Qy \equiv (y=z))]$'. If several descriptions occur in a sentence, it is a matter of indifference which of them is eliminated first with the help of D1; which is to say, eliminations in various orders lead to results that are L-equivalent.

While the sentences '$\sim(Qa)$' and '$(\sim Q)a$' are synonymous and L-equivalent (see D28-1a), the same cannot be said of the corresponding sentences obtained by replacing the individual constants by descriptions. For according to the theorem below (to T1b, in fact), the sentence '$\sim Q(\imath x)(Px)$' is L-equivalent to '$\sim[1(P) \,.\, (P \subset Q)]$' and hence to $\mathfrak{S}_1$: '$\sim 1(P) \vee \sim(P \subset Q)$'; on the other hand, by the same theorem the sentence '$(\sim Q)(\imath x)(Px)$' turns out L-equivalent to $\mathfrak{S}_2$: '$1(P) \,.\, (P \subset \sim Q)$'. Clearly, if the uniqueness condition '$1(P)$' is not satisfied (i.e. if there are either no individuals or else several with property P), then $\mathfrak{S}_1$ is true but $\mathfrak{S}_2$ false; $\mathfrak{S}_1$ and $\mathfrak{S}_2$ therefore cannot be L-equivalent. Thus descriptions require a treatment different from that of other individual expressions. In particular, a description may not simply be introduced in place of an individual variable. E.g. '$(y)(Qy)$' can hold (viz. each individual may have property Q) and still '$Q(\imath x)(Px)$' fail to hold because the uniqueness condition '$1(P)$' of the description is not satisfied. Hence '$Q(\imath x)(Px)$' is not L-implied by '$(y)(Qy)$' alone, but only by '$(y)(Qy)$' and '$1(P)$' together (see T1c below). Since the manipulation of descriptions demands special care, it is better to avoid their use when this does not lead to undue complications.

T35-1. The following sentential formulas are L-true:

+**a.** $G(\imath x)(Fx) \equiv (\exists y)[(x)(Fx \equiv (x=y)).Gy]$. (By D1a.)

b. $G(\imath x)(Fx) \equiv 1(F).(F \subset G)$.

c. $(y)(Gy).1(F) \supset G(\imath x)(Fx)$. (From (b).)

d. $G(\imath x)(Fx) \equiv 1(F).\exists(F.G)$.

e. $F(\imath x)(Fx) \equiv 1(F)$.

f. $[(\imath x)(Fx)=y] \equiv (x)(Fx \equiv (x=y))$. (By D1b.)

+**g.** $[(\imath x)(Fx)=y] \equiv 1(F).Fy$.

h. $[(\imath x)(Fx)=y] \equiv (F=\{y\})$.

i. $[(\imath x)(Fx)=(\imath y)(Gy)] \equiv (\exists z)[(x)(Fx \equiv (x=z)) \,.\, (y)(Gy \equiv (y=z))]$. (By D1c.)

j. $[(\imath x)(Fx)=(\imath y)(Gy)] \equiv 1(F).1(G).\exists(F.G)$.

k. $[(\imath x)(Fx)=(\imath y)(Gy)] \equiv 1(F).(G \subset F).\exists(G)$.

l. $[(\imath x)(Fx)=(\imath y)(Gy)] \equiv 1(F) \,{:}\, (F=G)$.

Exercises. 1. Give informal proofs of the following: a) T1a; b) T1c; c) T1g; d) T1l; e) '$a=(\imath x)(x=a)$'. — **2.** Give a derivation of '$\sim Q(\imath x)(Px)$' from 'Pa', 'Pb' and '$a \neq b$'. — **3.** Formulate in symbols and give an informal proof of the following sentences (a), (b) and (c) in the domain of natural numbers (state all assumptions explicitly): a) "It is not the case that the number greater than two is greater than two"; b) "The even prime number is even"; c) "It is not the case that the square number less than five is even"; d) Does c) imply "the square less than five is odd"? — **4.** Is '$(\imath x)Fx=(\imath x)Fx$' L-true? If so, give an informal proof. If not, state a sufficient assumption and show that it L-implies the negation of the formula mentioned by giving an informal derivation.

35b. Relational descriptions. Descriptions frequently have the form '$(\imath x)(Rxb)$', which means: "that individual which bears the relation R to b". The abbreviation 'R‘b' is used for '$(\imath x)(Rxb)$'. In these symbols, any two-place first-level predicate expression can stand in place of 'R' and any individual expression can stand in place of 'b'. Expressions like 'R‘b' are called *relational descriptions.* The restrictions previously noted on the manipulation of descriptions with $\imath$-operators apply equally to relational descriptions.

D35-2. H‘$y = (\imath x)(Hxy)$.

T35-2. The following sentential formulas are L-true:

a. $G(H$‘$y) \equiv (\exists z)[(x)(Hxy \equiv (x=z)).Gz]$. (By T1a.)
b. $G(H$‘$y) \equiv 1(H(-,y)).(H(-,y) \subset G)$. (By T1b.)
c. $Un_1(H).mem_2(H)y \supset 1(H(-,y))$.
d. $Un_1(H).mem_2(H)y.U(G) \supset G(H$‘$y)$. (By (c) and T1c.)

Descriptions are seldom used in language C. A common way of avoiding them is through the use of functors (provided conditions specified earlier (in **18b**) on the use of functors are satisfied). Descriptions of properties or relations of any level can always be avoided; they can be supplanted e.g. by full expressions of functor expressions, by compound predicate expressions, and by λ-expressions or expressions involving '–'. Thus, to illustrate, instead of the following expressions from [P.M.], viz. 'D‘R', 'C‘R', 'cnv‘R', 's‘κ', 'p‘κ', 'Cl‘κ', 'Rl‘κ', '$\overrightarrow{R}$‘b', '$\overleftarrow{R}$‘a', 'Nc‘α', 'Nr‘R', there appear in language C respectively the expressions '$mem_1(R)$', '$mem(R)$', 'R^{-1}', '$sm_1(M)$', '$pr_1(M)$', '$sub_1(M)$', '$sub_2(M)$', '$R(-,b)$', '$R(a,-)$', '$str_1(P)$', '$str_2(R)$'.

Suppose that the uniqueness condition for a given description is provable either on purely logical grounds or within a certain axiom system. In either case, the description can be treated as an individual constant. It can e.g. be admitted as an argument-expression of a functor, in contradistinction to the previous general restriction; and again, it can receive an individual constant as an abbreviation. Thus the rules governing the construction of admissible definitions may be extended to include the following: A sentence of the form $\mathfrak{a}_i = \mathfrak{A}_j$ with $\mathfrak{a}_i$ a new individual constant and $\mathfrak{A}_j$ a description can be accepted as a definition provided the uniqueness condition for $\mathfrak{A}_j$ is provable. Such so-called *definitions by description* are often convenient (see e.g. the remark under A2* in **44b**); nevertheless, to admit definition by description is to accept the disadvantage that the rules of formation for definitions thereby depend on the rules of transformation.

Exercises. 1. Translate the following sentences, using relational descriptions when possible: a) "The brother of *a* is a student". b) "The father of *a* is a friend of the father of *b*". c) "The successor of *x* is always greater than *x*" (do this in two ways: (i) using the two-place predicate '*Suc*' for "successor"; and (ii) using the functor '*suc*'); d) "The predecessor of *x* is always smaller than *x*" (Question: Can we here, as in exercise 1c, avoid the description by the use of a functor? Cf. **18b**.); e) "That number which is both prime and even is the predecessor of a prime" (with the help of the $\imath$-operator); f) "*a* is the father of *b*'s only brother"; g) "Anyone who is the father of the brother of his only daughter is also the father of the daughter of his only son". — **2.** Give informal proofs of the following: a) T2a; b) T2d; c) '$1(H(-,a)) \equiv (H(-,a)(H$‘$a))$'. — **3.** Taking the domain of individuals to be the natural numbers, which of the following are true?

a) '$(Suc`7)=8$'; b) '$\sim(Even(Gr`4))$'; c) '$Pred`2=Pred`(Suc`(Pred`2))$'; d) '$(x)[Pred`x=Pred`(Suc`(Pred`x))]$'; e) '$(x)[Even(x) \equiv Even(Sq`x)]$'.

The following systems in language C can now be read in Part Two (Application of symbolic logic) of this book: **52a,b**; **53a**.

36. HEREDITY AND ANCESTRAL RELATIONS

36a. Heredity. Ordinarily we say of a property (e.g. disposition to a certain disease, a proprietary interest, or the like) which always, or frequently, passes down from a man to his children that it is hereditary. In analogy to this let us say of a property P that it is *hereditary* with respect to a relation R or, for short, that it is R-hereditary (symbolically: '$Her(P,R)$') or that it is preserved under R, if the following condition is fulfilled: whenever an R-member has property P, then so do all the other members to which this R-member bears the relation R.

D36-1. $Her(F,H) \equiv (x)(y)(Fx.Hxy \supset Fy)$.

Examples. The property of being greater than 5 is hereditary with respect to the predecessor relation in the series of natural numbers. The structural properties of relations (D34-4) are those which are hereditary with respect to isomorphism.

Exercises. **1.** Taking the domain of individuals to be the natural numbers, give an example of a property which is hereditary with respect to each of the following relations: a) Immediate Successor; b) Divides; c) $(\lambda xy)[(y=x+2) \vee (x=y+2)]$. — **2.** Give for each of the following properties an example of a relation with respect to which it is hereditary: a) Even; b) Not Prime; c) $(\lambda x)[(\exists y)(x=5y+1)]$. — **3.** With respect to what relations are all properties hereditary?—**4.** What properties are hereditary with respect to all relations?

36b. Ancestral relations. Let us take '$Anc(a,b)$' to mean "a is an ancestor of b". How, then, might "ancestor" be explained in terms of "parent", i.e. how might 'Anc' be defined with the help of 'Par'? Speaking loosely, we would say '$Anc(a,b)$' amounts to '$Par(a,b) \vee Par^2(a,b) \vee Par^3(a,b) \vee$ etc.', i.e. the relation Anc holds between a and b provided some finite power of the relation Par holds between a and b. To make this loose characterization into a precise definition we must explicate the "etc.", i.e. the word "finite". But there is a difficulty here: we have not as yet defined the concept of a finite number. (Indeed, it is preferable that the concept of finite number be defined later, *in terms of* the ancestral relation being introduced here.) We consider the more general relation Anc', where '$Anc'(a,b)$' means "a is an ancestor of b, or a is the same as b". Now Anc' can be defined with the help of the concept of hereditary property treated just above; we can easily see that '$Anc'(a,b)$' holds just in case a is a Par-member and b has all the Par-hereditary properties that a has.

The following two considerations lead to the result just mentioned. 1. Suppose that $Anc'(a,b)$ holds. Then there is a certain number n, $n \geq 0$, such that one can proceed from a to b by n Par-steps: one step takes us to the children of a, two steps take us to the grandchildren of a, etc. Thus, from the assumption that a has a certain Par-hereditary property P it follows after n such Par-steps that b also has property P. 2. Conversely, suppose b has all the Par-hereditary properties that a does, and suppose a is a Par-member. If a is an ancestor of x or is the same as x, and if x is a parent of y, then evidently a is an ancestor of y. Hence the property of x to the effect that a bears the relation Anc' to x is itself a Par-hereditary property. Since a obviously has this same property, it follows from our original supposition that b has this property, i.e. that $Anc'(a,b)$ is the case.

Once 'Anc'' is defined, it is reasonable to define 'Anc' by '$Par|Anc'$'. Now let R be an arbitrary relation. The relation that is connected to R the way Anc' is connected to Par is called *the ancestral of R of the first kind* and is symbolized by '$R^{\geq 0}$'; the relation connected to R the way Anc is connected to Par is called *the ancestral of R of the second kind* and is symbolized by '$R^{>0}$'. [The corresponding symbols in [P.M.] are 'R_*' and 'R_{po}' respectively.] Thus the sentence '$R^{>0}(a,b)$' asserts that some finite positive power of R holds between a and b; the sentence '$R^{\geq 0}(a,b)$' asserts that either some finite positive power of R holds between a and b, or R^0 holds between them (i.e. a is the same R-member as b).

All these considerations lead to the following definitions:

D36-2. $H^{\geq 0}(x,y) \equiv mem(H)x.(F)[Her(F,H).Fx \supset Fy]$.

D36-3. $H^{>0}=(H|H^{\geq 0})$.

Examples. 1. If '$Pred$' designates the predecessor relation among natural numbers, then '$Pred^{>0}(a,b)$' says "a is less than b" and '$Pred^{\geq 0}(a,b)$' says "a is less than or equal to b". — **2.** The sentence '$Par^{>0}(a,b)$' reads "a is an ancestor of b", while '$Par^{\geq}(a,b)$' reads "a is an ancestor of, or the same as, b".

The theorems below summarize the main results about the ancestrals of a relation.

T36-1. The following sentential formulas are L-true.

a. $H^0 \subset H^{\geq 0}$.

b. $H \subset H^{\geq 0}$.

c. $H^2 \subset H^{\geq 0}$.

Etc.

d. $H \subset H^{>0}$.

e. $H^2 \subset H^{>0}$.

Etc.

f. $H^{>0} \subset H^{\geq 0}$.

g. $H^{>0} = H^{\geq 0}|H$.

h. $H^{\geq 0} = H^{>0} \vee H^0$.

i. $H^{>0}(x,y) \equiv (F)[Her(F,H).(z)(Hxz \supset Fz) \supset Fy]$.

j. (1) $Trans(H^{\geq 0})$; (2) $Trans(H^{>0})$.

Ancestrals of either kind are always transitive. (Thus the ancestral—of either kind—of a relation R is often a series or a partial order or a simple order, even though R itself is not.)

k. $Her(F,H) \equiv (H^{-1}``F) \subset F$.

We owe to Frege the idea of using the concept of hereditary property to explicate the "etc." in mathematics and to define the concept of a finite number (see [Begriffsschrift] 55 ff.; [Grundgesetze] I, 59 ff.; also [P.M.] I, 569 ff. and Russell [Introduction] Ch.3).

Exercises. **1.** Give informal proofs of the following: a) T1a; b) T1b; c) T1d; d) T1h; e) T1k. — **2.** Decide whether the following are L-true. If so, give an informal proof. If not, give a counter-example. a) '$H^0 \subset H^{\geqq 0}$'; b) '$\sim(H^0 \subset H^{>0})$'; c) '$Refl(H) \supset (H^{>0} = H^{\geq 0})$'; d) '$Sym(H) \supset (H^{>0} = H^{\geq 0})$'; e) '$Refl(H^{>0})$'.

36c. *R*-families. By the *R-posterity* of a we understand the class of all those R-members to which a bears the relation $R^{\geq 0}$, i.e. the R-posterity of a is the class $R^{\geq 0}(a,-)$. By the *R-ancestry* of a we understand the class of all those R-members which bear to a the relation $R^{\geq 0}$, i.e. the R-ancestry of a is the class $R^{\geq 0}(-,a)$. (These understandings entail that a is counted as belonging both to its own ancestry and its own posterity.) The union of a's R-ancestry and R-posterity, viz. $R^{\geq 0}(-,a) \vee R^{\geq 0}(a,-)$, is called the *R-family* of a and is designated symbolically by '$fam(R,a)$'. The *R-interval* between a and b, symbolized by '$int(R,a,b)$', is understood to be the intersection of a's R-posterity with b's R-ancestry, viz $R^{\geq 0}(a,-) \,.\, R^{\geq 0}(-,b)$. Definitions of the functors '*fam*' and '*int*' thus run as follows:

D36-4. $fam(H,x) = H^{\geq 0}(-,x) \vee H^{\geq 0}(x,-)$.
D36-5. $int(H,x,y) = H^{\geq 0}(x,-) \,.\, H^{\geq 0}(-,y)$.

Exercise. Symbolize and give an informal proof of the following: "If H is an equivalence relation (see **34a**), the H-family of x is the equivalence class of x with respect to H".

In Part II (Application of symbolic logic) of this book, the following systems in language C can now be read: **53b** and **54a**, **b**.

37. FINITE AND INFINITE

37a. Progressions. In the series of natural numbers the predecessor relation *Pred* has the following properties: (1) it is one-one; (2) it has exactly one initial member; (3) it has no terminal member; and (4) of any two distinct *Pred*-members, one can be reached from the other in finitely many *Pred*-steps, i.e. the relation $Pred^{>0}$ is connected. If an arbitrary relation R has these four properties, we say that R is a *progression* and write '$Prog(R)$'. Given two progressions, R and S, a correlator (see **19**) for them can be determined as follows: Let the initial member of R be coordinated with the initial member of S; and if a member x of R is coordinated with a member y of S, then let the R-successor of x be coordinated with the S-successor of y. Since any two progressions can thus be correlated, it is the case that any two progressions are isomorphic (see T1a below). It is clear, moreover, that any relation isomorphic to a progression is itself a progression (cf. T1b). Hence *Prog* is a (two-place) structure (cf. T1c).

We call a class P *denumerable*, and write '$\aleph_0(P)$', provided there is a progression whose members are the elements of P. The symbol '$\aleph_0$' is read "aleph-zero" (sometimes "aleph-null", due to a mistranslation from the German "Aleph-Null"). A correlator between two progressions being simultaneously a class-correlator between the fields of these progressions, it follows that there hold for $\aleph_0$ theorems analogous to those for *Prog* (cf. T1d, e, f). In particular, T1f says that $\aleph_0$ is a cardinal number; indeed, $\aleph_0$ is the smallest transfinite (i.e. non-finite) cardinal number.

On the basis of the discussion above we now proceed to state the definitions and theorems.

D37-1. $Prog(H) \equiv Un_{1,2}(H) \,.\, 1(init(H)) \,.\, 0(init(H^{-1})) \,.\, Connex(H^{>0})$.

D37-2. $\aleph_0(F) \equiv (\exists H)[Prog(H) \,.\, (F=mem(H))]$.

T37-1. The following sentential formulas are L-true:

a. $Prog(H).Prog(K) \supset Is_2(H,K)$.
b. $Prog(H).Is_2(H,K) \supset Prog(K)$.
+**c.** $Str_2(Prog)$. (By (a), (b), and T34-3c.)
d. $\aleph_0(F).\aleph_0(G) \supset Is_1(F,G)$. (By (a).)
e. $\aleph_0(F).Is_1(F,G) \supset \aleph_0(G)$. (By (b).)
f. $Str_1(\aleph_0)$. (By (d), (e) and T34-3c.)

Exercises. **1.** Give informal proofs of the following: a) T1a; b) T1b; c) T1e; d) '$Prog(H) \supset Ser(H^{>0})$'; e) '$Prog(H) \supset SOrd(H^{\geq 0})$'. — **2.** Is the converse of a progression a progression? If so, give an informal proof; if not, give a counter-example. — **3.** Which of the following are L-true? a) '$\aleph_0(F).(G \subset F).(G \neq F) \supset \aleph_0(F.\sim G)$'; b) '$Prog(H).mem(H)(x) \supset Prog(H \text{ in } H^{\geq 0}(x,-))$'; c) '$Prog(H).mem(H)(x) \supset \aleph_0(fam(H,x))$'.

37b. Sum and predecessor relation. If M_1 and M_2 are cardinal numbers (Str_1), we designate their arithmetic sum by '$sum(M_1,M_2)$'. (Our notation '$sum(M_1,M_2)$' supplants the more customary notation 'M_1+M_2'.) By '$sum(M_1,M_2)$' we mean the cardinal number of any class which can be partitioned into two subclasses with no elements in common and such that one of these subclasses has cardinal M_1 and the other, cardinal M_2. Again, if M_1 and M_2 are cardinal numbers we take '$Pred(M_1,M_2)$' to mean that M_1 is the *immediate predecessor* of M_2, i.e. that $M_1+1=M_2$. [In the definitions of '*sum*' and '*Pred*' below, arguments are not restricted to cardinal numbers but can be arbitrary classes (of at least the second level). However, since the use of '*sum*' and '*Pred*' is in practice confined to cardinal numbers, it is a matter of indifference what significance these signs have for other arguments.]

D37-3. $sum(N_1,N_2)(F) \equiv (\exists G_1)(\exists G_2)[(F=G_1 \vee G_2).\sim\exists(G_1.G_2).N_1(G_1).N_2(G_2)]$.

D37-4. $Pred(N_1,N_2) \equiv [sum(N_1,1)=N_2]$.

T37-2. The following sentential formulas are L-true:

a. $Str_1(N_1).Str_1(N_2) \supset Str_1(sum(N_1,N_2))$.
The sum of two cardinal numbers is again a cardinal number.

b. $Str_1(N_1).Pred(N_1,N_2) \supset Str_1(N_2)$. (By (a).)

c. $sum(0,1)=1$,
$sum(1,1)=2$,
$sum(2,1)=3$,
etc.

d. $Pred(0,1)$,
$Pred(1,2)$,
$Pred(2,3)$,
etc. (By (c).)

e. $sum(\aleph_0,1)=\aleph_0$.

Proof of (e). **1.** Let R be a progression, and P be the field of R. Thus P is denumerable, and has $\aleph_0$ for its cardinal number. Let Q be that subclass of P comprising all the elements of P except the initial member a of R. Let S be R confined to Q, i.e. let S be that subrelation of R comprising all the pairs of R except the first one. Clearly S is also a progression. Since Q is the field of S, Q is denumerable. Now P is also $Q \vee \{a\}$, and hence must have $sum(\aleph_0,1)$ for its cardinal number. — **2.** If $\aleph_0$ is empty, then $sum(\aleph_0, 1)$ is likewise empty. — Together, these considerations lead to (e) above.

Exercises. Give informal proofs of the following: a) T2a; b) T2c; c) '$sum(N_1,N_2)= sum(N_2,N_1)$'; d) '$sum(N_1,sum(N_2,N_3))=sum(sum(N_1,N_2),N_3)$'; e) '$\sim Pred(N,0)$'.

37c. Inductive cardinal numbers. There are two ways to explicate the difference between *finite* and *infinite* classes and, in connection with this, the difference between finite and infinite cardinal numbers. The first is explained here, the second in **37d**. The first way explicates the concept of the finite through the concept of inductive cardinal number. A cardinal number M is said to be an *inductive cardinal number* (symbolically: '$Str_1Induct(M)$') provided that either M is 0 or is attainable from 0 by finitely many additions of 1 (i.e. by finitely many *Pred*-steps); which is to say, M is an inductive cardinal number provided the relation $Pred^{\geq 0}$ holds between 0 and M. Similarly, a class P is called an *inductive class* (symbolically: '$ClsInduct(P)$') provided the cardinal of P is an inductive cardinal number.

D37-5. $Str_1Induct(N) \equiv Pred^{\geq 0}(0,N)$.

D37-6. $ClsInduct=sm_1(Str_1Induct)$.

The so-called *principle of mathematical induction* frequently used in arithmetical proofs runs as follows: "If something holds for the number 0 and, in case it holds for any number N, it holds also for $N+1$, then this something holds for every finite number". The word "finite" expresses an important limitation of this principle. It is not possible to say simply "... holds for every number" since e.g. the property expressed by '$N \neq N+1$'

attaches to 0 and also attaches to $N+1$ if it attaches to N, but nevertheless does *not* attach to $\aleph_0$ (cf. T2e). The explication above of "finite number" by way of "inductive cardinal number" amounts to characterizing finite numbers as those for which the principle of mathematical induction holds; for it follows from D5 that N is an inductive cardinal number if and only if the induction principle holds for N (see T3d below). The illustrative property just cited, viz. the one expressed by '$N \neq N+1$', shows that the inductive principle does not hold for $\aleph_0$, and further that $\aleph_0$ is not an inductive cardinal number (T3e). (The preceding remarks presuppose that $\aleph_0$ is not empty; compare **37e**.)

T37-3. The following sentential formulas are L-true:

a. $Str_1Induct(M)$,
where 'M' is any one of the predicates '0', '1', '2', etc., defined in accordance with D17-3. (By T2d.)

b. $Str_1Induct \subset Str_1$.

c. $Str_1Induct(N) \equiv (K)[Her(K,Pred).K(0) \supset K(N)]$. (By D36-2.)

d. $Str_1Induct(N_1) \equiv (K)[K(0).(N_2)[K(N_2) \supset K(sum(N_2,1))] \supset K(N_1)]$. (By (c), D36-1, and D4.)

e. $\exists(\aleph_0) \supset \sim Str_1Induct(\aleph_0)$. (By T2e.)

Exercises. **1.** Give informal proofs of the following: a) '$Str_1Induct(2)$' (do not use T3a); b) T3b; c) T3c; d) T3e; e) '$Str_1Induct(N_1).Str_1Induct(N_2) \supset Str_1Induct(sum(N_1,N_2))$'; f) '$Str_1Induct(N_1) \supset (\exists N_2)(Pred(N_1N_2).Str_1Induct(N_2))$'; g) '$Str_1Induct(N) \equiv Str_1Induct(sum(N,1))$'. — **2.** Translate the following into English and decide whether it is L-true: '$(M \subset Str_1Induct).\exists(M) \supset (\exists N_1)[M(N_1).(N_2)(M(N_2) \supset Pred^{\geq 0}(N_1,N_2))]$'.

37d. Reflexive classes. We saw earlier (in connection with the proof of T2e) that a certain subrelation of a progression R is also a progression, and hence that the field P of R is both denumerable and has a proper subclass which is denumerable. Thus P is isomorphic to a proper subclass of itself (cf. T1d). This last obviously cannot occur when the class in question is finite, since a proper subclass of a finite class must always have a smaller cardinal number. Here, then, is a second way to explicate the difference between the finite and the infinite, viz. to characterize infinite classes as precisely those classes which are isomorphic to proper subclasses of themselves. A class P that satisfies this condition is called a *reflexive class*; in symbols: '$ClsRefl(P)$'. (The notion of a reflexive class is, of course, not to be confused with that of a reflexive relation specified in D31-3a.) The cardinal number M of a reflexive class is called a *reflexive cardinal number*: '$Str_1Refl(M)$' (see D8 below). This concept of a reflexive cardinal number is here taken as an explicatum for the concept of an infinite cardinal number.

D37-7. $ClsRefl(F) \equiv (\exists G)[(G \subset F).(G \neq F).Is_1(G,F)]$.

D37-8. $Str_1 Refl = str_1$“$ClsRefl$.

A word, finally, about the contrasts between the classification of this section and of the preceding one. Inductiveness and reflexiveness are mutually exclusive (see T4b below). On the basis of the principle of choice (which appears as the primitive sentence P11 in the syntactical system B; see **22a** and the related discussion in **22b**), it can be shown that—apart from the improper null cardinal number Λ_1—each cardinal number is either inductive or reflexive, and hence that our two classifications coincide (cf. T4c). When it comes to classes, the two classifications agree without exception (T4d).

T37-4. The following sentential formulas are L-true:

a. $Str_1 Refl \subset Str_1$.
b. $Str_1 Refl \subset \sim Str_1 Induct$.
c. $Str_1(N).(N \neq \Lambda_1) \supset [Str_1 Refl(N) \equiv \sim Str_1 Induct(N)]$.
d. $ClsRefl = \sim ClsInduct$. (From (c).)
e. $\aleph_0 \subset ClsRefl$. (By T2e.)
The denumerable classes are reflexive.
f. $\exists(\aleph_0) \equiv Str_1 Refl(\aleph_0)$. (By (e) and T1f.)
If $\aleph_0$ is not empty, it is a reflexive cardinal number; and conversely.

Exercises. **1.** Give informal proofs of the following: a) T4b; b) T4e; c) T4f.

37e. Assumption of infinity. Some systems include in their bases an assumption to the effect that there are infinitely many individuals. Normally this assumption is included either as a primitive sentence of a syntactical system (in which case, the assumption is often called “axiom of infinity”; see the note to P12 of language B in **22a**, and comments in **22b** related thereto), or as a rule in a semantical system by which the assertion of infinity becomes L-true. [Whether it is justifiable to count this assertion as a purely logical one is, however, a contested question; cf. Carnap [Syntax E] § 38a.] Still other systems do not include this assumption in their bases, but use it only as a premiss from which other sentences are derived.

If it is desired to systematize the arithmetic of natural numbers in such a way that the familiar arithmetical theorems are provable within the system on the basis of the definition of inductive cardinal number (by which in turn the concept of natural number is explicated), then it is necessary to include the assumption of infinity in the basis of the system. While it is the case that all affirmative true sentences without variables, e.g. ‘$5+2=7$’, are provable *without* this assumption, the same is not so for certain negative true sentences, e.g. ‘$6 \neq 6+1$’ (in this connection, see T5e and the notes following T5). In T5 below we give various formulations of the assumption of infinity.

T37-5. The following sentences (a) through (i) are L-equivalent to each other; each of them says that the number of individuals is infinite. (If any one of these sentences is taken as primitive—i.e. as an "axiom of infinity", then each of the others is provable.)
[Here a superscript immediately to the left of a logical constant indicates the level of this constant in the sentence in question. E.g. '${}^2Prog$' designates a certain class of the second level, viz. the class of all progressions of the first level (progressions of individuals).]

a. $\exists({}^2\aleph_0)$.
There is a denumerable class of individuals.

+b. $\exists({}^2Prog)$.
There is a progression of individuals.

+c. $(N)[{}^3Str_1Induct(N) \supset \exists(N)]$.
For each inductive cardinal number N there is a class with N individuals.

d. $\sim {}^3Str_1Induct({}^2\Lambda_1)$.
The (second-level) empty class is not an inductive cardinal number.

e. $(N)[{}^3Str_1Induct(N) \supset (N \neq sum(N,1))]$.
For no inductive cardinal number N is it the case that $N = N+1$.

+f. $\exists({}^2ClsRefl)$.
There is a reflexive class of individuals.

g. ${}^3Str_1Refl({}^2\aleph_0)$. (From T4f.)
$\aleph_0$ is a reflexive cardinal number.

h. ${}^4Prog({}^3Pred\ in\ {}^3Str_1Induct)$.
The predecessor relation among inductive cardinal numbers is a progression.

i. ${}^4\aleph_0({}^3Str_1Induct)$.
The class of inductive cardinal numbers is denumerable.

To understand better these various formulations of the assumption of infinity, and the fact that certain sentences are provable only with the help of this assumption, it is helpful to see what follows if the *domain of individuals* is *finite*. Suppose e.g. the number of individuals is 5; then the following statements are readily shown to be true on the basis of our earlier definitions. (The corresponding sentences are *provable* if a sentence to the effect that the number of individuals is 5—e.g. '$5(V_1)$', where 'V_1' is a first-level predicate —is taken for our primitive sentence P12 in **22a**.) The cardinal numbers 0,1,2,3,4,5 are all different from each other and non-empty. Contrariwise, the inductive cardinals 6,7,8, etc., are all empty and hence identical with each other (cf. T29-3c). It is the case that $6=6+1=7$. Every class of individuals is an inductive class. It is the case that *Pred*(5,6) and *Pred*(6,7), and also that *Pred*(5,7) since $6=7$. Because $5\neq 6$, the relation *Pred* among inductive cardinal numbers is not one-many, and hence is not a progression. Although the number of classes increases from level to level, there is no finite level at which an infinite class or a progression appears; thus *Prog* and $\aleph_0$ are empty at every finite level.

Exercises. 1. Give informal proofs to show that each of the following is L-equivalent to some preceding sentence of T5: a) T5b; b) T5c; c) T5d; d) T5e. — **2.** How many different second level classes would there be if there were exactly one individual?

The following systems in language C may now be read in Part II (Application of symbolic logic): **44b, 46b, 51b**.

38. CONTINUITY

38a. Well-ordered relations, dense relations, rational orders. We say that an element *a* is a *minimum* of a class *P* with respect to a relation *R*, and write '$min(P,R)(a)$', provided *a* is an *R*-member which belongs to *P* but *no* other element of *P* bears the relation *R* to *a*. A minimum of *P* with respect to R^{-1} is counted a *maximum* of *P* with respect to *R*.

D38-1. $min(F,H)(x) \equiv Fx.mem(H)(x).\sim(\exists y)[(y \neq x).Fy.Hyx]$.

A relation *R* is called *well-ordered* or a well-ordering relation—we write '$WOrd(R)$'—if *R* is a simple order and every non-empty class of *R*-members has at least one minimum with respect to *R*. The structures of well-ordered relations are called *ordinal numbers*, and designated '*NO*' (from "*n*umerus *o*rdinalis").

D38-2. $WOrd(H) \equiv SOrd(H) \, . \, (F)[(F \subset mem(H)) \, . \, \exists(F) \supset \exists(min(F,H))]$.

D38-3. $NO = str_2``WOrd$.

To every ordinal number *M* there corresponds exactly one cardinal number *N*, viz. the cardinal number common to the fields of the relations which have the structure *M*. For inductive cardinal numbers, the converse holds also: each corresponds to exactly one ordinal number. Thus e.g. the cardinal number 1 corresponds to that ordinal number which is the class of all well-ordering relations having exactly one member, e.g. the relation $\{(a,a)\}$ (see the paragraph in small print in **31d**). [On the other hand, there is no series with exactly one member. Therefore, if the ordinal numbers are defined as structures of certain series, then there is no ordinal number One analogous to the other ordinal numbers.]

A relation *R* is called *dense* when with each two distinct members *x* and *y* such that *Rxy* there is a third ("intermediate") member *u* such that *Rxu* and *Ruy*. Thus "*R* is dense" is expressed by '$(R.J) \subset (R.J)^2$', and more simply by '$R \subset R^2$' if *R* is irreflexive.

A relation *R* is called a *rational order*, symbolically '$\eta(R)$', provided *R* is a simple order which is dense and whose field is denumerable.

D38-4. $\eta(H) \equiv SOrd(H) \, . \, [(H.J) \subset (H.J)^2] \, . \, \aleph_0(mem(H))$.

Rational orders can be divided into four kinds, separately designated with the help of subscripts: (1) rational orders which have no initial member and no terminal member (designation: 'η_{00}'); (2) rational orders which have a (one) initial member, but no terminal member ('η_{10}'); (3) rational orders which have no initial member, but do have a terminal member ('η_{01}'); and (4) rational orders which have both an initial member and a terminal member ('η_{11}'). Analogous distinctions will be made in connection with the concepts of **38b**. Rational orders of the same kind are isomorphic (see T1a below), and each of the four kinds is a structure (T1c).

Examples. The relation Smaller among the rational numbers between 2 and 3, but excluding 2 and 3, is a rational order of the η_{00} kind; including 2 but not 3, of the η_{10} kind; including 3 but not 2, of the η_{01} kind; including both 2 and 3, of the η_{11} kind.

Exercises. **1.** Taking the domain of individuals to be the natural numbers, what are the minima of the following classes with respect to the relation *Pred*? a) $\{2,3,4,5\}$; b) $\{2,3,5,6\}$; c) $\{2,5\}$; d) Is *Pred* a well-ordered relation? — **2.** Taking R to be a progression and S to be the converse of R, which of the following relations are well-ordered? a) R; b) S; c) $R^{>0}$; d) $S^{>0}$; e) $R^{\geq 0}$; f) $S^{\geq 0}$. — **3.** Does the class of rational numbers greater than two have a minimum with respect to the relation Smaller?

38b. Dedekind continuity and Cantor continuity. We say that R is a *Dedekind relation* and write '$Ded(R)$', provided: For each two classes F and G such that each element of F bears the relation R to every element of G, there is a z which "separates" F and G in the following sense: if x is any element of F different from z and y is any element of G different from z, then it is the case that both Rxz and Rzy. Precisely:

D38-5. $Ded(H) \equiv (F)(G)\big[(x)(y)\big(Fx.Gy \supset Hxy\big) \supset (\exists z)(x)(y)\big(Fx.(x \neq z).Gy.(y \neq z) \supset Hxz.Hzy\big)\big]$.

Let R be dense and a Dedekind relation. If, moreover, R is a series, R is called a *Dedekind series* or a series having *Dedekind continuity* (symbols: '$DedSer(R)$'). If, on the other hand, R is a simple order, R is called a *Dedekind order*, or an order having Dedekind continuity ('$DedOrd(R)$'):

D38-6. **a.** $DedSer(H) \equiv Ser(H).(H \subset H^2).Ded(H)$.
b. $DedOrd(H) \equiv SOrd(H).\big[(H.J) \subset (H.J)^2\big].Ded(H)$.

We say that P is a *median class* for the relation R provided P is such a subclass of the field of R that between any two distinct members of R for which R holds there is a third intermediate member which belongs to P:

D38-7. $Med(F,H) \equiv (x)(y)\big[Hxy.(x \neq y) \supset (\exists u)\big(Fu.(x \neq u).Hxu.(u \neq y).Huy\big)\big]$.

Let R be a series or a simple order. Then R is said to be a *continuous* series or order (more accurately: to have *Cantor continuity*) provided: R **is a** Dedekind series or order, and there is a denumerable median class for **R**.

That a relation R is a continuous series or order is symbolized by '$ContSer(R)$' or '$ContOrd(R)$' respectively. Cantor continuity implies Dedekind continuity, but the converse is far from being true.

D38-8. **a.** $ContSer(H) \equiv DedSer(H).(\exists F)[\aleph_0(F).Med(F,H)]$.
b. $ContOrd(H) \equiv DedOrd(H).(\exists F)[\aleph_0(F).Med(F,H)]$.

In analogy to our division of rational orders into four kinds η_{mn} (m is 0 or 1, n is 0 or 1), we divide Dedekind relations into four kinds, Ded_{mn}; Dedekind series into four kinds, $DedSer_{mn}$; $DedOrd$ into four kinds, $DedOrd_{mn}$; $ContSer$ into four kinds, $ContSer_{mn}$; and $ContOrd$ into four kinds, $ContOrd_{mn}$. Continuous series or orders of the same kind are isomorphic (see T1d below), and each kind of continuous series or continuous order is a structure (T1f and i). For these two reasons, the Cantor concept of continuity is preferred to the Dedekind one.

The relation Smaller among the real numbers of any interval is continuous, and the rational numbers in that interval constitute the denumerable median class. The relation Smaller among all real numbers is a continuous series of the kind $ContSer_{00}$.

T38-1. The following sentential formulas are L-true. [The subscript 'm' is to be supplanted by one of the two numerals '0', '1'; and similarly for the subscript 'n'.]

a. $\eta_{mn}(H).\eta_{mn}(K) \supset Is_2(H,K)$.
b. $\eta_{mn}(H).Is_2(H,K) \supset \eta_{mn}(K)$.
+c. $Str_2(\eta_{mn})$. (From (a), (b), and T34-3c.)
d. $ContSer_{mn}(H).ContSer_{mn}(K) \supset Is_2(H,K)$.
e. $ContSer_{mn}(H).Is_2(H,K) \supset ContSer_{mn}(K)$.
+f. $Str_2(ContSer_{mn})$. (From (d), (e), and T34-3c.)
g. $ContOrd_{mn}(H).ContOrd_{mn}(K) \supset Is_2(H,K)$.
h. $ContOrd_{mn}(H).Is_2(H,K) \supset ContOrd_{mn}(K)$.
+i. $Str_2(ContOrd_{mn})$. (From (g), (h), and T34-3c.)

The following systems in language C can now be read in Part II (Applications of symbolic logic): **45**; **48a, b, c**; **52c**.

Exercises. 1. Give informal proofs of the following: a) T1a; b) T1d; c) T1g; d) '$Ded(H) \equiv Ded(H^{-1})$'; e) '$Med(F,H) \equiv Med(F,H^{-1})$'; f) '$ContOrd(H) \equiv ContOrd(H^{-1})$'; g) "If H is dense, then the field of H is a median class for H"; h) "H is dense if and only if there is a median class for H".

PART TWO

APPLICATION OF SYMBOLIC LOGIC

Chapter D

Forms and methods of the construction of languages

Preliminary remarks. Part II of this book is devoted to showing how symbolic logic is used, be it in the symbolization of general languages or in the formulation of special axiom systems. Our demonstration will utilize the symbolic languages given in Part I, occasionally with some modifications (see e.g. **40**).

Chapter D sets forth several general considerations about forms and methods of the construction of languages. We begin in **39** with so-called thing languages without quantitative terms, languages which are formulable entirely within the framework of the language forms previously described. In contrast to these language forms, which contain designations of objects, we turn in **40** to language forms which contain designations of positions (numerical expressions as coordinates); we call these "coordinate languages". There follow in **41** certain general remarks about the formulation of quantitative concepts in thing languages and in coordinate languages; such formulations have for their main purpose the specification of the values of measurable magnitudes. Finally, in **42**, we discuss the method of axiom systems ("axiom system" is abbreviated "AS", "axiom systems" is abbreviated "ASs") and consider their relation to the procedure of symbolization and formalization.

Beginning with Chapter E, a series of particular axiom systems will be formulated symbolically.

39. THING LANGUAGES

39a. Things and their slices. In many branches of empirical science we have to do with the properties and relations of physical things. This happens whether we deal with inorganic things (e.g. rocks) or organic things (e.g. organisms and their parts; human beings). In any case a *thing* occupies a definite region of space at a definite instant of time, and a temporal series of spatial regions during the whole history of its existence. I.e. a thing occupies a region in the four-dimensional space-time continuum. A given thing at a given instant of time is, so to speak, a cross-section of the whole space-time region occupied by the thing. It is called a *slice* of the thing (or a thing-moment). We conceive a thing as the temporal series of its slices. The entire space-time region occupied by the thing is a class of particular space-time points which we speak of as "the space-time points of the thing".

Different language forms can be used in symbolizing sentences about things; what distinguishes these forms are the different types employed. The most significant questions respecting any language form so used are: (1) What do expressions of the individual type designate? (2) To what type do the designations of things belong?

In **39b** below we discuss various forms of the thing language. Before beginning that discussion it is helpful to identify several of the most important *relations* between space-time points or space-time regions, and to specify symbolic predicates for them (we will use these predicates later in examples). These predicates are either introduced into a particular language form as primitive predicates or defined therein on the basis of other predicates.

Among the most important relations between space-time *points* (regarded as individuals) are simultaneity and the time relation. Two space-time points x and y have the relation of simultaneity, and we write '$Sim(x,y)$', provided x and y are simultaneous, i.e. provided x and y have the same time instant. A space-time point x bears the time relation to a space-time point y, and we write 'Txy', provided x is earlier than y, i.e. provided x has an earlier time instant than y.

Among the most important relations between space-time *regions* (regarded as individuals) are simultaneity, the time relation, the part relation, and the slice-thing relation. Two space-time regions x and y have the relation of simultaneity, and here we write '$Simr(x,y)$', provided x is entirely simultaneous with y. A space-time region x bears the time relation to a space-time region y, and here we write '$Tr(x,y)$', provided x is entirely earlier than y. A space-time region x bears the part relation to a space-time region y, and we write 'Pxy', provided x is part of y (no new predicate is required if regions are conceived as classes of space-time points, rather than as individuals; in this case, the subclass relation suffices). Lastly, a space-time region x bears the slice-thing relation to a space-time region y, and we write '$Sli(x,y)$', provided x is a slice of the thing y.

Two remarks in closing. If a language form is adopted in which space-time regions are represented as classes, the same signs can be used for the relations named above—in this case, however, these signs must appear as predicates of a higher level, e.g. '$Simr(F,G)$'. Second, the relations named above occur in several of the axiom systems to be treated later, and may enter different systems in different ways; e.g. the system of **48** takes the relation *T* as a primitive concept and defines the relation *Sim*, the system of **49** takes both these relations as defined, the system of **52** takes '*Tr*' and '*P*' as primitive signs and '*Sli*' as a defined sign.

39b. Three forms of the thing language; language form I. We now divide thing languages into three main kinds, I, II, III and in each case make some further distinctions.

Language form I. Here the individuals are taken to be *space-time regions*, particularly things. We distinguish three subdivisions of this form, as follows:

Language form IA. Here only *four-dimensional space-time regions* are taken to be the individuals; here, therefore, things are individuals, but not thing slices. This choice is the simplest so long as sentences of the language are not expected to contain references to different time points. (Such is the case e.g. if assertions are to be made only about permanent properties of things, or if things are to be described only at a fixed instant of time or during a given interval of time within which changes are ignored.)

As examples of sentences in this language form we may take those illustrative sentences of Part One that employ thing predicates like '*Blue*', '*Stud*', '*Fa*', etc. (lists of such predicates appear in **2c**, under the heading of domain 1 and domain 2). The ASs in **54a, b** governing kinship relations likewise belong here.

Language form IB. Here the individuals are taken to be *space-time regions of definite but finite extent*; here, therefore, both things and thing slices count as individuals, but not space-time points. This language form is the most convenient when we are content to speak of small but definite space-time regions instead of space-time points, yet wish—here departing from IA—to distinguish between various instants of time. (Woodger's system belongs to this language form, see **52** and **53**; see also **55d**, Problems 26 and 27.) It is possible, within this language form, to represent space-time points as relations of individuals, viz. as sequences of regions converging to zero. (This representation is used e.g. by Whitehead in defining "point events" as "abstractive series" of "events"; see **55**, problem 22.)

Examples. To illustrate the present language form, as well as subsequent ones, let us agree now on two sentences which we propose to translate into the various language forms. Our two sentences are: 1. "Peter was once in Chicago and was later a student," and 2. "Peter was always happy when he was in Chicago at the same time Herbert was." The signs to be utilized in our translations are: for "Peter", the sign '*pe*' (used as an individual constant, as in forms IA, IB, IC) and the sign '*Pe*' (used as a predicate, as in forms II and III below; specifically, in forms IIAα and IIBα the sign '*Pe*' is a one-place predicate of the first level, in forms IIAβ and IIBβ it is a two-place predicate of the first level, in IIIα it is a one-place predicate of the first level, in IIIβ it is a one-place predicate of the second level, and in IIIγ it is a two-place predicate of the first level); for "Herbert", '*he*' and '*He*' in the senses just explained for "Peter"; for "Chicago", '*ch*' and '*Ch*' similarly; for "student", '*Stud*'; and for "happy", '*Hap*'.

Examples for language form IB. Translation of sentence 1: '$(\exists x)(\exists y)[Sli(x,pe).Sli(y,pe).Tr(x,y).P(x,ch).Stud(y)]$'. Translation of sentence 2: '$(x)(y)[Sli(x,pe).Sli(y,he).P(x,ch).P(y,ch).Simr(x,y) \supset Hap(x)]$'. Further, it is to be noted that the examples suggested for form IA are also examples for form IB.

Language form IC. Here *all space-time regions without exception*, including space-time points (the latter being defined as the smallest non-empty space-time regions), are taken as individuals. This language form is the simplest respecting matters of type, because in it space-time points and thing slices and things are all of the same type.

Examples. The examples cited for IA and IB serve here, too.

Exercises. **1.** Why could the translations for sentences 1 and 2 given in the examples not occur in language form IA? — **2.** For each of the following predicates (introduced in **39a**) decide to which of language forms IA, IB, IC they belong: a) '*Sim*'; b) '*T*'; c) '*Simr*'; d) '*Tr*'; e) '*P*'; f) '*Sli*'. — **3.** Translate the following into as many of language forms IA, IB, IC as possible: a) "Herbert was always happy when Peter was in Chicago"; b) "Herbert was a student after Peter was in Chicago".

39c. Language form II. In this form *space regions*, i.e. space-time regions with zero temporal extent, are taken as individuals; here, in particular, thing slices and slices of thing parts count as individuals. We distinguish two kinds of this form; and further, in each of these kinds, two subordinate kinds.

Language form IIA. Here only *space regions of finite spatial extent* are taken as individuals; hence space-time points are not regarded as individuals. (Such points can be represented here just as they were in form IB, viz. as convergent sequences.) Two subforms are distinguished on the basis of the representation of things.

IIAα. Here a thing is represented as the class of its slices.

IIAβ. Here a thing is represented as a relation of its slices, say as a temporal series of slices. (In this case, if R is a thing then 'Rab' reads: "a and b are slices of R and a is earlier than b.")

Examples. In form IIAα: 1. '$(\exists x)(\exists y)(\exists z)[Pe(x).Pe(y).Tr(x,y).Ch(z).Pxz.Stud(y)]$'. 2. '$(x)(y)(z)[Pe(x).He(y).Ch(z).Pxz.Pyz.Simr(x,y) \supset Hap(x)]$'. In form IIA$\beta$: 1. '$(\exists x)(\exists y)(\exists z)[Pe(x,y).mem(Ch)(z).Pxz.Stud(y)]$'. 2. '$(x)(y)(z)[mem(Pe)x.mem(He)y.mem(Ch)(z).Pxz.Pyz.Simr(x,y) \supset Hap(x)]$'. These translations show that Example 1 comes out simpler in form β than it does in form α, and that Example 2 comes out simpler in α than it does in β. Thus form β is to be preferred in cases that concern several slices of the same thing in their temporal order.

Language form IIB. Here *all space regions, including space-time points*, are taken as individuals. (Space-time points are defined as the smallest non-empty spatial regions.) Subforms IIBα and IIBβ are introduced here just as in IIA.

Examples. See those given for IIA.

Exercises. 1. Translate the following into language forms IIAα, IIAβ, IIBα, IIBβ: a) "Herbert was always happy when Peter was in Chicago"; b) "Herbert was a student after Peter was in Chicago".

39d. Language form III. This form takes as individuals just the *space-time points.* (The systems of space-time topology in **49** and **50** are of this form; the system of **48** is of a language form similar to III and to IIB, but with world-points—i.e. particle slices—as individuals instead of space-time points.) Here thing slices and slices of thing parts are represented as classes of space-time points. Three subforms are distinguished on the basis of the representation of things.

IIIα. A thing is represented as the class of its space-time points; here, therefore, a thing slice is a subclass of the thing.

IIIβ. A thing is represented as the class of its slices; thus here a thing slice is an element of the thing.

IIIγ. A thing is represented as a relation of its slices, say as a temporal series of slices (as in IIAβ); here, therefore, a thing slice is a member of the thing.

Examples. Again we furnish translations of our two prototype sentences, 1 and 2, in each of forms IIIα, IIIβ, IIIγ. (Our translations are formulated in the symbolism of language A of Part One; were language C used instead, we could formulate the subclass relation more concisely with the help of the sign '$\subset$' introduced in D28-3.) In form IIIα: 1. '$(\exists F)(\exists G)[Sli(F,Pe).Sli(G,Pe).Tr(F,G).(x)(Fx \supset Ch(x)).Stud(G)]$'. 2. '$(F)(G)[Sli(F,Pe).Sli(G,He).(x)(Fx \vee Gx \supset Ch(x)).Simr(F,G) \supset Hap(F)]$'. In form IIIβ: 1. '$(\exists F)(\exists G)(\exists H)[Pe(F).Pe(G).Tr(F,G).Ch(H).(x)(Fx \supset Hx).Stud(G)]$'. 2. '$(F)(G)(H)[Pe(F).He(G).Ch(H).(x)(Fx \vee Gx \supset Hx).Simr(F,G) \supset Hap(F)]$'. In form IIIγ: 1. $(\exists F)(\exists G)(\exists H)[Pe(F,G).mem(Ch)(H).(x)(Fx \supset Hx).Stud(G)]$'. 2. '$(F)(G)(H)[mem(Pe)(F).mem(He)(G).mem(Ch)(H).(x)(Fx \vee Gx \supset Hx).Simr(F,G) \supset Hap(F)]$'.

Exercises. 1. Translate the following into IIIα, IIIβ, IIIγ: a) "Herbert was always happy when he was in Chicago"; b) "Herbert was a student after Peter was in Chicago".

40. COORDINATE LANGUAGES

40a. Coordinate language with natural numbers. In many domains of individuals each individual is identified by its position in some appropriately ordered system. The basic ordering here may be a linear one (e.g. of people according to age), or a circular one (e.g. that of colors in a color wheel), or even a many-dimensional one (e.g. the three-dimensional ordering of points in space). By a coordinate language we understand a language in which the form of an individual expression indicates the position of that individual in the basic ordering system—this in contrast to an indication of position by means of sentences about relations between this individual and other individuals. Usually the order is represented by an association of positions with numbers, the numbers being viewed as "coordinates" of the position, in which case numerical expressions or n-tuples of such (when the basic ordering is n-dimensional) appear as individual expressions.

Now let us construct a particular coordinate language by supplementing language C of Part I in a certain way. Suppose the positions of the system in question have the order of a progression (recall **37a**), i.e. a one-dimensional discrete ordering with a *single* initial position and no terminal position. First, we introduce designations for the natural numbers. We agree that '0' designates the number Zero, and that the successor of a number a has the designation 'a''. Thus '0'' stands for the number One, '0''' for the number Two, etc. Next, the natural numbers are taken to be the values of the individual variables 'x', 'y', etc. Number expressions can then be used as indirect references to the positions of the progression, through the device of "coordinates": the number 0 is taken as the coordinate of the initial position of the system; the number 1, designated '0'', is taken as the coordinate of the next position, etc. Thereupon e.g. '*Blue*(0'')' may be read: "The position with coordinate 2 is blue". Strictly, '0''' designates only the pure number Two; reference to the position does not belong to the significance of '0''', but to that of the predicate '*Blue*', whose significance is "The position having ... as coordinate is blue." It is convenient, however, to speak as if the individual expressions designate not only numbers, but coordinated positions of the system as well. For this reason we often call such positions (be they space points, time points, or space-time points) the individuals of the coordinate language in question.

An important new means of expression in this coordinate language is the *K-operator.* This operator is only used with numerical variables; we take '$(Kx)(...x...)$ to mean: "the smallest natural number x satisfying the condition '...x...', or Zero in case no natural number satisfies that condition". Accordingly, a K-expression (i.e. a full expression of the K-operator, comprising both operator and operand; as e.g. '$(Kx)(Px)$') is not a sentence, but an individual expression and so a numerical expression. Thus K-expressions are in contrast to full expressions of the universal and existential quantifiers, but are in analogy with full expressions of the $\imath$-operator (recall **35a**). However, K-expressions have a distinct advantage over $\imath$-descriptions, viz. they always designate precisely one number; for this reason precautions and restrictive rules of the sort that hedge the use of $\imath$-descriptions are not needed for K-expressions.

The formation and transformation rules comprising the syntax for a coordinate language of the present form are taken to be the same basically as those specified earlier for language B (see **21** and **22**). We *add* only the following:

Additions to the rules of formation. We add '0', ''' and 'K' to the stock of *primitive signs*. These signs are logical signs, since they serve in the formulation of arithmetic (i.e. in the formulation of logical assertions about numbers). Thus the individual expressions '0'', '0''', etc., are also **logical.** (Under certain circumstances, however, a ***K*-expression** is descriptive, viz. when a descriptive sign occurs in its operand.) The type of the individual expressions includes: the individual variables, the constant '0', certain defined constants, the expression $\mathfrak{A}_i'$ in case $\mathfrak{A}_i$ is an individual expression, expressions of the form $(K\mathfrak{v}_i)(\mathfrak{S}_j)$ with $\mathfrak{v}_i$ an individual variable, and full expressions of certain functors (e.g. see D4 and D5 below).

Additions to the rules of transformation. The following are to be added to the stock of *primitive sentences*:

1. $(x)(0 \neq x')$.
2. $(x)(y)[x' = y' \supset x = y]$.
3. $(F)[F(0).(x)(Fx \supset Fx') \supset (x)Fx]$.
4. $(G)(F)[G(Kx)(Fx) \equiv [\sim(\exists x)(Fx).G(0)] \vee (\exists x)[Fx.(z)((H)[Hz'.$ $(u)(Hu \supset Hu') \supset Hx] \supset \sim Fz).Gx]]$.
   (In case use is made of restricted universal quantifiers (see below), the second component of the disjunction above can be written more simply as '$(\exists x)[(z)x(Fz \equiv (z = x)).Gx]$'.)

Sentence (1) says that 0 is not the successor of any individual. Sentence (2) says that different individuals do not have the same successor. Sentence (3) expresses the principle of mathematical induction (cf. **37c**). Sentence (4) serves as a definition of the K-operator and, in accordance with our previous discussion, says the following: $(Kx)(Fx)$ has property G if and only if either no individual has property F and 0 is G, or if there is an individual x such that x is F and no z smaller than x is F (in this connection, see T36-1i and D2 below) and x is G.

So-called *restricted operators* (including quantifiers) are useful in many connections. The restriction is imparted thusly: between the operator and its operand there is inserted a number expression, and this number expression is understood to limit the domain to which the operator refers. E.g. '$(x)0'''(Px)$' says "every number up through 3 (i.e. 0,1,2,3) is P"; again, '$(\exists x)0''''(Qx)$' says "there is a number not beyond 4 which is Q"; and '$(Kx)0'''(Px)$' denotes: the smallest number not beyond 3 which is P, and 0 in case there is no such.

Primitive sentences for the kinds of restricted operator just exemplified may be found in [Syntax E] §30, PS II 7,8,9, 14. In [Syntax] Chapter I there is presented a language form I which employs *only* restricted operators. The only variables there are individual variables whose values are natural numbers. That language form provides a way to formulate unrestricted universal propositions about numbers, viz. with the help of open sentential formulas that are admitted as sentences. However, that language form cannot provide a formulation of unrestricted existential propositions. Recursive definitions are admissible (see e.g. D4 and D5 below). Each closed logical sentence $\mathfrak{S}_i$ of that language is decidable, i.e. exactly one of $\mathfrak{S}_i$ and $\sim\mathfrak{S}_i$ is provable and there is a (decision) procedure for discovering the proof. Each closed logical numerical expression $\mathfrak{A}_i$ is computable, i.e. there is a procedure for discovering a numerical expression $\mathfrak{A}_j$ in normal form ('0',

'0″', etc.) such that $\mathfrak{A}_i = \mathfrak{A}_j$ is provable. The language form under discussion here agrees with certain philosophical views sometimes called "finitism" or "constructivism." According to these views it is the case e.g. that unrestricted existential quantifiers with respect to infinite domains give rise to meaningless sentences, and that predicates and functors are meaningful only if there is a fixed procedure by which their applicability in any concrete case can be decided.

Exercises. **1.** Let R be such that $R = (\lambda x,y)(x' = y)$; give an informal proof (with the help of the four primitive sentences) that R is a progression (D37-1). — **2.** Write L-true sentences of the following forms with one of the expressions '0', '0′', '0″', etc., substituted for 'z', and give informal proofs of them: a) '$(Kx)(Prime(x).Even(x)) = z$'; b) '$(Kx)(\exists y)(Gr(y,x).\sim Prime(y)) = z$'; c) '$(Kx)(y)(Gr(y,x) \supset \sim Prime(y)) = z$'.

40b. Recursive definitions. In a coordinate language of the form set forth in **40a** it is possible to define arithmetical concepts quite simply. We give several examples to suggest how this can be done. It is useful in this connection to permit *recursive definitions* of first-level predicates and functors, a procedure which is customary in arithmetic—especially for functors.

A recursive definition comprises two sentential formulas; the first formula specifies the value at zero of the functor being defined (or the truth-value at zero of the predicate being defined), and the second formula specifies the value at x' in terms of the value at x. (E.g. definitions D4 and D5 below are recursive.)

The following list contains definitions of predicates for the relations Predecessor ('*Pred*'), Smaller ('*Sm*'), and Greater ('*Gr*'); of functors for the functions Sum ('*sum*') and Product ('*prod*'); of predicates for the properties Divisible ('*Div*') and Prime number ('*Prime*'); and of some of the usual symbols.

D40-1. $Pred(x,y) \equiv (x' = y)$.

D40-2. $Sm = Pred^{>0}$. (See **36b**, example 1.)

D40-3. $Gr = Sm^{-1}$. (See D30-3.)

D40-4. (1) $sum(0,y) = y$;
(2) $sum(x',y) = sum(x,y)'$.

D40-5. (1) $prod(0,y) = 0$;
(2) $prod(x',y) = sum(prod(x,y),y)$.

D40-6. $Div(x,y) \equiv (\exists u)(x = prod(y,u))$.

D40-7. $Prim(x) \equiv [(x \neq 0).(x \neq 0').(u)((u = 0') \vee (u = x) \vee \sim Div(x,u))]$.

D40-8. **a.** $1 = 0'$.
b. $2 = 1'$.
c. $3 = 2'$.
Etc.

Exercises. **1.** Give informal proofs of the following: a) '$Sm(1,3)$'; b) '$prod(1,3) = 3$'; c) '$(x)[prod(1,x) = x]$'; d) '$(x)[prod(x,1) = x]$'; e) '$(x)[Div(x,x)]$'.

40c. Coordinate language with integers. A coordinate language similar to that of **40a** (a coordinate language with natural numbers) can be constructed with integers as the individuals. [Integers comprise the positive and negative whole numbers, and zero.] As before, '0' designates a certain basic individual, i.e. the number zero; and 'a'' designates the successor of a, i.e. the number $a+1$. It is also convenient now to have a symbol for the predecessor of a; we agree that '$'a$' designates the predecessor of a, i.e. the number $a-1$. In accordance with this agreement, '$'0$' denotes -1, '$''0$' denotes -2, etc.

Our previous interpretation of the K-operator for natural numbers (in **40a**) cannot be carried over unmodified to integers. Unlike the domain of natural numbers, the domain of integers may provide a number with, say, the property P, but no smallest such number; this happens e.g. when there are arbitrarily "small" negative numbers with P. Let us agree that in cases of this sort our K-expression also denotes the number zero. Thus when the domain of individuals is the class of integers, '$(Kx)(Px)$' denotes: the smallest integer x with property P, or zero in case either there is no integer with P or there is no smallest integer with P.

Finally, let us symbolize "integer a is smaller than, or equal to, integer b" by '$SmEq(a,b)$', and agree to take '$SmEq$' as a primitive sign. Thus the primitive signs of our present language form are those established in **40a**, together with '$SmEq$'. [Actually '$SmEq$' can be treated as a defined sign; e.g. the primitive sentence (2) below could be taken as a definition of '$SmEq$'. However, by taking this sign to be primitive we simplify our formulations of the primitive sentences (3) and (4) below.]

In place of the primitive sentences added in **40a**, let us add the following to our regular stock (see **22**) of *primitive sentences*:

1. a. $(x)[{}'(x')=x]$.

b. $(x)[('x)'=x]$.

2. $(x)(y)[SmEq(x,y) \equiv (F)(Fx\,.\,(u)[Fu \supset Fu'] \supset Fy)]$.

3. $(x)[SmEq(x,0) \vee SmEq(0,x)]$.

4. $(G)(F)[G(Kx)(Fx) \equiv ([(\sim(\exists x)(Fx) \vee (x)(\exists y)[SmEq(y,x)\,.\,Fy])\,.\,G(0)] \vee (\exists x)[(y)(SmEq(y,x) \supset [Fy \equiv (y=x)])\,.\,Gx])]$.

Sentence (1a) says that the predecessor of the successor of x is always x itself; and (1b) says that the successor of the predecessor of x is always x. (In short, the predecessor relation is one-one.) Sentence (2) says that the relation $SmEq$ holds between x and y if and only if y has every hereditary property of x (recall **36**), i.e. if y is either the same as x or is attainable from x in finitely many steps. Sentence (3) says respecting any number that between it and 0 the relation $SmEq$ holds either in one direction or in the other; the effect of this primitive sentence is to restrict the domain of individuals to finite integers. Sentence (4) expresses our earlier explanation of the K-operator respecting the domain of integers.

Taken together, (2) and (3) yield a generalization of the principle of mathematical induction to the domain of integers; this generalization calls for mathematical induction in the usual way respecting positive integers and in the reverse direction respecting negative integers.

With respect to the domain of integers, a recursive definition comprises three sentential formulas—the first about 0, the second about x' in terms of x with $x \geq 0$, and the third about $'x$ in terms of x with $x \leq 0$ (e.g. definitions D11, D12 and D14 below are recursive definitions of this kind).

The list which follows presents several examples of definitions respecting integers. Note that '$opp(a)$' designates the number opposite to a, and that '$diff(a,b)$' designates the difference $a-b$. The remaining predicates and functors defined below have meanings corresponding to their earlier ones (in **40b**). In D15 and D16 we introduce the customary notations for several integers.

D40-9. $Pred(x,y) \equiv (x' = y)$.

D40-10. $Sm(x,y) \equiv SmEq(x,y).(x \neq y)$.

D40-11. (1) $sum(0,y) = y$.
(2) $SmEq(0,x) \supset (sum(x',y) = sum(x,y)')$.
(3) $SmEq(x,0) \supset (sum('x,y) = {}'sum(x,y))$.

D40-12. (1) $opp(0) = 0$.
(2) $SmEq(0,x) \supset (opp(x') = {}'opp(x))$.
(3) $SmEq(x,0) \supset (opp('x) = opp(x)')$.

D40-13. $diff(x,y) = sum(x,opp(y))$.

D40-14. (1) $prod(0,y) = 0$.
(2) $SmEq(0,x) \supset (prod(x',y) = sum(prod(x,y),y))$.
(3) $SmEq(x,0) \supset (prod('x,y) = diff(prod(x,y),y))$.

D40-15. **a.** $+1 = 0'$.
b. $+2 = +1'$.
c. $+3 = +2'$.
Etc.

D40-16. **a.** $-1 = {}'0$.
b. $-2 = {}'-1$.
c. $-3 = {}'-2$.
Etc.

The language form presented in this section is used later in defining the concept of the dimension number (see **46c**).

Exercises. **1.** Give informal proofs for the following, using '*Pred*' as defined in D40-9: a) "*Pred* is one-one" (recall D31-11); b) "*Pred* has no initial member" (recall D32-8); c) "*Pred* has no terminal member" (recall **32c**); d) "$Pred^{>0}$ is connected" (recall D31-4, D36-3); e) '$Pred^{\geq 0} = SmEq$' (recall D36-2). — **2.** Write L-true sentences of the following forms with one of the expressions '0', '0′', '′0', '0″', '″0', etc., substituted for 'z' and give informal proofs of them (note that the K-operator now ranges over integers and not just natural numbers): a) '$(Kx)(SmEq(x,0).Gr(x,-2))=z$'; b) '$(Kx)(\exists y)(Gr(y,x))=z$'; c) '$(Kx)(y)(x=prod(0,y))=z$'.

40d. Real numbers. We now have at our disposal two different procedures for introducing further kinds of numbers, in particular the real numbers. For the sake of definiteness, let us talk here about the introduction of real numbers. We can construct them from the natural numbers, or from the integers; or we can establish a fresh basis by taking the real numbers as the individuals of an entirely new coordinate language form.

Consider the first procedure, that of constructing the real numbers on a previously-prepared basis. Our basis can be either the natural numbers (given in **40a**) or the integers (given in **40c**); the choice here depends on whether we want to confine our use of the subsequent kinds of numbers to the positive domain, or allow their use in the whole positive and negative domain. In any event, the first step in the construction is to introduce rational numbers as pairs of natural numbers (or of integers), i.e. denote a rational number by an expression of the form '(a,b)'. And the second step in the construction is to introduce real numbers as classes or functions of rational numbers. (We remark parenthetically that, once real numbers are in hand, complex numbers can be constructed as pairs of real numbers.)

Concerning the step-by-step introduction of additional kinds of numbers beginning with the natural numbers, cf. Russell [Introduction] Chap. 7; Waismann [Math. Thought]; [Syntax] §39; Cooley [Logic] §37.

Consider now the second procedure, which is to construct a language form in which the real numbers enter as individuals. This procedure utilizes the following additional primitive signs: '0' and '1' (with their familiar signification), the functors '*sum*' and '*prod*', and the two-place predicate '*Sm*' (the relation Smaller-than). The construction itself is similar to that in **45** respecting Tarski's axiom system for real numbers. (Since here only the real numbers appear as individuals, Tarski's predicate 'R' is superfluous.) Sixteen primitive sentences are added, viz. all axioms of **45** with the exception of A5, A12, A10 and A18. (Of course, all free variables in these primitive sentences are covered by universal quantifiers; and further, the components involving '*R*' are struck out of A15.)

Expressions for real numbers are of special importance in the construction of a *language of physics*. This construction first sees the association with each space-time point of four real numbers as its coordinates—three of them as spatial coordinates, one as a temporal coordinate. Thereafter the designation of properties of space-time points, or of relations between such points, or of physical state-magnitudes, is accomplished with the help of predicates and functors having one or more quadruples of real number expressions as their argument-expressions (cf. **41c**).

41. QUANTITATIVE CONCEPTS

41a. Quantitative concepts in thing languages. Progress in the different areas of science discloses an ever-increasing use of quantitative numerical concepts in the description of things and processes. This quantitative method of description has essential advantages over non-quantitative or purely qualitative methods. First, it permits a more exact description of the separate facts. And second, it makes possible the elaboration of decidedly more effective general laws expressing connections between the values of various quantitative concepts with the help of mathematical functions.

Quantitative concepts, e.g. length, weight, temperature, price, degree of attention, etc., are also called "measurable magnitudes" because the procedure for establishing their value is that of measurement. Such concepts are most conveniently designated by means of functors; their value expressions are considered of greatest general usefulness when they are real number expressions. (In certain circumstances it is possible to simplify the language form by using expressions for rational numbers, or even integers, instead of real numbers; however, this results in important limitations on the construction of laws.)

Let us first discuss the use of quantitative concepts in thing languages; later (in **41c**) we shall discuss their use in coordinate languages.

For the thing languages, the language forms explained in **39** are of chief importance. In these forms, functors have as their argument-expressions mostly expressions for things, for thing slices, and for space-time points. Now the most important kind of measurable magnitudes—of frequent occurrence not only in physics but in any branch of empirical science (including psychology and social science) that operates quantitatively—are the magnitudes which ascribe a real number to a definite space region at a definite time (e.g. a thing slice). Examples of magnitudes that are representable in this form are: temperature, energy, mass, weight, intelligence, performance in mathematics (or in chess, tennis, etc.), life expectancy, and so on. If a measurement or a battery of experimental tests indicates that today Mr. Smith has such and such a weight, or such and such a blood pressure, or can jump so and so high, or can multiply so and so fast, or can concentrate to such and such a degree, etc., the result is expressed in each case by ascribing to a thing slice of Mr. Smith a definite number as value of some particular measurable magnitude.

41b. Formulation of laws. In the terminology customarily employed by physicists (a terminology, by the way, which is not entirely clear) measurable magnitudes like length, pressure, current intensity, etc., are sometimes termed "variables". According to the terminology of modern logic, however, it is signs and *not* their designata that are divided into variables and constants. Each concept, therefore, is to be designated by a constant, not a variable; and in particular, measurable magnitudes are to be designated by functor constants.

Nevertheless, physical laws are intended to refer to arbitrary space-time points or regions, hence (at least when completely formulated) must exhibit variables as well as functor constants. It is usual in physics, however, to give not the complete formulation of a physical law, but an abbreviated formulation in which the variables have been omitted. Also the specific conditions under which the law holds are ordinarily omitted from this abbreviated symbolic formulation (at most, these conditions are explained in the verbal text accompanying the formulation). Consider e.g. the so-called perfect gas law. The usual formulation of this law in physics is '$p \cdot V = R \cdot T$'. If we use 'P' to designate the conditions which a system x at time t (a thing slice of a body of gas) must satisfy before the perfect gas law applies to it, then the complete formulation of this law runs '$(x)(t)$ $[Pxt \supset (p(x,t) \cdot V(x,t) = R(x) \cdot T(x,t))]$'. The full form of the law makes clear that 'p', 'V' and 'T' are functors, and indeed constants (viz. for pressure, volume and temperature of the body x at time t respectively), and 'R' is a functor (for a characteristic of the body x independent of time). Of course, the usual abbreviated formulation has important advantages; and it is well to note that the suppression of variables here bears some analogy to our own practice of writing predicates without argument-expressions. The functor character of the symbols that survive in the abbreviation should, however, not be overlooked.

The preceding paragraph suggests one kind of completion of the abbreviated formulations found in physics. For another kind, in which e.g. the signs 'p', 'V', etc., are taken as variables and their interpretation as values (for pressure, volume, etc., respectively) is incorporated into the antecedent of the completed law, see Carnap [Foundations] §23, Axiom A1.

Another question deserves attention here. Values of a measurable magnitude are expressed in terms of some *unit* of measure (e.g. a centimeter or an inch, a second or a day, a shilling or a dollar); where and how should this unit be specified? Ordinary practice here is to add to the number expression for the value of the magnitude a sign indicating the unit of measure, e.g. "the length of rod a is 5 cm", "the price of a is \$5". Strictly speaking, however, the specification of the unit is part of the definition of the functor; the value of the functor is always a pure number. Should an explicit indication of the unit be wanted in the symbolization of the measurable magnitude (perhaps because the same body of text makes references to various units), this indication must be achieved by way of an inseparable part of the functor sign, e.g. by a subscript. For example, in the matter of length we might write "$lg_{cm}(a) = 5$" or "$lg_{inch}(a) = 2$"; each of 'lg_{cm}' and 'lg_{inch}' is to be regarded as one sign, and each designates a different magnitude.

Examples. We propose to translate each of the following two sentences into the various language forms of **39**: 1. "Peter was (or is, or will be) at one time heavier than Herbert"; 2. "The energy of an isolated system remains constant." Our translations utilize the following additional signs (with variations in type, according to the arguments): a. Functors—'*wt*' designates weight, '*energ*' designates energy; b. Predicates—'*Gr*' designates Greater (respecting real numbers; cf. **40d**), '*Isol*' designates Isolated System.

Translations into language forms IB and IC (language form IA is not appropriate for these examples): 1. '$(\exists x)(\exists y)[Sli(x,pe).Sli(y,he).Simr(x,y).Gr(wt(x),wt(y))]$'.— 2. '$(x)(y)(z)[Isol(x).Sli(y,x).Sli(z,x) \supset (energ(y)=energ(z))]$'.

Translations into language forms IIAα and IIBα (the change into form β is analogous to that illustrated in the examples of **39c**): 1. '$(\exists x)(\exists y)[Pe(x).He(y).Simr(x,y).Gr(wt(x),wt(y))]$'. 2. '$(F)(y)(z)[Isol(F).Fy.Fz \supset (energ(y)=energ(z))]$'.

Translations into language form IIIβ (language form IIIα is not appropriate for these examples; the change into form IIIγ is analogous to that of the examples in **39d**): 1. '$(\exists F)(\exists G)[Pe(F).He(G).Simr(F,G).Gr(wt(F),wt(G))]$'. 2. '$(N)(F)(G)[Isol(N).N(F).N(G) \supset (energ(F)=energ(G))]$'.

Exercises. **1.** Translate the following into language forms IB, IC, IIAα, IIBα, IIIβ: a) "At no time is a thing heavier than itself"; b) "If at one time Peter was heavier than Herbert and at a later time Herbert was heavier than Peter, then at some intermediate time they had the same weight"; c) "If the energy of x remains constant, then x is an isolated system".

41c. Quantitative concepts in coordinate languages. The use of measurable magnitudes in coordinate languages does not differ essentially from their use in thing languages. Thus e.g. magnitudes used in coordinate languages are also designated by functors. Here, however, the form and type of argument-expressions are different. A quadruple of real numbers corresponds to a space-time point; the question what type these number expressions are depends on the particular language form (see **40d**). Slices of things, of thing parts, and of other systems are representable as classes of space-time points. Such representation is suited to language form III of **39d**, and hence here there are further alternative subforms analogous to the special forms α, β, γ.

Of these, the α version of language form III may be the most useful. In this form, things and other physical systems, as well as their slices, are represented as classes of space-time points, i.e. are denoted by predicates which take quadruples of real number expressions as their argument-expressions. The functors of measurable magnitudes then have as their argument-expressions either predicates of the sort just described or else quadruples of the sort mentioned, according as the values of the magnitudes in question are counted as ascribed to a space-time region or a space-time point. It is convenient to admit in a coordinate language also compound expressions as value-expressions of functors—these compound expressions consisting of several real number expressions. While the values of certain physical magnitudes are real numbers (such magnitudes are called "scalar magnitudes"), certain others such as space vectors have values that are triples of real numbers, and still others such as space-time vectors have values that are quadruples of real numbers, etc.

42. THE AXIOMATIC METHOD

42a. Axioms and theorems. By an *axiom system* (abbreviation: AS) we understand the representation of a theory in such a way that certain sentences of this theory (the *axioms*) are placed at the beginning, and from them further sentences (the *theorems*) are derived by means of logical deduction.

There is a traditional view of an AS—current in Euclid's time, and continuing into our own—that requires its axioms to be self-evident, i.e. immediately clear to the intuition and hence in no need of proof. (Even today, common usage tends to attribute this meaning to the word "axiom".) The modern conception of an AS does not include this requirement; arbitrary sentences may be selected as axioms.

For the formulation of an AS we need to choose or construct a language L, the so-called *basic language of the AS*. Usually this basic language contains only logical signs. The axioms and theorems of the AS contain certain constants not occurring in language L, called the *axiomatic constants of the AS*. Some of them are given without definitions; they are called *the axiomatic primitive constants of the AS*. All other axiomatic constants are introduced by definitions on the basis of the primitives. The language L′ obtained from the basic language L by adding the axiomatic constants is called the axiomatic language.

In the modern conception of an AS, the derivations of theorems must be a matter of purely logical deduction. Nothing may be referred to the intuition—this in contradistinction to derivations found in Euclid's system —and no knowledge of the objects of the theory may be utilized except that which the axioms pronounce. Since derivation in this sense is purely logical, it is open to formalization. E.g. we may have a syntactical system for L, specifying primitive sentences in L and rules of inference for L, and may extend this system to a syntactical system for L′ with the axioms as additional primitive sentences; the theorems of the AS are then those sentences which are provable in L′ but not in L.

Two ways of treating an AS are now at hand. 1. The way just described; here the axioms are counted as primitive sentences, and the theorems are obtained by proofs, i.e. without premisses. 2. The axioms are not counted as primitive sentences, and the theorems are obtained by derivations in which the axioms appear as premisses. There is no essential difference between these treatments, beyond that of presentation. A third way of treating an AS is discussed below (in the first of the two notes concluding **42d**).

42b. Formalization and symbolization; interpretations and models. In connection with the construction of the language L′ in which an AS is formulated, the following additional procedures may be applied. The language L′ can be *formalized*, i.e. a syntactical system with explicit formal rules for L′ may be constructed as indicated above; see **21, 22**. Also, the language L′ can be *symbolized*, i.e. artificial symbols used in place of the words of the natural language.

Neither of these procedures is absolutely required. And indeed, neither of them is used in the majority of published presentations of ASs, including those conceived in the modern sense. For the most part, these ASs are formulated in words, and rules of transformation are not specified. The rules of the basic language are, so to speak, tacitly presupposed, i.e. ways of deducing common in the word language are usually assumed to be familiar. Further, there is tacitly presupposed a particular interpretation of the basic language L, viz. the usual interpretation of the logical words of the word language; only the interpretation of the axiomatic constants in L′ is deliberately kept open.

It is also to be observed that the two procedures of formalization and symbolization described above are independent of each other. A word language can be formalized by introducing transformation rules phrased with the logical words "and", "or", "not", "every", "some", etc., instead of the symbols corresponding to these words. On the other hand, all or part of the language can be symbolized without also formalizing it, i.e. without explicitly laying down syntactical transformation rules; our treatment of language A in Chapter A was of this nature.

When an AS is stated, the basic language used is assumed to be understood. Usually its *interpretation* is tacitly presupposed; only in special cases is it explicitly specified, e.g. by semantical rules. On the other hand, the interpretation of the axiomatic constants is not supposed to be fixed. The author of an AS often specifies a certain interpretation, i.e., an assignment of meanings to the axiomatic primitives, based on a specified domain *D* of individuals. He usually does this informally; it may also be done in a semantical system by rules of designation (cf. **25b**). In either case, the statement of the interpretation is not to be regarded as part of the description of the AS. When an interpretation of the primitives is given, the remaining axiomatic constants straightway receive an interpretation through their definitions, and thereupon all sentences of L′ have an interpretation, including the axioms and theorems. An interpretation of an AS is called a true interpretation if under it all axioms are true; and, moreover, an L-true interpretation, if all axioms are L-true. One of the essential characteristics of axiomatization in the modern sense consists in the fact that the deduction of the theorems makes no use of any interpretation of the axiomatic constants. Each theorem is L-implied by the axioms. Therefore under any true interpretation all theorems are true; and under any L-true interpretation they are L-true. In this way, the same AS may serve as a representation of many different theories.

We say an interpretation of an AS is a *logical interpretation* provided all axiomatic primitive constants are interpreted as logical constants, otherwise a *descriptive interpretation*. Thus an interpretation of an AS is a descriptive interpretation provided at least one axiomatic primitive is interpreted as a descriptive constant.

By a *model* (more specifically, a logical or mathematical model) for the axiomatic primitive constants of a given AS with respect to a given domain D of individuals we mean a value assignment VA (**25a**) to these primitives such that both D and VA are specified without the use of descriptive constants. A model is said to be a *model of the AS* provided it satisfies all the axioms. D may, for example, be the class of numbers of a certain kind, or of ordered k-tuples of such numbers, or the like. VA assigns to each primitive an extension of the corresponding type with respect to D, e.g., to an individual constant an element of D, to a one-place predicate of first level a subclass of D, etc. [The study of models is simpler than that of interpretations, since it deals with extensions, not intensions; e.g., with classes, not properties. Logical interpretations are essentially the same as models. Therefore, if we are only interested in possible applications of a given AS within the field of mathematics, the investigation of models is sufficient. For this reason, some mathematical books use the terms "interpretation" and "model" as synonyms. However, if we are interested in the use of a given AS in fields of empirical science, e.g., physics, economics, etc., or in the construction of an AS as a formal representation of a given scientific theory, then we have to consider descriptive interpretations.]

According to our definition of L-implication (**6a**), the following holds:

(1) The sentence $\mathfrak{S}_i$ is L-implied by one or more other sentences if and only if every model satisfying these sentences satisfies $\mathfrak{S}_i$ also.
(2) If we can construct a model satisfying the other sentences but not $\mathfrak{S}_i$, we have shown that $\mathfrak{S}_i$ is not L-implied by those sentences.

42c. Consistency, completeness, monomorphism. Now let us explain certain properties of ASs which are important in the critical examination of any given AS. An AS is said to be *inconsistent* provided that among its theorems is one of the form $\mathfrak{S}_i$ and another of the form $\sim\mathfrak{S}_i$. An AS is said to be *consistent* provided it is not inconsistent. In view of T6-15, any sentence of the language is derivable from $\mathfrak{S}_i$ and $\sim\mathfrak{S}_i$ together; the theorems of an inconsistent AS therefore include all the sentences of the language L′, and the AS in consequence is trivial and useless for practical purposes. Consistency is thus an obvious requisite of any non-trivial AS. The consistency of any particular AS is established by constructing a model of the AS.

An AS is said to be (deductively) *complete* provided it is the case for any sentence $\mathfrak{S}_i$ in L′ that either $\mathfrak{S}_i$ itself or $\sim\mathfrak{S}_i$ is a theorem. The incompleteness of a given AS can, according to (2) above, be shown by constructing two models M_1 and M_2 of the AS and a sentence $\mathfrak{S}_i$ in L′ such that M_1 satisfies $\mathfrak{S}_i$ and M_2 satisfies $\sim\mathfrak{S}_i$. Suppose the language L′ has means of expression sufficient to permit a formulation of the arithmetic of natural numbers up through general statements about numbers; then it follows from Gödel's result (see the end of **26**) that the AS cannot be complete. For this reason the concept of completeness is frequently inapplicable, and the weaker concept of monomorphism (to be defined below) becomes of some interest because it represents a kind of completeness.

An AS is said to be *monomorphic* (or categorical) provided it is consistent and all its models (with respect to a given domain of individuals for which it has models) are isomorphic to each other; and to be *polymorphic* in case it has non-isomorphic models. Isomorphism of models being a more comprehensive concept than isomorphism of classes or relations (this last was defined earlier, in **19**), an explanation of the concept is in order here. Suppose a model M of an AS consists of the extensions B_1, B_2, ..., B_n corresponding respectively to the n axiomatic primitive constants of the AS, and suppose that another such model M' similarly consists of B_1', B_2', ..., B_n'. We say that M is isomorphic to M' provided there is a correlator between the individuals of M and those of M' and, on the basis of this correlator, B_p is isomorphic (in the sense of **19**) to B_p' for each p from 1 to n.

An AS that is monomorphic thus specifies all the structural properties of its possible models; it is this sense of completeness that can be imputed to a monomorphic AS.

Examples of monomorphic ASs: Peano's AS for the natural numbers (see **44**; all models of this AS are progressions and hence, by T37-1a, isomorphic to one another); Tarski's AS for the real numbers (see **45**; all models of this AS are essentially continuous series having the structure $ContSer_{00}$ and hence, by T38-1d, isomorphic to one another); and several modern ASs for Euclidean geometry, e.g. Hilbert's (in his *Foundations of geometry*) and E. Roth's (see **47**).

Consider an AS containing n axioms A_1, ..., A_n. If neither the axiom A_i nor its negation is deducible from the remaining axioms, A_i is said to be *independent* in the AS. This can be shown by constructing a model of the AS and another model which satisfies the other axioms but not A_i. The AS itself is called independent provided every axiom of it is independent in it.

Example. To illustrate the concepts just defined, we consider an extremely simple AS. As the axiomatic primitive constants we take three one-place first-level predicates 'P', 'Q', 'R'. There are two axioms: A1 is '$(x)(Px \supset \sim Qx)$'; A2 is '$(x)(Rx \supset Px)$'. Example of a theorem: '$(x)(Rx \supset \sim Qx)$'. To give an example of an interpretation, we take as the domain D the material bodies of a specified space-time region, and as the designata of the predicates 'P', 'Q', and 'R' the properties Red, Blue, and Cherry, respectively. This is a true but not L-true interpretation, since under it the two axioms are true but not L-true. This interpretation is descriptive since the three properties assigned are non-logical. For the construction of models, we shall use the domain D' of natural numbers. Let M_1 be the model which assigns to 'P' the class $\{1,3,8\}$ (i.e., the class whose elements are 1, 3, and 8, cf. **32e**), to 'Q' $\{4,10,15\}$, and to 'R' $\{1,8\}$. Obviously it can be proved in a purely logical way that this model satisfies both axioms; thereby it is shown that the AS is consistent. Let the model M_2 be like M_1 except that 'R' has the extension $\{3\}$ instead of $\{1,8\}$; M_2 likewise satisfies both axioms. Since the two classes just mentioned are non-isomorphic, the two models are non-isomorphic, and thus the AS is polymorphic. Now we can easily show that the AS is not complete. Let $\mathfrak{S}_1$ be the sentence '$(\exists y)(x)(Rx \equiv (x=y))$', which says in effect that exactly one individual has the property R (**35a**). M_2 satisfies $\mathfrak{S}_1$, but M_1 does not. Therefore the AS is incomplete.

42d. The explicit concept. Given an AS with n axiomatic primitive constants $\mathfrak{a}_{i_1}$, $\mathfrak{a}_{i_2}$, ..., $\mathfrak{a}_{i_n}$, it can be transformed into a statement about the n primitive concepts. The transformation is accomplished as follows: First, we eliminate all defined axiomatic constants. Next, we form in L′ the conjunction $\mathfrak{S}_i$ of all the axioms (each axiom component of $\mathfrak{S}_i$ is a sentence; if the initial form of an axiom is that of an open formula, this axiom is first converted into a sentence by prefixing a universal quantifier for each free variable of the axiom). Noting that, besides logical constants and variables, only the n primitive constants occur in $\mathfrak{S}_i$, we now can abbreviate $\mathfrak{S}_i$ into a sentence of the form $\mathfrak{a}_k(\mathfrak{a}_{i_1},\mathfrak{a}_{i_2},...,\mathfrak{a}_{i_n})$. The sign $\mathfrak{a}_k$ is an n-place predicate called the *explicit predicate* of the AS, and the n-place attribute designated by $\mathfrak{a}_k$ is called the *explicit concept* of the AS. This explicit concept is an n-place attribute which holds for an n-tuple of concepts if and only if this n-tuple of concepts satisfies the AS.

The definition of $\mathfrak{a}_k$ can evidently be constructed in the basic language L in the form $\mathfrak{a}_k(\mathfrak{v}_{i_1},\mathfrak{v}_{i_2},...,\mathfrak{v}_{i_n}) \equiv \mathfrak{S}_j$, where $\mathfrak{S}_j$ is obtained from $\mathfrak{S}_i$ by replacing each occurrence of the primitive constant $\mathfrak{a}_{i_p}$ in $\mathfrak{S}_i$ by a variable $\mathfrak{v}_{i_p}$ having the same type, this for each p from 1 to n. Since $\mathfrak{S}_j$ consists entirely of signs of the basic language L (all axiomatic constants having been excluded), it appears that $\mathfrak{a}_k$ must be a constant of the basic language L. And if (as is commonly the case) the basic language is interpreted as a logical language, $\mathfrak{a}_k$ is in fact a logical constant.

For an example of a definition of an explicit concept, see D2* in **44b**; it is shown there that the explicit concept of the Peano AS for natural numbers (formulated with a *single* primitive) is the class of progressions (D37-1). Further examples of explicit predicates of ASs: '*ZF*' in **43b**; '*Hausd*' in **46c**.

If we similarly transform a theorem $\mathfrak{S}_k$ of the AS—viz. if we bind its free variables, eliminate its defined signs, replace its axiomatic primitive constants by the corresponding variables—there results an open sentential formula $\mathfrak{S}_0$. Now $\mathfrak{S}_k$ is derivable from the above $\mathfrak{S}_i$ in L′. Hence the universal sentence $(\mathfrak{v}_{i_1})(\mathfrak{v}_{i_2})...(\mathfrak{v}_{i_n})[\mathfrak{S}_j \supset \mathfrak{S}_0]$ is provable in L, and is logically true under the usual interpretation of this language.

In the note at the end of **42a** mention was made of two ways to formulate an AS. To these two presentations we now add a third, viz. the axioms and theorems of the AS are not presented as sentences in L′, but as open sentential formulas in L constructed in the fashion indicated above. In place of the derivation (or of the proof, as may be) of a theorem in L′ there now appears a proof in L of the universal conditional just specified. This mode of presentation of the AS uses no signs except those of the basic language L. In place of the axiomatic primitive constants in L′ we now have axiomatic variables in L. In place of the interpretation of the uninterpreted axiomatic constants we now have a substitution of n constants, which represent the interpretation, for the n axiomatic variables.

Concerning the *axiomatic method* see Hilbert, "Das axiomatische Denken", *Math. Ann.*, 78, 405 ff., 1918; Fraenkel [Einleitung] §18 (where additional copious references may be found); Russell [Principles]; Woodger [Biology] and [Theory construction] (these two are especially concerned with applications of the method in the empirical sciences); Tarski [Logic] Chap. VI "Deductive method"; Copi [Logic] Chap. VI; Wilder [Foundations] Chaps. I and II; Carnap [Foundations].

42e. Concerning the axiom systems (ASs) in Part Two of this book. In the chapters which follow we present various ASs. Our presentation makes use of the symbolic languages explained in Part One of the book. In this connection it is possible to utilize the syntactical rules for language B explained earlier (in **21** and **22**); such an AS is then not only symbolized, but formalized as well. The logical constants of the basic language are presumed known; only the axiomatic primitive constants are specified, definitions being constructed for the remaining axiomatic constants.

For certain ASs some theorems are given by way of illustration. [These theorems belong to the object language and hence are to be carefully distinguished from the theorems established in Part One, which belong to the metalanguage—viz. either to semantics or to syntax.]

The ASs are arranged in four fields on the basis of their specified interpretations; of course, the impression should not be gained from this that the specified interpretation of an AS is the only one possible.

In the case of many axioms, *two formulations* are given: a formulation marked 'A', which belongs to the simple language A; and a formulation marked 'C', which belongs to the extended language C. If neither of the marks 'A' and 'C' appears, the formulation belongs to language A. A few axioms are formulated only in language C, since the formulation in A would be tedious.

For the sake of brevity, universal quantifiers which refer to the whole formula are omitted from axioms and theorems.

The order of appearance of the various ASs is not that of increasing difficulty, but follows the order of the four fields. It is therefore advisable for the reader wishing to examine systems in formulation C that he choose them in accordance with the necessary prerequisites from Chapter C, viz. **44a** and **46a** can be read after the study of **32**; **47** and **51a** after **33**; **52a** and **b**, and **53a** after **35**; **53b** and **54a** and **b** after **36**; **44b, 46b** and **51b** after **37**; and **45**, **48a** and **b** and **c**, and **52c** after **38**. In **46c** use is made of the coordinate language explained in **40**; and **48d**, **49** and **50** utilize several logical concepts introduced in **46**.

Chapter E

Axiom systems (ASs) for set theory and arithmetic

43. AS FOR SET THEORY

The following AS is a modification of Fraenkel's system ([Grundlegung]; see [Einleitung] §16 and [Set Theory]) which in turn is based on the system of Ernst Zermelo (Math. Annalen, 65, 1908). (Axiom A9 was proposed later by Zermelo, Fund. Math., 16, 1930.) In Fraenkel's system the following is the case: (1) sets are not classes, but individuals; (2) every element of a set is itself a set; (3) there are no individuals other than sets. Our modification of this system consists in retaining (1) and (2), but abolishing (3); the modification permits a clearer formulation of the axiom of restriction (axiom A10 in **43b** below).

A set in set theory is, in practice, essentially the same as a class in logic. The logical rules for the two concepts differ, however, since in the AS now to be considered (as well as in the majority of ASs of set theory) no distinctions of type are made between sets: the same variables (e.g. 'x', 'y', etc.) are used for sets, for sets of sets, etc. This is the meaning of statement (1) above that sets are individuals of the system. Sometimes we also speak of a property of sets (notice e.g. the variable 'F' in A5); in this connection it is to be noted that such a property of sets does *not* necessarily correspond to another set (say, the set of those sets having the property in question)—the question whether a set of a certain kind exists is to be settled in each case solely by appeal to the axioms. Observe, finally, that our axioms are for the most part existence statements; they assert that under certain circumstances there is a set which satisfies certain conditions.

Among other ASs of set theory are those of: J. von Neumann, "Eine Axiomatisierung der Mengenlehre", *Jour. reine u. ang. Math.* 154, 1925, and "Die Axiomatisierung der Mengenlehre", *Math. Zeitschr.* 27, 1928; P. Bernays, "A system of axiomatic set theory", *Jour. Symbolic Logic* 2, 1937, and subsequent volumes; and K. Gödel, "The consistency of the axiom of choice, etc.", *Annals of Mathematics Studies*, No. 3, Princeton, 1940. A survey of the various forms of ASs for set theory is given in: Hao Wang and R. McNaughton, *Les systèmes axiomatiques de la théorie des ensembles*, Paris and Louvain, 1953.

43a. The Zermelo-Fraenkel AS. This AS features a single primitive sign, 'E'; the expression 'Exy' may be read "The set x is an element of the set y" (the customary notation is '$x \in y$'). The axioms and definitions here and in **43b** and **43c** are formulated only in language A (because of the transformation to be described later, in **43c**). The AS can be read after the study of Chapter A.

Sets are the members of relation E:

D1. $Sx \equiv mem(E)x$.

Subset (analogous to subclass):

D2. $Ss(x,y) \equiv Sx.Sy.(z)(Ezx \supset Ezy)$.

Sets with the same elements are identical:

A1. $Ss(x,y).Ss(y,x) \supset (x=y)$.

A set x is a pair set comprising y and z provided y and z are the only elements of x:

D3. $Prs(x,y,z) \equiv Sy.Sz.(u)(Eux \equiv (u=y) \vee (u=z))$.

Existence of a pair set comprising two given sets:

A2. $Sy.Sz.(y \neq z) \supset (\exists x)Prs(x,y,z)$.

A set x is a (the) union set of y provided the elements of x are the elements of the elements of y:

D4. $Us(x,y) \equiv Sx.Sy.(u)[Eux \equiv (\exists z)(Euz.Ezy)]$.

Existence of a union set of a given set:

A3. $Sy \supset (\exists x)Us(x,y)$.

A set x is a (the) power set of y provided the elements of x are the subsets of y:

D5. $Ps(x,y) \equiv Sy.(u)[Eux \equiv Ss(u,y)]$.

Existence of a power set:

A4. $Sy \supset (\exists x)Ps(x,y)$.

Axiom of comprehension (in simple form; Fraenkel's axiom V). Given any set y and any property F, there exists a comprehension set x of y respecting F, i.e. a set x whose elements are those elements of y that have property F:

A5. $(y)(F)[Sy \supset (\exists x)[Sx.(u)(Eux \equiv Euy.Fu)]]$.

A set x is a selection set for y provided x is a subset of a (the) union set of y and x has exactly one element in common with each set that is an element of y.

D6. $Sls(x,y) \equiv (\exists w)(Us(w,y).Ss(x,w)).(z)[Ezy \supset (\exists u)(v)(Evz.Evx \equiv (v=u))]$.

Axiom of choice (or selection). If y is a set whose elements are non-empty and mutually exclusive, then there is at least one selection set for y:

A6. $Sy.(z)(Ezy \supset (\exists u)Euz).(v)(w)(u)[Evy.Ewy.Euv.Euw \supset (v=w)] \supset (\exists x)Sls(x,y)$.

Set x is a (the) unit set of y:

D7. $Uts(x,y) \equiv Prs(x,y,y)$.

Axiom of infinity. Axiom A7 below says there is a set z such that: (1) every empty set belongs to z, and (2) if v belongs to z, then so does every unit set of v also. It is a consequence of the other axioms that, if there is any set, then there is exactly one empty set and that for each set there is exactly one unit set; hence A7 guarantees a set z containing a progression of elements, viz. the empty set, the unit set of the empty set, the unit set of this last, etc. Hence z is an infinite set.

A7. $(\exists z)[Sz.(y)[Sy.\sim(\exists x)(Exy) \supset Eyz].(v)(w)(Evz.Uts(w,v) \supset Ewz)]$.

Axiom of replacement (A8): Given any set x and any function from sets to sets, there exists a set y comprising those elements which the function associates with the elements of x. (In the present system we designate such set functions not by functors, but by two-place predicates for one-many relations (the variable 'K')).

A8. $(x)(K)[Sx.(v)(w)(Kvw \supset Sv).(u)(v)(w)(Kuw.Kvw \supset (u=v)) \supset (\exists y)[Sy.(v)(Evy \equiv (\exists w)(Ewx.Kvw))]]$.

Axiom of regularity (A9): For any non-empty set x, there is an element y of x such that y and x have no element in common:

A9. $Eux \supset (\exists y)[Eyx.\sim(\exists z)(Ezy.Ezx)]$.

With the help of this axiom (which was proposed by Zermelo in 1930) it can be shown that the relation E is irreflexive (i.e., no set is an element of itself) and asymmetric (i.e., no two sets are elements of each other).

43b. The axiom of restriction. It can easily be seen that the axioms in **43a** leave open certain questions concerning the existence of sets. Therefore Fraenkel considered a further axiom which should restrict the system of sets as much as possible under the previous axioms. He formulated tentatively this axiom of restriction ("Axiom der Beschränktheit") as follows: "No sets exist beyond those which are required by the previous axioms". He remarked, however, that it was extremely doubtful whether this or any similar axiom was meaningful. For this reason he did not include an axiom of this kind in his AS.

It will now be shown that Fraenkel's doubts were not justified and that an axiom of the kind he intended can be stated in an unobjectionable form. The above-mentioned formulation of the axiom contains a reference to the previous axioms. Taken literally, such a reference can be formulated only in the metalanguage. However, this difficulty can be overcome by allowing the new axiom to contain an open sentential formula that corresponds to the conjunction of the previous axioms, but with a variable 'H' in place of the primitive axiomatic constant 'E'. Then the axiom can be formulated in the symbolic object language. To avoid writing out this long formula, we shall make use of the predicate of second level, '*ZF*', to be defined as the explicit predicate (**42d**) for the Zermelo-Fraenkel AS or, more specifically, for the seven axioms A1, A3, A4, A6, A7, A8, and A9 of **43a**. (We include here Zermelo's axiom A9 although it did not belong to Fraenkel's AS; on the other hand, we omit A2 and A5 because they are redundant, i.e. derivable from the other axioms.) Then '$ZF(E)$' is an abbreviation for the conjunction of the seven axioms, and '$ZF(H)$' is an abbreviation for the corresponding open sentential formula. The definition of '*ZF*' is readily obtained by the procedure outlined in **42d**; in accordance with this procedure, '*E*' is replaced by the variable '*H*'. The definition being very long, we give only the beginning of it here:

D8. $ZF(H) \equiv (x)(y)\big[mem(H)x \,.\, mem(H)y \,.\, (z)\big(Hzx \equiv Hzy\big) \supset (x=y)\big] \,.$ $(y)\big[mem(H)y \supset (\exists x)\big(mem(H)(x).(u)[Eux,$ etc.

Now the meaning of Fraenkel's axiom is this: "For the system of sets ordered by the relation *E*, there is no subsystem of a different structure (i.e., not isomorphic to the original system) which likewise fulfills the previous axioms". Thus, in terms of the explicit predicate '*ZF*', the axiom of restriction (A10) can now be formulated in this way: "Every subrelation *H* of *E* with property *ZF* is isomorphic to *E*". (Our statement of A10 makes use of 'Is_2'; for this symbol, recall D19-5.)

A10. $(H)\big[(x)(y)\big(Hxy \supset Exy\big).ZF(H) \supset Is_2(H,E)\big].$

If to a given polymorphic AS (recall **42c**) an axiom of this form, containing the explicit predicate with respect to the AS, is added, the effect is to restrict the admitted model structures to the minimal structures (i.e., those which have no other admitted structures as parts). Therefore we call axioms of this kind "minimal-structure axioms". This is one of four kinds of so-called extremal axioms, whose nature and general method of application is explained in Carnap-Bachmann [Extrem.]. The addition of an extremal axiom to a given polymorphic AS often yields a monomorphic (or categorical) AS (**42c**). Whether the addition of A10 has this result for the AS now under consideration is not known. But the result is obtained by each of the following axioms: by the axiom A4* in Peano's AS in the form **44b**, when reformulated as a minimal-structure axiom (see [Extrem.] p. 179); and by Hilbert's axioms of completeness in his AS of Euclidean geometry and in his AS for real numbers (see the reference in **45**), each reformulated as a so-called maximum-model axiom.

43c. A modified version of the AS in an elementary basic language. The AS stated in **43a** makes use of predicate variables, viz. '*F*' in A5 and '*K*' in A8. For certain purposes, however, it seems desirable to have an AS for set theory with a more elementary basic language (**42a**) containing only individual variables but no predicate variables. Especially is this so if set theory is constructed for the purpose of serving as the logical theory of abstract concepts (classes, relations, functions, etc.), for then we should avoid a basic language that already contains a logic of classes, etc.

Let L_i be a basic language with individual variables as the only variables; the primitive constants and the defined logical constants may be those of language A (as far as their definitions do not make use of variables other than individual variables). Let L_i' be obtained from L_i by adding the axiomatic primitive predicate '*E*'. Then we take instead of the one axiom A5 an infinite class of axioms A5* containing just those sentences of the language L_i' which result from A5 by deleting the quantifier '(*F*)' and substituting for '*Fu*' any sentential formula of the language L_i' in accordance with the rules for formula substitution (see **12c**). (According to these rules, the formula to be substituted must not contain '*u*' in a quantifier and must not contain '*x*' or '*y*'. If the formula contains still other variables as free variables, they must be bound by universal quantifiers placed at the beginning of the axiom.) Analogously, instead of the axiom A8 we take an infinite class of axioms A8* obtained by deleting '(*K*)' and making any formula substitution for '*Kvw*'. (Here, the substitutum must not contain '*v*' or '*w*' in a quantifier and must not contain '*x*', '*y*', or '*u*'.) Each of the axiom classes A5* and A8* could, of course, be specified by an axiom schema in the metalanguage (analogous to the primitive sentence schemata in **22a**).

It should be noticed that the class of axioms A5*, although infinite, is weaker than the one axiom A5. The latter refers, by the use of the variable '*F*', to *all* properties of sets without regard to expressibility in any given language, while the axioms of the class A5* refer only to those properties which are expressible by sentential formulas in language L_i'. Likewise, the class of axioms A8* is weaker than the axiom A8.

44. PEANO'S AS FOR THE NATURAL NUMBERS

44a. The first version: the original form. For the original account, see Peano [Formulaire] II, §2: Arithmétique, 1898, pp. 1 ff. For another account, see Russell [Introduction] Ch. I. Our formulation A may be read after **18**, formulation C after **32**. The AS features *three primitive signs*: '*ze*', '*N*', '*sc*'. The sign '*ze*' is an individual constant, '*N*' a one-place predicate, and '*sc*' a one-place functor. The usual interpretation is: '*ze*' denotes the number 0; '*Nx*' reads "*x* is a (natural) number"; and '*sc*(*x*)' reads "the successor of *x*" or "the (natural) number following *x*".

Zero is a number:

A1. $N(ze)$.

The successor of a number is a number:

A2. **A.** $Nx \supset N(sc(x))$.
C. $sc``N \subset N$.

Numbers with the same successor are identical:

A3. $Nx.Ny.\big(sc(x)=sc(y)\big) \supset (x=y)$.

Zero is not the successor of any number:

A4. **A.** $Nx \supset \big(sc(x)\neq ze\big)$.
C. $\sim(sc``N)(ze)$.

Axiom A5 is the Principle of Mathematical Induction ("complete" induction); recall **37c**. Every number is F if the property F satisfies the two conditions: (1) zero is F; and (2) if any individual is F, then so is its successor:

A5. **A.** $(F)\big[F(ze).(x)\big(Fx\supset F(sc(x))\big) \supset (y)\big(Ny\supset Fy\big)\big]$.
B. $(F)\big[F(ze).\big(sc``F\subset F\big) \supset \big(N\subset F\big)\big]$.

44b. The second version: just one primitive sign. The single primitive sign here is the two-place predicate 'Pr'; its customary interpretation: immediate predecessor in the series of natural numbers. For discussions, see Russell [Introduction] Ch. I and [P.M.] II, 245. Formulation A may be read after **18**, formulation C after **37**.

The (natural) numbers are the members of Pr:

D1*. **A.** $Nx \equiv mem(Pr)x$.
C. $N=mem(Pr)$.

The relation Pr is one-one:

A1*. **A.** $\big(Pr(x,z).Pr(y,z) \supset x=y\big).\big(Pr(x,y).Pr(x,z) \supset y=z\big)$.
C. $Un_{1,2}(Pr)$.

The relation Pr has exactly one initial member:

A2*. **A.** $(\exists x)(y)\big[Ny.\sim(\exists z)\big(Pr(z,y)\big) \equiv (y=x)\big]$.
C. $1(init(Pr))$.

If definitions by description are admitted into language C (recall again the note at the end of **35**), axiom A2* provides a basis on which the number zero can be defined by D3*(C): '$ze=(\iota x)(init(Pr)x)$'.

Every number is the predecessor of something, i.e. Pr has no terminal member:

A3*. **A.** $Nx \supset (\exists y)Pr(x,y)$.
C. $N \subset mem_1(Pr)$.

Each member of Pr can be reached from an initial member in finitely many Pr-steps, i.e. every member of Pr possesses all the Pr-hereditary properties (**36a**) of any initial member of Pr:

A4*. **A.** $(x)(y)(F)\big[Nx.\sim(\exists z)\big(Pr(z,x)\big).Ny.Fx.(u)(v)\big(Fu.Pr(u,v) \supset Fv\big) \supset Fy\big]$.
C. **a.** $init(Pr)x.Ny \supset Pr^{\geq 0}(x,y)$; or
b. $N \subset (Pr^{\geq 0})^{-1}``(init(Pr))$.

Definition of the *explicit concept M* of this AS (in formulation C). We form the definiens in accordance with the procedure outlined in **42d**, viz. we eliminate '*N*' from the axioms by means of D1*, replace the primitive constant '*Pr*' by the variable '*H*', and form the conjunction:

D2*. **C.** $M(H) \equiv Un_{1,2}(H) . 1(init(H)) . \big(mem(H) \subset mem_1(H)\big) . \big(mem(H) \subset (H^{\geq 0})^{-1}``(init(H))\big)$.

In view of this definition, *M* is the class of relations that satisfy axioms A1* through A4*.

The definiens of '*M*' can readily be transformed into that of '*Prog*' (recall D37-1). Thus '*M*' and '*Prog*' are synonymous, the models of the AS now under consideration are the progressions, and the explicit concept of this AS is the class of the progressions.

Finally, let us cite a simple example of a theorem in this AS, viz. *Pr* is asymmetric:

T1*. **C.** $As(Pr)$.

This theorem corresponds to the open sentential formula '$As(H)$' and so to the universal conditional sentence '$(H)[M(H) \supset As(H)]$' that says: every relation that satisfies the four axioms is asymmetric. This sentence is provable in language C (which here serves as our basic language); in the usual interpretation of this language, the sentence is L-true.

45. AS FOR THE REAL NUMBERS

This AS stems from Tarski [Logic] §63. An account of it may also be found in Cooley [Logic] §36. The AS has six primitive signs, viz. two predicates: '*R*' (Real number) and '*S*' (relation Smaller); two two-place functors: '*su*' (sum) and '*prod*' (product); and two individual constants (numerals): '0' and '1'. Tarski mentions the fact that the axioms are not mutually independent (i.e. several of them are derivable from the others and hence are superfluous, theoretically speaking); he also gives another AS (it is in §61 of the book cited above) which is distinctly shorter but which makes the derivation of theorems far more complicated. (Formulation A of the following AS can be read after **18**, formulation C after **38**.)

The first AS for real numbers was given by Hilbert ("On the number concept", orig. 1900, later published in the appendix of his book *The Foundations of geometry*, 1902).

Of two different numbers, one is smaller than the other:

A1 **A.** $(x \neq y) \supset Sxy \vee Syx$.
C. $Connex(S)$. (See **31b**.)

The relation *S* is asymmetric:

A2. **A.** $Sxy \supset \sim Syx$.
C. $As(S)$.

The relation S is transitive:

A3. **A.** $Sxy . Syz \supset Sxz$.
C. $Trans(S)$.

The relation S is a Dedekind relation (**38b**):

A4. **A.** $(x)(y)[Fx . Gy \supset Sxy] \supset (\exists z)(x)(y)[Fx . (x \neq z) . Gy . (y \neq z) \supset Sxz . Szy]$.
C. $Ded(S)$.

It follows from A4 and other axioms that S has no initial member and no terminal member, and hence that S belongs to the kind Ded_{00}.

The sum of two numbers is a number:

A5. $Rx . Ry \supset R(su(x,y))$.

The sum is commutative:

A6. $su(x,y) = su(y,x)$.

The sum is associative:

A7. $su(x,su(y,z)) = su(su(x,y),z)$.

Existence of the difference of two numbers:

A8. $Rx . Ry \supset (\exists z)\big(Rz . (x = su(y,z))\big)$.

[Here Cooley takes the simpler axiom: '$Rx \supset (\exists z)\big(Rz . (su(x,z) = 0)\big)$'.]

Monotony of the sum:

A9. $Syz \supset S(su(x,y),su(x,z))$.

It is the case that 0 is a number:

A10. $R(0)$.

It is the case that $x + 0 = x$:

A11. $su(x,0) = x$.

The product of two numbers is a number:

A12. $Rx . Ry \supset R(prod(x,y))$.

The product is commutative:

A13. $prod(x,y) = prod(y,x)$.

The product is associative:

A14. $prod(x,prod(y,z)) = prod(prod(x,y),z)$.

Existence of the quotient:

A15. $Rx . Ry . (y \neq 0) \supset (\exists z)\big(Rz . (x = prod(y,z))\big)$.

Monotony of the product:

A16. $S(0,x) . S(y,z) \supset S(prod(x,y),prod(x,z))$.

The distributive law:

A17. $prod(x,su(y,z)) = su(prod(x,y),prod(x,z))$.

It is the case that 1 is a number:

A18. $R(1)$.

It is the case that $x \cdot 1 = x$:

A19. $prod(x,1) = x$.

It is the case that 0 is distinct from 1:

A20. $0 \neq 1$.

Chapter F

Axiom systems (ASs) for geometry

46. AS FOR TOPOLOGY (NEIGHBORHOOD AXIOMS)

The AS below is constructed following Hausdorff [Grundzüge] 213 ff. (compare also Rosser [Logic] Ch. IX, sec. 8; and H. F. Bohnenblust, *Theory of functions of real variables*, Princeton 1937). With the elements, called points, certain classes of points are associated as neighborhoods. Such a neighborhood system forms a topological space.

46a. The first version. Here the only primitive sign is the predicate '*Nb*'. The expression '$Nb(F,x)$' reads "the class F (of points) is a neighborhood of (the point) x". (Formulation A can be read after **19**; formulation C, after **32**.)

The points are the second-place members of *Nb*:

D1. **A.** $Px \equiv mem_2(Nb)x$.
C. $P = mem_2(Nb)$.

The neighborhoods ('*Nbh*') are the first-place members of *Nb*:

D2. **A.** $Nbh(F) \equiv mem_1(Nb)(F)$.
C. $Nbh = mem_1(Nb)$.

The point classes:

D3. **A.** $PC(F) \equiv (z)(Fz \supset Pz)$.
C. $PC = sub_1(P)$.

Each neighborhood is a class of points:

A1. **A.** $Nbh(F) \supset PC(F)$.
C. $Nbh \subset PC$.

Every neighborhood of x contains x:

A2. $Nb(F,x) \supset Fx$.

If F_1 and F_2 are neighborhoods of x, then there is a neighborhood of x which is a subclass both of F_1 and of F_2:

A3. **A.** $Nb(F_1,x) \,.\, Nb(F_2,x) \supset (\exists G)[Nb(G,x) \,.\, (y)(Gy \supset F_1 y \,.\, F_2 y)]$.
C. $Nb(F_1,x) \,.\, Nb(F_2,x) \supset (\exists G)[Nb(G,x) \,.\, (G \subset F_1 \,.\, F_2)]$.

If y belongs to the neighborhood F of x, then there is a neighborhood G of y such that G is a subclass of F:

A4. **A.** $Nbh(F) \,.\, Fy \supset (\exists G)[Nb(G,y) \,.\, (z)(Gz \supset Fz)]$.
C. $Nbh(F) \,.\, Fy \supset (\exists G)[Nb(G,y) \,.\, (G \subset F)]$.

Two different points have neighborhoods with no points in common:

A5. **A.** $Px.Py.(x \neq y) \supset (\exists F)(\exists G)[Nb(F,x).Nb(G,y).\sim(\exists z)(Fz.Gz)]$.
C. $Px.Py.(x \neq y) \supset (\exists F)(\exists G)[Nb(F,x).Nb(G,y).\sim\exists(F.G)]$.

46b. The second version. Here the sole primitive sign is the predicate '*Nbh*' (of second level); *Nbh* is the class of all neighborhoods. (Formulation A can be read after **17**; formulation C, after **37**.) We take any class which belongs to *Nbh* as a neighborhood of any of its points. (This is a simplified version of Rosser's AS, p. 273, which uses two primitives and essentially three axioms.)

D1*. **A.** $Px \equiv (\exists F)[Nbh(F).Fx]$.
C. $P = sm_1(Nbh)$.

The following axioms A1* and A2* correspond to A3 and A5, respectively.

A1*. **A.** $Nbh(F_1).Nbh(F_2).F_1x.F_2x \supset (\exists G)[Nbh(G).Gx.(y)(Gy \supset F_1y.F_2y)]$.
C. $Nbh(F_1).Nbh(F_2).F_1x.F_2x \supset (\exists G)[Nbh(G).Gx.(G \subset F_1.F_2)]$.

A2*. **A.** $Px.Py.(x \neq y) \supset (\exists F)(\exists G)[Nbh(F).Fx.Nbh(G).Gx.\sim(\exists z)(Fz.Gz)]$.
C. $Px.Py.(x \neq y) \supset (\exists F)(\exists G)[Nbh(F).Fx.Nbh(G).Gx.\sim\exists(F.G)]$.

We now define the two-place predicate '*Nb*' so that it corresponds to the primitive predicate '*Nb*' of the first version:

D2*. $Nb(F,x) \equiv Nbh(F).Fx$.

D3*. For '*PC*', as in D3.

It can easily be shown that on the basis of these axioms and definitions, the five axioms of the first version are derivable.

The additional concepts of topology (point set theory) can be defined on the basis of '*Nbh*'. Some examples follow below:

A point x of the class F is called an "inner" point of F provided there is a subclass of F which is a neighborhood of x:

D4*. **A.** $Inn(x,F) \equiv PC(F).(\exists G)[Nb(G,x).(z)(Gz \supset Fz)]$.
C. $Inn(x,F) \equiv PC(F).\exists(Nb(-,x).sub_1(F))$.

A point class is called "open" ('*OPC*'), if all its points are inner points:

D5*. **A.** $OPC(F) \equiv PC(F).(x)(Fx \supset Inn(x,F))$.
C. $OPC(F) \equiv PC(F).(F \subset Inn(-,F))$.

By the complement of F we understand the class of all those points which do not belong to F (note that '*cpl*' is a functor of the second level):

D6*. **A.** $cpl(F)x \equiv Px.\sim Fx$.
C. $cpl(F) = P.\sim F$.

We say that x is a limit point of F ('$Lim(x,F)$') provided F is a point class and x is a point (not necessarily belonging to F) such that every open point class containing x also contains a point of F different from x:

D7*. $Lim(x,F) \equiv PC(F).Px.(G)[OPC(G).Gx \supset (\exists y)(y \neq x.Fy.Gy)]$.

A point class is called "closed" if it contains all its limit points:

D8*. **A.** $Clos(F) \equiv PC(F).(x)[Lim(x,F) \supset Fx]$.
C. $Clos(F) \equiv PC(F).[Lim(-,F) \subset F]$.

The closure of F is defined as the union of F and the class of the limit points of F. It is denoted by '$clos(F)$', where '$clos$' is a functor of second level:

D9*. **A.** $clos(F)(x) \equiv Fx \vee Lim(x,F)$.
C. $clos(F) = (F \vee Lim(-,F))$.

A point x is said to be a point of accumulation of F ('$Acc(x,F)$') provided every neighborhood of x contains infinitely many points of F:

D10*. **A.** $Acc(x,F) \equiv (G_1)[Nb(G_1,x) \supset (\exists G_2)(\exists G_3)[(z)(G_3z \supset G_2z).$ $(\exists y)(G_2y. \sim G_3y).Is_1(G_3,G_2).(z)(G_2z \supset G_1z.Fz)]]$.
C. $Acc(x,F) \equiv (G)[Nb(G,x) \supset ClsRefl(F.G)]$.

Theorems. The whole space, i.e., the class of all points, is both open and closed:

T1. **a.** $OPC(P)$. **b.** $Clos(P)$.

Every neighborhood is open:

T2. **A.** $Nbh(F) \supset OPC(F)$.
C. $Nbh \subset OPC$.

The closure of any point class is closed:

T3. $PC(F) \supset Clos(clos(F))$.

A point class is closed if and only if it is identical with its closure:

T4. **A.** $Clos(F) \equiv (x)[Fx \equiv clos(F)x]$.
C. $Clos(F) \equiv [F = clos(F)]$.

A point class is closed if and only if its complement is open:

T5. $Clos(F) \equiv OPC(cpl(F))$.

46c. Definition of logical concepts. What follows is given in formulation C, and may be read after **40**. We begin by defining the explicit concept (recall **42d**) for the Hausdorff AS in its second version (**46b**); in this connection we use the symbol '$Hausd(M)$', which reads "The class M (of the second level) satisfies the Hausdorff AS" or "M is a (Hausdorff) neighborhood system." Thereafter we list definitions of additional logical concepts, culminating in the concept of the *dimension number* (see Karl Menger,

Dimensionstheorie, 1928, pp. 77 ff.; see also his "What is dimension?", Amer. Math. Mon., 50, 1943). All these definitions are formulated just in language C; their formulation in language A is too long and complicated.

D11*. C. $Hausd(M) \equiv (F_1)(F_2)(x)[M(F_1) . M(F_2) . F_1x . F_2x \supset (\exists G)$
$[M(G) . Gx . (G \subset F_1 . F_2)]] . (x)(y)[sm_1(M)x . sm_1(M)y . (x \neq y)$
$\supset (\exists F)(\exists G)[M(F) . Fx . M(G) . Gx . \sim \exists(F . G)]]$.

To the axiomatic predicate '*Acc*' (recall D10*) there corresponds the logical predicate '*Acp*'; the sentence '*Acp(x,F,M)*' says "*x* is a point of accumulation of *F* with respect to neighborhood system *M*" (all the other concepts below similarly refer to a neighborhood system *M*):

D12*. C. $Acp(x,F,M) \equiv Hausd(M) . (G)[M(G) . Gx \supset ClsRefl(F . G)]$.

The boundary of a class *F* with respect to *M* (symbols: '*bd(F,M)*') is the class of those accumulation points of *F* respecting *M* which are not points of *F*:

D13*. C. $bd(F,M)x \equiv Acp(x,F,M) . \sim Fx$.

For subsequent definitions we enlarge our language by adding a second type of individuals. (Thus we are establishing a two-sorted language, in the sense of **21c**.) There is already at hand the type of the objects; points are of this type, as are the variables '*x*', '*y*', etc. Besides that type we now include the type of the integers, and take the variables '*m*', '*n*', etc., to be of this type. We use the language form explained in **40c**, with its additional primitive signs '0', ''', '*K*' and '*SmEq*'.

Our definition D14* below is a three-part recursive definition; excepting the fact that it defines a predicate, this definition is analogous in form to D40-11. Definition D14* is our initial step towards a definition of dimension number in the fashion of Menger. In defining this latter concept we have departed from Menger so as to avoid the appearance of a vicious circle and to represent the concept exactly.

We treat first the preliminary concept "A dimension number at most *n* is possessed by *F* at point *x* with respect to neighborhood system *M*". We want this concept to conform to the following rules: (1) A dimension number at most -1 is to be possessed by the empty class at every point *x*; (2) A dimension number at most $n+1$ is to be possessed by a class *F* at any one of its own points *x* if and only if there is an arbitrarily small neighborhood G_2 of *x* (which is to say, provided there is within each neighborhood G_1 of *x* another neighborhood G_2 of *x*) such that a dimension number at most *n* is possessed by the intersection of *F* and $bd(G_2,M)$ at each point of this intersection. A final remark: So that our recursive definition may have 0 rather than -1 as initial argument, we define in D14* the auxiliary predicate '*Di*' such that '*Di(n,F,x,M)*' reads "A dimension number at most $n-1$ is possessed by *F* at point *x* with respect to neighborhood system *M*".

D14*. **C. 1.** $Di(0,F,x,M) \equiv Hausd(M).\sim\exists(F)$.

2. $SmEq(0,n) \supset [Di(n',F,x,M) \equiv Hausd(M).(F \subset sm_1(M)).Fx.(G_1)[M(G_1).G_1x \supset (\exists G_2)[M(G_2).G_2x.(G_2 \subset G_1).(y)(Fy.bd(G_2,M)y \supset Di(n,F.bd(G_2,M),y,M))]]]$.

3. $SmEq(n,0) \supset [Di('n,F,x,M) \equiv (x \neq x)]$.

Parts 1 and 2 of this definition express respectively the two rules (1) and (2) set forth above. Part 3 is added simply to establish that numbers less than 0 are not first-place members of *Di*, i.e. that numbers less than -1 do not occur as dimension numbers (notice that the definiens is L-false).

By the dimension number of class F at point x respecting neighborhood system M (symbolically: '$dimp(F,x,M)$') we understand the smallest number n such that $Di(n',F,x,M)$, i.e. the smallest number n such that a dimension number at most n is possessed by F at x respecting M:

D15*. **C.** $dimp(F,x,M) = (Kn)(Di(n',F,x,M))$.

Omitting the reference to a point, we say the *dimension number* of class F respecting neighborhood system M (symbolically: '$dim(F,M)$') is n provided: either F is empty and $n=-1$; or else F is not empty, the dimension number of F at each of its points does not exceed n, and the dimension number of F at one at least of its points is n. Thus:

D16*. **C.** $dim(F,M) = (Kn)[(\sim\exists(F).(n=-1)) \vee [(x)(Fx \supset SmEq(dimp(F,x,M),n)).(\exists y)(Fy.dimp(F,y,M)=n.)]]$

We say that F has the *homogeneous dimension number n* provided either F is empty and $n=-1$, or else F is non-empty and has dimension number n at each of its points:

D17*. **C.** $Dimhom(n,F,M) \equiv Hausd(M).[(\sim\exists(F).(n=-1)) \vee [\exists(F).(x)(Fx \supset dimp(F,x,M)=n)]]$.

The concepts defined above, especially the logical predicate '*Dimhom*', are utilized in **48d, 49** and **50.**

47. ASs OF PROJECTIVE, OF AFFINE AND OF METRIC GEOMETRY

This system follows on the whole that of Roth [Axiomat.]; our formulation A may be read after **18**, formulation C after **33**. Our program is as follows: first we set up an AS of projective geometry (**47a**); this system is then enlarged through the addition of a new primitive sign (and, under certain circumstances, of new axioms) to an AS of affine geometry (**47b**); and finally, this last system is similarly extended to an AS of metric (Euclidean) geometry (**47c**).

The first modern AS of Euclidean geometry is due to Hilbert (*Foundations of geometry*, 1899). A modified form of Hilbert's system has been formulated symbolically by O. Helmer (*Axiomatischer Aufbau der Geometrie in formalisierter Darstellung*, Diss. Berlin 1934; *Schriften des Math. Seminars der Universität Berlin*, 2, 1935).

47a. AS of projective geometry: A1–A20. The primitive signs are three predicates of the first level: 'O', 'In', and 'S'. The sentence 'Oxu' reads "the point x lies on the line u"; the sentence '$In(x,r)$' reads "the point x lies in the plane r"; and the sentence '$Sxyvw$' reads "the points x and y separate the points v and w on a line". In connection with this last reading we remark that projective lines are closed and thus the points of any such line have a cyclic order.

We distinguish three types of individuals, viz. points, lines and planes. Thus we require a three-sorted language (recall **21c**) and three kinds of individual variables. We agree to use the variables 'x', 'y', 'z', 'v', 'w' for points, 't' and 'u' for lines, and 'r' and 's' for planes.

[Were we to employ a one-sorted language, we would need to introduce three additional primitive signs, viz. three one-place predicates denoting respectively the class of points, the class of lines, and the class of planes. Moreover, we would require eight new axioms: three axioms to the effect that the three classes just mentioned are mutually exclusive, and five other axioms stipulating to which of these three classes the members of 'O', of 'In', and of 'S' belong. Further, the additional axioms we take would frequently call for extra conditions to the effect that x and y are points, or the like; a three-sorted language dispenses with this, since the sort of the individual is conveyed by the shape of the variable.]

Axioms A1 through A10 are called *axioms of connection* (see Roth: I, 1–8).

For any two distinct points, there is at least one line (A1) and at most one line (A2) on which they both lie:

A1. **A.** $(x \neq y) \supset (\exists u)(Oxu . Oyu)$.
C. $J \subset O | O^{-1}$.

A2. **A.** $(x \neq y) . Oxu . Oyu . Oxt . Oyt \supset (u = t)$.
C. $2_m(O(-, u) . O(-, t)) \supset (u = t)$.

On each line there are at least two distinct points:

A3. **A.** $(\exists x)(\exists y)(Oxu . Oyu . (x \neq y))$.
C. $2_m(O(-, u))$.

Three points are said to be collinear provided they lie on the same line (D1); and similarly for four points (D2):

D1. $Coll_3(x,y,z) \equiv (\exists u)(Oxu . Oyu . Ozu)$.

D2. $Coll_4(x,y,z,w) \equiv (\exists u)(Oxu . Oyu . Ozu . Owu)$.

There are three non-collinear points:

A4. **A.** $(\exists x)(\exists y)(\exists z)(\sim Coll_3(x,y,z))$.
C. $\exists(\sim Coll_3)$.

Every three non-collinear points lie in a plane:

A5. $\sim Coll_3(x,y,z) \supset (\exists r)(In(x,r).In(y,r).In(z,r))$.

In each plane there is at least one point:

A6. $(\exists x)In(x,r)$.

For every three distinct non-collinear points there is at most one plane in which they lie:

A7. **A.** $\sim Coll_3(x,y,z).(x\neq y).(x\neq z).(y\neq z).In(x,r).In(y,r).In(z,r).In(x,s).In(y,s).In(z,s) \supset (r=s)$.
C. $\exists[(\sim Coll_3.J_3)in(In(-,r).In(-,s))]\supset(r=s)$.

The line u is said to *lie in* the plane r (symbolically: '$LinIn(u,r)$' provided all points which lie on u also lie in r:

D3. $LinIn(u,r) \equiv (z)(Ozu \supset In(z,r))$.

If each of two distinct points of a line lie in one plane, then the entire line lies in that plane:

A8. **A.** $Oxu.Oyu.In(x,r).In(y,r).(x\neq y) \supset LinIn(u,r)$.
C. $2_m(O(-,u).In(-,r)) \supset LinIn(u,r)$.

If two planes have a point in common, they also have a second different point in common:

A9. **A.** $In(x,r).In(x,s) \supset (\exists y)((y\neq x).In(y,r).In(y,s))$.
C. $1_m(In(-,r).In(-,s)) \supset 2_m(In(-,r).In(-,s))$.

There are four points which lie in no one plane:

A10. $(\exists x)(\exists y)(\exists z)(\exists w)\sim(\exists r)[In(x,r).In(y,r).In(z,r).In(w,r)]$.

Axioms A11 through A19 are called *axioms of order* (see Roth: II, 1–8).

If points x, y separate points v, w, then points x, y, v, w are distinct and collinear:

A11. **A.** $Sxyvw \supset Coll_4(x,y,v,w).(x\neq y).(x\neq v).(x\neq w).(y\neq v).(y\neq w).(v\neq w)$.
C. $S \subset (Coll_4.J_4)$.

If x,y separate v,w, then x,y separate w,v:

A12. $Sxyvw \supset Sxywv$.

If x,y separate v,w, then v,w separate x,y:

A13. $Sxyvw \supset Svwxy$.

If x,y,v are distinct collinear points, then there is a point w such that x,y separate v,w:

A14. **A.** $Coll_3(x,y,v).(x\neq y).(x\neq v).(y\neq v) \supset (\exists w)Sxyvw$.
C. $(Coll_3.J_3)xyv \supset \exists(S(x,y,v,-))$.

If x,y,v,w are distinct collinear points, then either x,y separate v,w; or x,v separate y,w; or y,v separate x,w:

A15. **A.** $Coll_4(x,y,v,w).(x \neq y).(x \neq v).(x \neq w).(y \neq v).(y \neq w).(v \neq w) \supset Sxyvw \vee Sxvyw \vee Syvxw.$

C. $(Coll_4.J_4)xyvw \supset Sxyvw \vee Sxvyw \vee Syvxw.$

If x,y separate v,w, then x,v do not separate y,w:

A16. $Sxyvw \supset \sim Sxvyw.$

If x,y separate z,v, if z,v,w are collinear, and if w is distinct from x and from y, then x,y separate z,w if and only if x,y do not separate v,w:

A17. $Sxyzv.Coll_3(z,v,w).(w \neq x).(w \neq y) \supset (Sxyzw \equiv \sim Sxyvw).$

Axiom A18 is the axiom of Pasch. Suppose that three non-collinear points x,y,z and also all the points of line u and of line t lie in the same plane r, but that none of x,y,z lies on either of u and t; and suppose further that v is a point on u, that w is a point on t, and that x,y separate v,w; then there is a point v on u and a point w on t such that v,w separate either y,z or x,z:

A18. **A.** $In(x,r).In(y,r).In(z,r).\sim Coll_3(x,y,z).LinIn(u,r).LinIn(t,r).\sim Oxu.\sim Oyu.\sim Ozu.\sim Oxt.\sim Oyt.\sim Ozt.(\exists v)(\exists w)(Ovu.Owt.Sxyvw) \supset (\exists v)(\exists w)[Ovu.Owt.(Svwyz \vee Svwxz)].$

C. $\sim Coll_3(x,y,z).LinIn(u,r).LinIn(t,r).[\{x,y,z\} \subset (In(-,r).\sim O(-,u).\sim O(-,t))].(\exists v)(\exists w)(Ovu.Owt.Sxyvw) \supset (\exists v)(\exists w)[Ovu.Owt.(Svwyz \vee Svwxz)].$

We say that point w *belongs* to *segment* x,y,z (and write: '$Segm(w,x,y,z)$') provided x,y,z are three distinct points on a line u, w lies on u, and w,y do not separate x,z:

D4. **A.** $Segm(w,x,y,z) \equiv (\exists u)(Oxu.Oyu.Ozu.Owu).(x \neq y).(x \neq z).(y \neq z).\sim Swyxz.$

C. $Segm(w,x,y,z) \equiv (J_3 \text{ in } O(-,u))xyz.Owu.\sim Swyxz.$

We say that w is an *inner point* of segment x,y,z (and write: '$ISegm(w,x,y,z)$') provided w belongs to segment x,y,z and is distinct from x and from z:

D5. $ISegm(w,x,y,z) \equiv Segm(w,x,y,z).(w \neq x).(w \neq z).$

Axiom A19 is the *axiom of continuity*. If F is a subclass of a segment and has at least two points, then there exist three points x_1,y_1,z_1 such that F is contained in segment x_1,y_1,z_1 and each segment having either x_1 or z_1 as an inner point also has an inner point that belongs to F:

A19. $(\exists x)(\exists y)(\exists z)(v)(Fv \supset Segm(v,x,y,z)).(\exists v)(\exists w)(Fv.Fw.(v \neq w)) \supset (\exists x_1)(\exists y_1)(\exists z_1)[(v)(Fv \supset Segm(v,x_1,y_1,z_1)).(x_2)(y_2)(z_2)(ISegm(x_1,x_2,y_2,z_2) \vee ISegm(z_1,x_2,y_2,z_2)) \supset (\exists w)(ISegm(w,x_2,y_2,z_2).Fw)].$

Axiom A20 is the *projective axiom* (see Roth: III): Two lines in a plane always have at least one point in common:

A20. A. $LinIn(u,r).LinIn(t,r) \supset (\exists z)(Ozu.Ozt)$.
C. $(LinIn|LinIn^{-1}) \subset (O^{-1}|O)$.

47b. AS of affine geometry. We obtain affine geometry from projective geometry by singling out one (projective) plane and giving it a particular role in the system. This plane can be selected arbitrarily from all the available projective planes. Once selected, the plane is called the *improper plane* (sometimes the *ideal* plane) and is designated by the sign '*improp*'. Hence our AS of affine geometry requires an *additional primitive sign* '*improp*'. The sign '*improp*' is an individual constant of the third sort in the three-sorted language (points, lines, planes) we have developed for our ASs of geometry. The role of this improper plane is indicated in definition D9 below. Our AS of affine geometry thus comprises the four primitive terms '*O*', '*In*', '*S*' and '*improp*', the axioms and definitions already laid down for projective geometry, and the four definitions now to be given.

Points and lines lying in the improper plane are called respectively *improper points* and *improper lines.* All other planes, points and lines are called respectively *proper planes* ('*PPl*'), *proper points* ('*PPo*') and *proper lines* ('*PLi*').

D6. $PPl(r) \equiv (r \neq improp)$.
D7. $PPo(x) \equiv \sim In(x,improp)$.
D8. $PLi(u) \equiv \sim LinIn(u,improp)$.

Two proper lines are said to be *parallel* ('*Par*') provided they have an improper point in common:

D9. $Par(u,t) \equiv PLi(u).PLi(t).(\exists x)(In(x,improp).Oxu.Oxt)$.

This form of the system requires no additional axioms for affine geometry. A formation rule lays it down that '*improp*' is an individual constant of the third sort, i.e. of the same sort as the variable '*r*'. And from this it follows that '$(\exists r)(r = improp)$' is provable ("There is a plane *improp*").

[There are alternative routes to affine geometry besides that of introducing '*improp*' as a new primitive sign. Of these we mention the following two: (1) We may take as a primitive sign the predicate '*IPl*' designating the class of improper planes, and then lay it down as an axiom that one and only one plane belongs to this class; again (2) we may take '*Par*' ("parallel") as a primitive sign, lay down suitable axioms for it, and then with the help of '*Par*' define '*IPl*' and if desired '*improp*'. This last constant is introduced by descriptional definition (recall **35b**), once '$1(IPl)$' has been proved.]

47c. AS of metric Euclidean geometry: A1–A32. Euclidean geometry is obtained from affine geometry by omitting from the latter the improper elements—more exactly, by the introduction of concepts which refer only to the proper elements. The *additional primitive sign* here is '*Perp*'; the sentence '*Perp(u,r)*' is read "the proper line *u* is perpendicular to the proper plane *r*". Axioms A21 through A32 below are called *axioms of orthogonality* (Roth: V, 1–3).

The first-place members of *Perp* are proper lines (A21), and the second-place members thereof are proper planes (A22):

A21. **A.** $mem_1(Perp)u \supset PLi(u)$.
C. $mem_1(Perp) \subset PLi$.

A22. **A.** $mem_2(Perp)r \supset PPl(r)$.
C. $mem_2(Perp) \subset PPl$.

If the proper point *x* lies in the (proper) plane *r*, then there is at least (A23) and at most (A24) one line through *x* perpendicular to *r*:

A23. $PPo(x).In(x,r) \supset (\exists u)(Oxu.Perp(u,r))$.

A24. **A.** $PPo(x).In(x,r).Oxu.Perp(u,r).Oxt.Perp(t,r) \supset (u=t)$.
C. $PPo(x).In(x,r) \supset \sim 2_m(O(x,-).Perp(-,r))$.

If the proper point *x* lies on the line *u*, then there is at least (A25) and at most (A26) one plane *r* in which *x* lies and to which *u* is perpendicular:

A25. $PPo(x).Oxu \supset (\exists r)(In(x,r).Perp(u,r))$.

A26. **A.** $PPo(x).Oxu.In(x,r).Perp(u,r).In(x,s).Perp(u,s) \supset (r=s)$.
C. $PPo(x).Oxu \supset \sim 2_m(In(x,-).Perp(u,-))$.

If the proper point *x* lies on the (proper) line *u*, then there is at least (A27) and at most (A28) one plane *s* such that *x* lies in *s* and the following is the case: if *u* lies in the plane *r* and line *t* is perpendicular to *r* and *x* lies on *t*, then all points of *t* lie in *s*:

A27. $PPo(x).Oxu \supset (\exists s)[In(x,s).(r)(t)(LinIn(u,r).Perp(t,r).Oxt \supset LinIn(t,s))]$.

A28. $PPo(x).Oxu.In(x,s_1).In(x,s_2).(r)(t)[LinIn(u,r).Perp(t,r).Oxt \supset LinIn(t,s_1).LinIn(t,s_2)] \supset (s_1=s_2)$.

If the proper point *x* lies on the lines *u* and *t* and in the planes *s* and *r*, if *u* is perpendicular to *s* and *t* to *r*, and if *u* lies in *r*, then *t* lies in *s*:

A29. $PPo(x).Oxu.Oxt.In(x,s).In(x,r).Perp(u,s).Perp(t,r).LinIn(u,r) \supset LinIn(t,s)$.

If the proper point *x* lies on the line *u* and in the plane *r*, and if *u* is perpendicular to *r*, then *u* does not lie in *r*:

A30. $PPo(x).Oxu.In(x,r).Perp(u,r) \supset \sim LinIn(u,r)$.

If lines u and t are perpendicular to a plane, then there is a plane r in which both u and t lie:

A31. **A.** $Perp(u,s).Perp(t,s) \supset (\exists r)(LinIn(u,r).LinIn(t,r))$.
C. $(Perp|Perp^{-1}) \subset (LinIn|LinIn^{-1})$.

If two different lines u and t are perpendicular to the same plane, then u and t have no proper point in common:

A32. $Perp(u,s).Perp(t,s),PPo(x).Oxu.Oxt \supset (u=t)$.

Chapter G

ASs of physics

48. ASs OF SPACE-TIME TOPOLOGY: 1. THE *C-T* SYSTEM

48a. General remarks. The topological structure of the physical world is independent of measurable magnitudes. However, the method ordinarily employed in physics to treat topological properties of space and time makes use of measurable magnitudes, viz. of coordinate systems. Such a coordinate system associates with each space-time point a quadruple of real numbers; while this association is based on certain arbitrary conventions, the arbitrariness is subsequently eliminated in topology through the device of considering only those properties which are invariant under any one of a certain class of transformations from one coordinate system to another. This usual procedure is convenient mathematically because it utilizes the familiar and effective means of real numbers and real functions; nevertheless it is, so to speak, methodologically impure.

The question thus arises whether it is possible to treat topological properties of space and time by a *purely topological method*, i.e. a method which makes no use of conceptual means—such as e.g. real numbers and coordinate systems—that have a metric (non-topological) character. Such a method is possible on the basis of the logic of relations; indeed, this is true for topological problems generally, and not simply for topological problems concerning space and time that arise in physics. The AS presented herewith is intended to illustrate how the logic of relations makes possible a treatment of topological questions by purely topological means. The AS is based on the conception of space and time found in Einstein's general theory of relativity; a knowledge of this theory is, of course, not presupposed.

For more detailed discussions of the concepts here employed from relativity theory, see e.g. Reichenbach, *Axiomatik der relativistischen Raum-Zeit-Lehre*, 1924. Concerning the *C-T* system used here (as presented by me earlier in [Abriss]) and related systems stemming from Robb, Reichenbach and Russell, see H. Mehlberg, "Essai sur la théorie causale du temps", *Studia Philosophica* I (1935) and II (1937). A similar system, making reference to Reichenbach and to the present system, is given by K. Schnell, *Eine Topologie der Zeit in logistischer Darstellung*, Diss., Münster i. W., 1938. Concerning the philosophical significance of the present AS, cf. Carnap, "Über die Abhängigkeit der Eigenschaften des Raumes von denen der Zeit", *Kantstudien*, 30, 1925.

The present *C-T* system treats the motions and coincidences of physical particles. No assumptions are made concerning the physical nature of these particles (they may be thought of as particles proper, e.g. electrons;

again, they may be thought of as the smallest elements of electromagnetic radiation); they are regarded as idealized, i.e. unextended.

As *individuals* we take moments or slices of particles. Following Minkowski, we call the moments of a particle its world-points; the class of all world-points of a particle we call the world-line of the particle. Each world-point is assigned to a space-time point, i.e. is associated with a position in the space-time continuum.

Suppose a_1, b_1, c_1, ... are world-points of a certain particle, and similarly a_2, b_2, c_2, ... are world-points of another particle. Now if, say, b_1 and b_2 are assigned to the same space-time point, we take this to mean that at the instant in question both particles are in the same place, i.e. they touch or coincide. For this relation of coincidence let us now introduce the sign 'C', the first primitive sign of our system. Thus the state of affairs just described is formulated in the sentence 'Cb_1b_2'.

[We observe here parenthetically that our second and third forms of the present system (in **49** and **50**, respectively) proceed differently, viz. world-points are *identified* with space-time points. For the case referred to above, this entails taking b_1 and b_2 not as coincident but as identical: '$b_1=b_2$'.]

Following Kurt Lewin, we say that world-points of the same particle are *genidentical.* In the example above, a_1 and b_1 are genidentical; so likewise are a_2 and c_2; but b_1 and b_2 are not genidentical even when they coincide.

The second primitive sign 'T' of our system denotes the relation Earlier between two genidentical world-points. Relation T thus represents only a local time order (the *Eigenzeit* of relativity theory), and not a temporal relation between remote processes. E.g. assuming the world-points a_1,b_1,c_1 of a particle to occur in this temporal order, each of the following sentences holds: 'Ta_1b_1', 'Tb_1c_1', 'Ta_1c_1'. It is to be noted that T is a topological concept, not a metrical one; which is to say, statements about T presuppose a comparison of earlier and later, but no measurement of temporal durations. It is often remarked that all observational statements of physics can be referred back to observational statements about coincidences. This claim is imprecise. Observational statements about coincidences need to be supplemented by observational statements about the *Eigenzeit* relation, for by observation of coincidences alone it is not possible to establish the temporal order of the processes—viz. coincidences with other particles—involving a single particle.

The construction of the present system (of which only the chief features are indicated in what follows) shows that our relations C and T suffice to express not only the topological structure of temporal order, but that of spatial order as well.

48b. *C*, *T*, and world-lines. The first form of the system, presented in this section, contains two primitive signs, viz. 'C' and 'T'. [The two leading subsections, **48b** and **48c**, may be read in formulation A (which omits A13 and the theorems) after **18**; in formulation C, after **38**. Several axioms are given alternative formulations in language C. For the sake of brevity, we formulate the theorems in language C only.]

Relation C is symmetric (A1) and transitive (A2), thus C is an equivalence relation (see **34a**):

A1. **A.** $Cxy \supset Cyx$.
C. $C \subset C^{-1}$. [Alternatively: $Sym(C)$.]

A2. **A.** $Cxy.Cyz \supset Cxz$.
C. $C^2 \subset C$. [Alternatively: $Trans(C)$.]

Every individual coincides with something:

A3. **A.** $(x)(\exists y)Cxy$.
C. $mem_1(C)$. [Abbreviation for '$U(mem_1(C))$'; see **28b**.]

Theorem. Every individual coincides with itself, i.e. relation C is totally reflexive:

T1. **C.** $I \subset C$. [Alternatively: $Reflex(C)$.] [By A1, A2, A3 and T31-1(d) and (c).]

Relation T is transitive (A4), irreflexive (A5), and dense (A6; see **38a**):

A4. **A.** $Txy.Tyz \supset Txz$.
C. $T^2 \subset T$. [Alternatively: $Trans(T)$.]

A5. **A.** $\sim Txx$.
C. $T \subset J$. [Alternatively: $Irr(T)$.]

A6. **A.** $Txy \supset (\exists u)(Txu.Tuy)$.
C. $T \subset T^2$.

Every individual is a first member (A7) and a second member (A8) of T:

A7. **A.** $(x)(\exists y)Txy$.
C. $mem_1(T)$.

A8. **A.** $(y)(\exists x)Txy$.
C. $mem_2(T)$.

Theorems. Relation T is asymmetric (T2); T has no initial member (T3) and no terminal member (T4):

T2. **C.** $T \subset \sim(T^{-1})$. [Also: $As(T)$.] [From A4, A5, and T31-1g.]

T3. **C.** $\sim\exists(init(T))$. [From A8, D32-8.]

T4. **C.** $\sim\exists(init(T^{-1}))$. [From A7.]

Axiom A9 leads to the theorem (T5) that C and T are mutually exclusive:

A9. **A.** $Cxy.(x \neq y) \supset \sim Txy$.
C. $C.J \subset \sim T$.

T5. **C.** $C \subset \sim T$. [By A9 and A5.]

World-points x and y are genidentical provided x is identical with y, or the relation T holds between them in one direction or the other:

D1. **A.** $Gen(x,y) \equiv Txy \vee Tyx \vee (x=y)$.
C. $Gen = (T \vee T^{-1} \vee I)$.

Theorems. Relation *Gen* is symmetric (T6) and totally reflexive (T7):

T6. **C.** $Sym(Gen)$. [From D1.]

T7. **C.** $Reflex(Gen)$. [From D1.]

A world-line never branches into two parts, either in the direction of the past (A10) or in the direction of the future (A11):

A10. **A.** $Txz.Tyz \supset Gen(x,y)$.
C. $(T|T^{-1}) \subset Gen$.

A11. **A.** $Tux.Tuy \supset Gen(x,y)$.
C. $(T^{-1}|T) \subset Gen$.

Theorem. Relation *Gen* is transitive:

T8. **C.** $Trans(Gen)$. [From A4, A10, and A11.]

It follows from T6 and T8 that *Gen* is an equivalence relation. World-lines are the non-empty equivalence classes of *Gen*, i.e. a world-line is the class of world-points genidentical with some world-point:

D2. **A.** $Wl(F) \equiv (\exists x)[(y)(Fy \equiv Gen(y,x))]$.
C. $Wl(F) \equiv (\exists x)[F{=}Gen(-,x)]$.

The world-points of each world-line are ordered into a series by means of a subrelation of T; these series relations we call "world-line series" ('*Wlin*'):

D3. **A.** $Wlin(H) \equiv (\exists F)[Wl(F).(x)(y)(Hxy \equiv Txy.Fx.Fy)]$.
C. $Wlin(H) \equiv (\exists F)[Wl(F).(H = (T \; in \; F))]$.

Theorems. The world-line series are transitive (T11; from A4 and T32-2c), irreflexive (T12; by A5), asymmetric (T13; from T2) and connected (T14; from A4, A10, A11); hence they are properly series (T15; from T11, T12, T14); moreover, they are dense (T16; by A6):

T11. **C.** $Wlin \subset Trans$.

T12. **C.** $Wlin \subset Irr$.

T13. **C.** $Wlin \subset As$.

T14. **C.** $Wlin \subset Connex$.

T15. **C.** $Wlin \subset Ser$.

T16. **C.** $Wlin(H) \supset (H \subset H^2)$.

Every world-line series is a Dedekind relation (A12; recall **38b**) and hence is a series of Dedekind continuity without initial or terminal members (T17):

A12. **A.** $Wlin(H) \supset (F)(G)[(x)(y)(Fx.Gy \supset Hxy) \supset (\exists z)(x)(y)(Fx.(x \neq z).Gy.(y \neq z) \supset Hxz.Hzy)]$.
C. $Wlin \subset Ded$.

T17. **C.** $Wlin \subset DedSer_{00}$.

The following axiom A13, formulated only in language C, may be passed over inasmuch as it is not used hereafter. It says that every world-line series has a denumerable class of intermediate members (recall D38-7). Hence such a series also has Cantor continuity (T18; from T17 and A13). This topological structural property of these series makes possible a transition to a metric, viz. it permits a one-one association of real numbers with world-points of a world-line.

A13. **C.** $Wlin(H) \supset (\exists F)(\aleph_0(F).Med(F,H))$.

T18. **C.** $Wlin \subset ContSer_{00}$.

48c. The signal relation. [Here formulation A (axioms only) can be read after **18**; formulation C, after **38**.] An effect reaches from a world-point x to a world-point y if and only if x is connected to y by a signal. The simplest case of such a connection sees x coincident with the world-point u of a particle which so moves that a later world-point v of the same particle coincides with y. (According as this mediating particle is a material particle or a particle of radiant energy, we have to do with a material signal or a radiation signal, e.g. a light signal.) In other cases the signal is not by a single particle, but by a chain of particles: x and y are joined by a linkage consisting of segments of world-lines, the linkage being such that the end of each constituent segment is joined to the beginning of the next by a coincidence. Figure 4 illustrates how world-point b_1 could be joined to world-point e_3 by the signal chain: Tb_1c_1, Cc_1c_2, Tc_2d_2, Cd_2d_3, Td_3e_3.

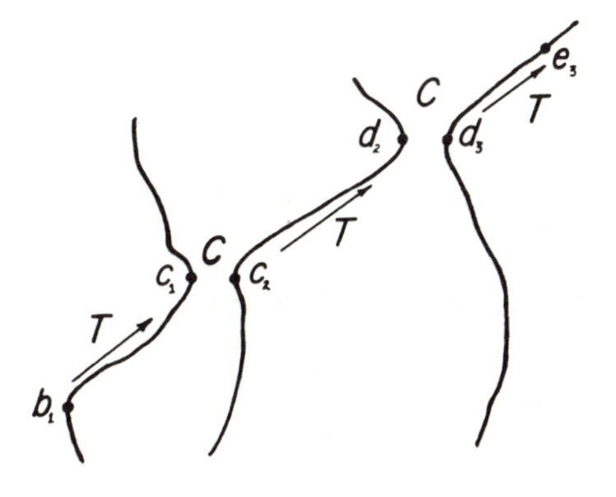

FIG. 4. Signal chain

Since identical world-points are also regarded as coincident (T1), our explication of the concept of the signal chain loses no generality if we require that every such chain begins with C and ends with C. If there is a chain of this kind, we say that the *signal relation* ('S') exists between the initial member and terminal member of the chain. E.g. adding 'Cb_0b_1' and 'Ce_3e_4' to the chain pictured in Figure 4, we obtain the signal chain: Cb_0b_1, Tb_1c_1, Cc_1c_2, Te_2d_2, Cd_2d_3, Td_3e_3, Ce_3e_4; in view of this chain we see that Sb_0e_4 is the case. Of course, it is also the case that Sb_1e_3, since C is totally reflexive and hence both 'Cb_1b_1' and 'Ce_3e_3' hold. Thus 'S' represents the same thing as '$C|T|C|T|C|T...|C$'.

On the basis of the considerations above we construct for 'S' the definition D4. [From this point on we give the definitions as well as the theorems only in formulation C; formulation A becomes quite complicated beginning with D4.]

D4. **C.** $S = (C|T)^{>0}|C$.

Theorems. If T holds, so also does S (T19; by T1); and S is transitive (T20; by A2):

T19. **A.** $T \subset S$.

T20. **C.** $Trans(S)$.

The following axiom A14 serves to establish the irreflexivity of S (T21):

A14. **A.** $Sxy . \sim Txy \supset (x \neq y)$.
C. $(S . \sim T) \subset J$.

T21. **C.** $Irr(S)$. [From A14, A5.]

Relation S is asymmetric (T22; from T20, T21), and S and C are mutually exclusive (T23; from A1, A2, T21):

T22. **C.** $As(S)$.

T23. **C.** $S \subset \sim C$.

Two further axioms are to the following effect. Suppose x bears the relation S to y, and either lies outside the world-line of y or else on this world-line but not before y. Then, first, there is a world-point u before y on the world-line of y which is so early that no signal (i.e. S-relation) from x reaches it (A15); and, second, there is a world-point v after x on the world-line of x which is so late that no signal from it reaches y (A16). From these assumptions it follows that the same also holds for arbitrary world-points x and y (T24: from A15, T19, T20, A8; and T25: from A16, T19, T20, A7). This in turn entails that on each world-line there are arbitrarily early and arbitrarily late world-points.

A15. **A.** $Sxy . \sim Txy \supset (\exists u)(\sim Sxu . Tuy)$.
C. $(S . \sim T) \subset ((\sim S)|T)$.

T24. **C.** $(\sim S)|T$. [Abbreviation for '$U((\sim S)|T)$'; see **28b**.]

A16. A. $Sxy.\sim Txy \supset (\exists v)(Txv.\sim Svy)$.
C. $(S.\sim T) \subset (T|\sim S)$.

T25. C. $T|\sim S$.

Axiom A17 concerns the *finite limiting velocity*. If there were an infinite signal velocity, there could be two non-coincident points x and y with a signal from x to y and also a signal from y to x; but this is impossible because of the asymmetry of S (T22). However, it might still be the case that there are signal velocities of any arbitrary finite value. Were this last to be so, it could happen that from each point before x on the world-line of x—if not from x itself—a signal could go to y, and from y a signal to each point after x on the world-line of x. Axiom A17 leads to T26 (with the help of T22) and thereby excludes this possibility, in accordance with relativity theory.

A17. A. $(u)(Tux \supset Suy).(z)(Txz \supset Syz) \supset (Sxy.Syx) \vee Cxy$.
C. $(T(-,x) \subset S(-,y)).(T(x,-) \subset S(y,-)) \supset (Sxy.Syx) \vee Cxy$.

T26. C. $(T(-,x) \subset S(-,y)).(T(x,-) \subset S(y,-)) \supset Cxy$.

48d. The structure of space. [From here on everything, including axioms, is formulated in language C only; the material can be read after **38** of Part One and **40** and **46** of Part Two.] We say two world-points x and y are *simultaneous* (and write '$Sim(x,y)$') provided the signal relation fails to hold between x and y, and likewise between y and x. This definition is in agreement with that feature of relativity theory according to which there is an admissible coordinate system furnishing the same value to the time coordinate of both x and y when and only when it is impossible that a signal go from x to y or from y to x. (Cf. Reichenbach, *Philosophie der Raum-Zeit-Lehre*, Berlin, 1928, p. 171; or its English translation, *Philosophy of space and time*, 1958.)

D5. C. $Sim = (\sim S.\sim S^{-1})$.

The class $S(-,a)$ of world-points that bear the signal relation S to the world-point a we call (following Minkowski) the *prior cone of a* (see Figure 5). The class $S(a,-)$ of world-points to which a bears the signal relation S we call the *posterior cone of a*. In view of the finite limiting velocity (A17), there exists between the prior cone and the posterior cone the so-called *intermediate region of a*; this intermediate region of a is the class $Sim(-,a)$ of world-points simultaneous with a. A world-line F having no coincidence with a has a whole segment in common with the intermediate region of a (in Figure 5, this segment is labelled '$F.Sim(-,a)$'). Such a world-line F has not simply *one* world-point simultaneous with a, but many (indeed, infinitely many; see T34 below). While these last-mentioned world-points of F are all simultaneous with a, they are not simultaneous with each other, i.e. *Sim* is not a transitive relation (in contradistinction to simultaneity with reference to a fixed coordinate system).

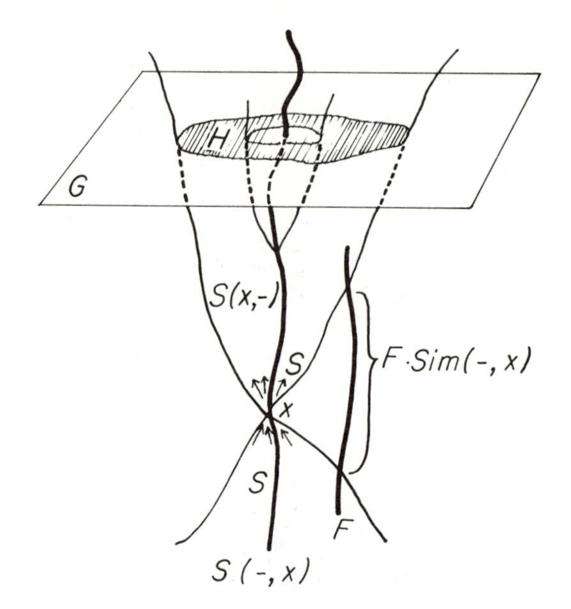

FIG. 5. The shaded area represents the *effected region* H of the point x in the space G

Theorems regarding *Sim*. *Sim* is totally reflexive (T27; by T21) and symmetric (T28). Coincident world-points are simultaneous (T29; from T23 and A1); simultaneous genidentical points are identical (T30; by T19); *Sim* and S are mutually exclusive, hence *Sim* and T are likewise (T31 and T32, by T19).

T27. **C.** *Reflex*(*Sim*).
T28. **C.** *Sym*(*Sim*).
T29. **C.** $C \subset Sim$.
T30. **C.** $(Sim.Gen) \subset I$.
T31. **C.** $Sim \subset \sim S$.
T32. **C.** $Sim \subset \sim T$.

Additional theorems. For each world-point x there is on each world-line F a world-point simultaneous with x (T33), and even infinitely many world-points simultaneous with x provided no world-point of F coincides with x (T34):

T33. **C.** $(x)[Wl(F) \supset \exists(F.Sim(-,x))]$.

T34. **C.** $Wl(F).\sim(C``F)(x) \supset ClsRefl(F.Sim(-,x))$.

Outlines of *proofs* for T33 and T34. We distinguish two cases; T33 refers to both, T34 to the second only. (1) Suppose x coincides with a point of world-line F; T33 then follows with the help of T29—and in the special event that x belongs to F, with the help of T27. (2) Suppose x does not coincide with a point of F. Let $F_1 = F.S(-,x)$, i.e. F_1 is the class of those world-points of F which bear the relation S to x; and let $F_2 = F.S(x,-)$, i.e. F_2 is the class of those world-points of F to which x bears the relation S. Because of the Dedekind continuity of T in F (recall T17) there is an upper limit, say y_1, for F_1 (i.e. a world-point of F which separates the class F_1 and its complement in F; cf. **38b**), and a lower limit, say y_2, for F_2 (i.e. a world-point of F which separates F_2 and its complement in F). In accordance with the axiom of the finite limiting velocity (A17), world-points y_1 and y_2 are different; indeed, it is the case that Ty_1y_2. By A6 there are infinitely many points of F between y_1 and y_2. All of these intervening points are simultaneous with x (cf. Figure 5).

A spatial class, or *space* for short, is so to speak a three-dimensional cross section of the four-dimensional space-time world, the sectioning being done across the time direction—i.e. across all world-lines. Thus our definition runs as follows: A space is a class of world-points which are simultaneous with each other, the class itself being such that it has in common with each world-line at least one world-point.

D6. **C.** $Sp(G) \equiv (x)(y)[Gx.Gy \supset Sim(x,y)].(F)[Wl(F) \supset \exists(G.F)]$.

In view of definition D6 it is the case that every space has exactly one point in common with each world line (T35; by T32):

T35. **C.** $Sp(G).Wl(F) \supset 1(G.F)$.

Axiom A18 is adopted to assure that for each world-point there is a space to which it belongs (T36). To weaken our formulation of A18 we add to it the condition that every world-line not containing a point coincident with x has infinitely many points simultaneous with x. The condition can be omitted from T36, because by T34 it is already satisfied. Finally, T37 says that points coincident with points of a space also belong to the space.

A18. **C.** $(F)[Wl(F).\sim(C``F)x \supset ClsRefl(F.Sim(-,x))] \supset (\exists G)(Sp(G).Gx)$.

T36. **C.** $sm_1(Sp)x$.

T37. **C.** $Sp(G) \supset (C``G \subset G)$.

What the primitive concepts C and T furnish directly is a topological order for time alone. The question arises whether it is possible on this same basis, i.e. without additional new primitive concepts, also to determine a topological order in each space. This question can be answered affirmatively with the help of the concept of *effected region*. We say (D7; see

Figure 5) a class H is the effected region in space G of world-point x, and write '$Effreg(H,x,G)$', provided H is non-empty and is the class of all points z of G to which a signal (i.e. the S-relation) leads from some point y later than x (so to speak, H is the intersection of G with the class of interior points of the posterior cone of x):

D7. C. $Effreg(H,x,G) \equiv Sp(G).\big[H = ((T|S)(x,-).G)\big].\exists(H)$.

Coincident world-points have the same spatial position. Hence we take as elements of our space order, i.e. as space-points ('SpP'), not world-points but classes of world-points coincident among themselves—which is to say, we count as space-points the non-empty equivalence classes respecting the relation C (recall T34-1b):

D8. C. $SpP(F) \equiv (\exists x)\big(F = C(-,x)\big)$.

The nearer x lies to space G, the smaller is the effected region of x in G. Thus each G contains arbitrarily small effected regions. On the other hand, an arbitrary world-point of G can be reached by a signal from a given world-line provided only the signal emanates from a sufficiently early world-point x of this given world-line. Hence G also contains arbitrarily large effected regions. These considerations suggest that effected regions—more precisely: space-point classes that correspond to effected regions—be taken as neighborhoods within space G. We shall do this; '$Nbd(N,G)$' is to mean "the class N (of space points) is a neighborhood in space G" (D9). Then we shall regard such a class N as a neighborhood of each space-point F in G which belongs to it (cf. **46b**):

D9. C. $Nbd(N,G) \equiv (\exists x)(\exists H)\big[Effreg(H,x,G).(N \subset SpP).(sm_1(N) = H)\big]$.

To show (T40) that in each space the neighborhoods just defined constitute a Hausdorff neighborhood system (recall **46b**), we require axioms A19 and A20. Axiom A19 says: If y and v are two non-coincident points of space G, then there is an x preceding y and a u preceding v such that no point of G can be reached both by a signal from a point after x and by a signal from a point after u. It follows from this axiom A19 that there are in G neighborhoods of the space-points corresponding to world-points y and v such that these neighborhoods have no points in common (T38)—viz., the neighborhoods corresponding to the effective regions of x and of u in G.

A19. C. $Sp(G).Gy.Gv.\sim Cyv \supset (\exists x)(\exists u)\big[Txy.Tuv.\sim(\exists t)\big[Gt.(T|S)(x,t).(T|S)(u,t)\big]\big]$.

T38. C. $Sp(G).SpP(F_1).SpP(F_2).(F_1 \subset G).(F_2 \subset G).(F_1 \neq F_2) \supset (\exists N_1)(\exists N_2)\big[Nbd(N_1,G).N_1(F_1).Nbd(N_2,G).N_2(F_2).\sim\exists(N_1.N_2)\big]$.

Axiom A20 says: If there is a point z in space G which receives a signal from a point later than x and also a signal from a point later than y, then there is also a point u of which the same is true and which is such that from a later point (i.e. a point later than u) there is a signal that leads to z. I.e. if the effective regions F_1 and F_2 in G of x and of y respectively have a point z in common, then in the intersection of F_1 and F_2 there is also an effective region, viz. that of u. From this axiom we obtain the following result (T39): If N_1 and N_2 are neighborhoods of F in G, then there is a neighborhood N_3 of F in G such that N_3 is a subclass of N_1 and a subclass of N_2.

A20. C. $Sp(G).Gz.(T|S)(x,z).(T|S)(y,z) \supset (\exists u)[(T|S)(x,u).(T|S)(y,u).(T|S)(u,z)]$.

T39. C. $Sp(G).Nbd(N_1,G).N_1(F).Nbd(N_2,G).N_2(F) \supset (\exists N_3)[Nbd(N_3,G).N_3(F).(N_3 \subset N_1.N_2)]$.

That the two neighborhood axioms A1* and A2* in **46b** hold is shown by T39 and of T38, respectively. (Notice that A2* would not hold for two different but coincident world-points; it is for this reason that we use space-points, rather than world-points, as the elements of the neighborhoods of our system.) Thus in each space the classes of space-points defined here (by D9) as neighborhoods constitute a Hausdorff system of neighborhoods (T40). (Recall that '*Hausd*' is a logical constant; see D11* in **46c**).

T40. C. $Sp(G) \supset Hausd(Nbd(-,G))$.

The foundation just laid enables us to employ all the topological concepts defined earlier (in **46c**) with respect to neighborhood systems. Thus, a description of any of the topological properties of space can be formulated in the signs of our AS—and this means in terms of C and T ultimately. E.g. we can now construct an axiom (it is A21) stipulating that each space is three-dimensional. Axiom A21 says: If any space G is such that it carries a Hausdorff system of neighborhoods, then the class of space-points of G has the homogeneous dimension number 3 respecting the neighborhood system in G (recall D17* in **46c**). Theorem T41 says the same thing without the restrictive condition involving the neighborhood system in G, for in view of T40 this condition holds in any case.

A21. C. $Sp(G).Hausd(Nbd(-,G)) \supset Dimhom(3,SpP.sub_1(G),Nbd(-,G))$.

T41. C. $Sp(G) \supset Dimhom(3,SpP.sub_1(G),Nbd(-,G))$.

49. ASs OF SPACE-TIME TOPOLOGY: 2. THE *Wlin*-SYSTEM

The present *second form* is called the *Wlin*-system. Its single *primitive sign* is '*Wlin*'. This sign designates the class of time relations (in previous terms: world-line series) on world-lines; recall D3 in **48b**. In the present system, world-points are again taken as *individuals*—however, world-points not as particle slices, but as the space-time points corresponding thereto. Here, therefore, coincident world-points are identical, and hence the relation C is superfluous. On the other hand, discrimination between the various world-lines now requires the class $Wlin$ of relations rather than the relation T. The present form of the system makes especially clear how the axioms ascribe topological properties to the time order, while also permitting a representation of the nature of space order. [The present system, as well as that given in **50**, are formulated in language C only; both systems may be read after **38** of Chapter C, together with **46** of Chapter F.]

Axioms A1 through A6 say that each of the time relations $Wlin$ is irreflexive, transitive, devoid of initial member, devoid of terminal member, dense and connected:

A1. **C.** $Wlin \subset Irr$.

A2. **C.** $Wlin \subset Trans$.

A3. **C.** $Wlin(H) \supset (mem(H) \subset mem_2(H))$.

A4. **C.** $Wlin(H) \supset (mem(H) \subset mem_1(H))$.

A5. **C.** $Wlin(H) \supset (H \subset H^2)$.

A6. **C.** $Wlin \subset Connex$.

It follows from A1, A2 and A6 that the relations comprised by $Wlin$ are series:

T1. **C.** $Wlin \subset Ser$.

We can now introduce a sign 'T' with roughly the same meaning as that imputed to the primitive sign 'T' of the first form (**48b**). Here, however, the relation T is not transitive; if the present T holds between x and y and between y and z, and if x and z belong to different world-lines, then this T does not hold between x and z.

D1. **C.** $T = sm_2(Wlin)$.

It follows from A1 that T is irreflexive (T2), and from A5 that T is dense (T3):

T2. **C.** $Irr(T)$.

T3. **C.** $T \subset T^2$.

The signs defined next below correspond to the same signs given in the first form (**48b, c**): 'Wl' denotes the class of world-lines, i.e. of the fields of the relations constituting $Wlin$; 'Gen' denotes genidentity; and 'S' the signal relation.

D2. **C.** $Wl = mem``Wlin$.

D3. **C.** $Gen(x,y) \equiv (\exists F)(Wl(F).Fx.Fy)$.

D4. **C.** $S = T^{>0}$.

Axioms A7 through A9 of the present system are identical in appearance with axioms A12 through A14 of the previous system; for that reason we do not list them here explicitly.

Axioms A10, A11, and A12 below are similar to our preceding axioms A15, A16, and A17 respectively:

A10. C. $Wl(F) \supset (S``F . \sim F \subset (\sim S)``F)$.
A11. C. $Wl(F) \supset (S^{-1}``F . \sim F \subset (\sim S^{-1})``F)$.
A12. C. $(T(-,x) \subset S(-,y)).(T(x,-) \subset S(y,-)) \supset (Sxy . Syx) \vee (x=y)$.

From the axioms given to date there follow theorems with the same phrasing as our earlier theorems T17 through T22, T24, and T25; we do not repeat these theorems here.

Again, our present D5 for '*Sim*', D6 for '*Sp*', and D7 for '*Effreg*' run like D5, D6 and D7 of our first form (**48d**), and are not repeated here.

From this point our present system continues in a fashion analogous to the previous one, though in some respects it is markedly simpler. Since here coincident points are identical, we need not distinguish between world-points and space-points. Further, neighborhoods can be defined directly as the effected region themselves:

D8. C. $Nbd(F,G) \equiv (\exists x)Effreg(F,x,G)$.

Additional axioms A13 through A15 are to be constructed in analogy with axioms A18 through A20 of the first form. Thereupon there follows a theorem with the same wording as our earlier T40.

Axiom A17 stipulates that each space has the homogeneous dimension number 3; the formulation of this axiom is somewhat simpler than that of the corresponding axiom (A21) in the first system.

A17. C. $Sp(G) . Hausd(Nbd(-,G)) \supset Dimhom(3,G,Nbd(-,G))$.

And finally, T41 here is analogous to T41 in the first form:

T41. C. $Sp(G) \supset Dimhom(3,G,Nbd(-,G))$.

50. ASs OF SPACE-TIME TOPOLOGY: 3. THE S-SYSTEM

We now turn to the *third form*, called the S-system. The single *primitive sign* of this system is 'S', standing for the signal relation. Here, as in the second form (**49**), we regard coincident points as identical. However, the concepts of genidentity and of world-line do not appear in the present system. From a certain point of view, this omission is an advantageous feature of the third form because the use e.g. of the concept of genidentity is questionable in some cases—notably, in the matter-free electromagnetic field and for particles in quantum theory. (The formulations that follow are given in language C alone.)

The first axioms say that S is transitive, irreflexive, dense, and devoid of initial and of terminal members. Subsequent axioms are analogous to some of the first form; however, a smaller number of axioms suffices for this system. We shall not state the axioms here, but give only the definitions.

Definition D1 for '*Sim*' reads like D5 of the first form (**48d**).

The present form poses a difficulty in connection with the definition of '*Sp*' ("space"). In order that each space be sufficiently comprehensive, our earlier definition (D6 in **48d**) required a space to have a point in common with every world-line. Our difficulty here stems from the fact that the concept of world-line does not appear in the system. However, we can avoid this difficulty and reach the same goal with the help of the concept of signal line (*Sln*). A signal line is a series which is contained in S and which—this being the essential thing so far as the definition of "space" is concerned—is as comprehensive as possible, i.e. neither the ends nor the middle lack a piece. Our definition of '*Sln*' (D2) exhibits this requirement in the form of a condition that a signal line not be extensible, i.e. not be a proper subrelation of a relation which itself is a series and is contained in S.

D2. C. $Sln(H_1) \equiv Ser(H_1).(H_1 \subset S).(H_2)[Ser(H_2).(H_2 \subset S).(H_1 \subset H_2) \supset (H_1 = H_2)]$.

D3. C. $Sp(G) \equiv (x)(y)[Gx.Gy \supset Sim(x,y)].(H)[Sln(H) \supset \exists(G.mem(H))]$

Our definition of the effected region is analogous to that given for the first form (D7 in **48d**), but simpler:

D4. C. $Effreg(F,x,G) \equiv Sp(G).(F = G.S(x,-)).\exists(F)$.

The present definition of '*Nbd*' (D5) reads like that of the second form (D8 in **49**).

The axiom relating to three-dimensionality runs here exactly as it did in the second system (A17 in **49**). From this axiom follows the theorem about the homogeneous three-dimensionality of each space; this theorem has exactly the same phrasing as T41 in **49**:

T1. C. $Sp(G) \supset Dimhom(3, G, Nbd(-,G))$.

If, with the help of the definitions so far given, we eliminate from T1 all the defined axiomatic signs and simplify the result slightly, we obtain theorem T2 below. Besides logical constants and variables, this theorem contains only 'S' as the single axiomatic sign; hence the theorem expresses the three-dimensionality of spaces as a property of S:

T2. C. $(V_2 \; in \; G \subset \sim S).(H_1)[Ser(H_1).(H_1 \subset S).(H_2)[Ser(H_2).(H_2 \subset S).(H_1 \subset H_2) \supset (H_1 = H_2)] \supset \exists(G.mem(H_1))] \supset Dimhom(3,G,(\lambda F)[(\exists x)(F = G.S(x,-)).\exists(F)])$.

Further, every other topological property of space order can similarly be expressed as a property of the signal relation. In a certain sense, therefore, it is possible to say that space order is the order among simultaneous points determined by the signal relation.

51. DETERMINATION AND CAUSALITY

51a. The general concept of determination. (Formulation A may be read after **19**, formulation C after **33**.) There are two *primitive signs*: '*Magn*' and '*Pos*'. The sentence '$Magn(f)$' says "f is a state magnitude"; this means that f is a function and that to each position of the domain in question f associates either a quantity (recall **41a**), say a real number or an n-tuple of real numbers, or else a quality. The sentence '$Pos(H)$' says "H is a two-place positional relation between positions"; the positional relations determine the order of the positions, but not their nature.

We take positions as individuals (or as individuals of the first sort, in case the values of the state magnitudes—e.g. real numbers—are taken as individuals of the second sort in a two-sorted language). Individual variables 'x', etc., thus refer to positions.

The relation H is called a *positional correlator* between classes F and G, and we write '$PosCorr(H,F,G)$', provided: If K_1 is any positional relation, and K_2 and K_3 are the subrelations of K_1 for the elements of F and of G respectively, then H is a correlator between K_2 and K_3.

D1. A. $PosCorr(H,F,G) \equiv (K_1)(K_2)(K_3)[Pos(K_1).(x)(y)(K_2xy \equiv K_1xy.Fx.Fy).(x)(y)(K_3xy \equiv K_1xy.Gx.Gy) \supset Corr_2(H,K_2,K_3)]$.

C. $PosCorr(H,F,G) \equiv (K)[Pos(K) \supset Corr_2(H,K \text{ in } F,K \text{ in } G)]$.

A positional correlator between F and G is a *magnitude correlator* between F and G with respect to the class N of state magnitudes (we write: '$MagnCorr(H,F,G,N)$') provided each state magnitude of class N has at each position of F the same value that it does at the position of G corresponding thereto under H.

D2. $MagnCorr(H,F,G,N) \equiv PosCorr(H,F,G).(f)(x)(y)[N(f).Hxy.Fx.Gy \supset Magn(f).(f(x)=f(y))]$.

The class F of positions is called a *determining class* of position x with respect to the class N of state magnitudes ('$Det(F,x,N)$') provided it is the case that the values of the state magnitudes of N at x are determined by their values at the positions of F (more precisely: if, on the basis of a positional correlator H, a position y has the same positional relations to a class G of positions as position x does to class F, and if H is also a state magnitude correlator between F and G with respect to N, then the state magnitudes of N at y have the same values that they do at x):

D3. A. $Det(F,x,N) \equiv (f)(N(f) \supset Magn(f)).(F_2)(G_1)(G_2)(y)(H)(f)[(u)(F_2u \equiv Fu \vee (u=x)).(u)(G_2u \equiv G_1u \vee (u=y)).PosCorr(H,F_2,G_2).Hxy.MagnCorr(H,F,G,N).N(f) \supset (f(x)=f(y))]$.

C. $Det(F,x,N) \equiv (N \subset Magn).(G)(y)(H)(f)[PosCorr(H,F \vee \{x\},G \vee \{y\}).Hxy.MagnCorr(H,F,G,N).N(f) \supset (f(x)=f(y))]$.

51b. The principle of causality. (What follows is phrased in language C only; it may be read after **37**.) With the help of present concepts, and some earlier ones from **48–50**, we can now formulate various versions of the principle of causality. We assume at the outset the following interpretations of present concepts: individuals (positions) are space-time points, i.e. we employ language form III explained in **39d**; *Pos* is the class of geometric relations between space-time points (e.g. distance of 3 cm.); and *Magn* is the class of physical magnitudes (e.g. temperature).

Version 1. "There is a non-empty finite class N of state magnitudes such that the state at every space-time point x with respect to N is determined by the state with respect to N at a class F of space-time points not including x":

CP_1. C. $(\exists N)(x)(\exists F)[\exists(N) . ClsInduct(N) . \sim Fx . Det(F,x,N)]$.

Version 2. Suppose some physical state magntiudes are specified, and M is defined as the class of these specified magnitudes. The causality principle with respect to M runs as follows: "The state at every space-time point x with respect to M is determined by the state respecting M at a class F which does not include x":

CP_2. $(x)(\exists F)[\sim Fx . Det(F,x,M)]$.

Version 3. Appealing to the signal relation (**48c**), we can express the temporal relation between a point x and a determining class F, whether in respect to an unspecified finite class N of state magnitudes or in respect to a class M defined by enumeration. We choose the second route (as in CP_2), for the sake of simplicity. "The state at any space-time point x with respect to M is determined by the state respecting M at a class F of points which temporally precede x, i.e. which belong to the prior cone $S(-,x)$":

CP_3. C. $(x)(\exists F)[(F \subset S(-,x)) . Det(F,x,M))$.

Version 4. A stronger assertion is the following one. "The state at x with respect to M is determined by the state respecting M at an arbitrary spatial cross-section F through the prior cone of x" (regarding '*Sp*', see **48d**):

CP_4. C. $(x)(F)(G)[Sp(G) . (F = G . S(-,x)) . \exists(F) \supset Det(F,x,M)]$.

A similar assertion of still greater strength makes the same claim for any spatial cross-section through the prior cone or through the posterior cone; i.e. in this case—the case of classical physics—determinism is assumed in both directions. To formulate this assertion, we simply replace '$S(-,x)$' by '$(S(-,x) \vee S(x,-))$' in CP_4.

Chapter H

ASs of biology

52. AS OF THINGS AND THEIR PARTS

52a. Things and their parts. In **52** and **53** there is constructed an AS which is a small portion (slightly modified) of the AS set up by Woodger [Biology] for certain basic concepts of biology, notably of genetics. The present section contains a *preliminary part* concerned with *things in general*, without specialization to biology. This AS can therefore serve as a basis for other fields besides biology. The next section enlarges this AS into an AS with certain primitive concepts of a biological character. (The formulations of **52a** and **52b** given in language A can be read after **17**; those given in language C, after **35**.)

The present AS treats part-relations and time-relations between space-time regions. These regions are taken as individuals, i.e. we employ language form I explained in **39b**. The *primitive signs* of this AS are: 'P', 'Tr', 'Th'. (Woodger uses 'P', 'T',— instead.) Interpretations of the first two agree with those given in **39a**: 'Pxy' is read "x is a (spatial, or temporal, or spatio-temporal) part of y", and '$Tr(x,y)$' is read "x is temporally earlier than y—more exactly: every part of x is temporally earlier than every part of y". Our interpretation of the third primitive sign runs: '$Th(x)$' means "x is a thing".

Relation P is transitive:

A1. **A.** $Pxy.Pyz \supset Pxz.$
C. $Trans(P).$

We say that x is the *sum* of the class F, and write '$Su(x,F)$', provided the elements of F are parts of x and for each part y of x there is an element z of F such that y and z have at least one part in common:

D1. **A.** $Su(x,F) \equiv (u)(Fu \supset Pux).(y)[Pyx \supset (\exists z)(\exists w)(Fz.Pwy.Pwz)].$
C. $Su(x,F) \equiv (F \subset P(-,x)).(y)[Pyx \supset (\exists z)(Fz.(P^{-1}|P)yz)].$

Every non-empty class has exactly one sum:

A2. **A.** $(\exists u)(Fu) \supset (\exists x)(y)(Su(y,F) \equiv (y=x)).$
C. $\exists(F) \supset 1(Su(-,F)).$

[Axiom A2 shows that 'Su' is designed so that any description of the form 'Su‘Q' (see D35-2), for Q a non-empty class, satisfies the uniqueness condition. Instead of the two-place predicate '*Su*', therefore, we could just as well take a one-place functor '*su*' as a primitive sign (recall **18b**); in this case we would have to take as the (improper) sum of the empty class ($su(\Lambda)$) some fixed region, e.g. the empty region.]

Of the theorems which follow from A1 and A2 we give two. The first says that relation P is totally reflexive:

T1. **A.** Pxx.
C. $Reflex(P)$.

The second theorem runs as follows: If x and y are parts of each other, then they are identical (i.e. between two different individuals the relation P holds in at most one direction):

T2. **A.** $Pxy.Pyx \supset (x=y)$.
C. $(P.P^{-1}) \subset I$.

The time relation Tr is asymmetric:

A3. **A.** $Tr(x,y) \supset \sim Tr(y,x)$.
C. $As(Tr)$.

If a (the) sum of F is earlier (Tr) than a (the) sum of G, then F and G are not empty and every element of F is earlier than every element of G; and conversely:

A4. **A.** $(\exists u)(\exists v)[Su(u,F).Su(v,G).Tr(u,v)] \equiv (\exists x)(Fx).(\exists x)(Gx).(x)(y)(Fx.Gy \supset Tr(x,y))$.
C. $Tr(Su\text{‘}F,Su\text{‘}G) \equiv \exists!(F).\exists!(G).(x)(y)(Fx.Gy \supset Tr(x,y))$.

If no part of x is later than y, then every individual later than y is also later than x:

A5. **A.** $(u)(Pux \supset \sim Tr(y,u)) \supset (v)(Tr(y,v) \supset Tr(x,v))$.
C. $(P(-,x) \subset \sim Tr(y,-)) \supset (Tr(y,-) \subset Tr(x,-))$.

If no part of x is earlier than y, then every individual earlier than y is also earlier than x:

A6. **A.** $(u)(Pux \supset \sim Tr(u,y)) \supset (v)(Tr(v,y) \supset Tr(v,x))$.
C. $(P(-,x) \subset \sim Tr(-,y)) \supset (Tr(-,y) \subset Tr(-,x))$.

Theorems. Relation Tr is transitive:

T3. **A.** $Tr(x,y).Tr(y,z) \supset Tr(x,z)$.
C. $Trans(Tr)$.

If x is earlier than y, then x is earlier than every part of y:

T4. **A.** $Tr(x,y).Pzy \supset Tr(x,z)$.
C. $(Tr|P^{-1}) \subset Tr$.

If x is a part of something which is earlier than z, then x itself is earlier than z:

T5. **A.** $Pxy.Tr(y,z) \supset Tr(x,z)$.
C. $(P|Tr) \subset Tr$.

If x is earlier than y, then any part of x is earlier than every part of y:

T6. **A.** $Tr(x,y).Pux.Pvy \supset Tr(u,v)$.
C. $(P|Tr|P^{-1}) \subset Tr$.

If w is earlier than x and x is a part of y and y is earlier than z, then w is earlier than z:

T7. **A.** $Tr(w,x).Pxy.Tr(y,z) \supset Tr(w,z)$.
C. $(Tr|P|Tr) \subset Tr$.

Relations Tr and P are mutually exclusive:

T8. **A.** $Tr(x,y) \supset \sim Pxy$.
C. $Tr \subset \sim P$.

52b. The slices of things. A space-time region x is said to be *momentary* provided no part of x is earlier than any other part of x:

D2. **A.** $Mom(x) \equiv (u)(v)(Pux.Pvx \supset \sim Tr(u,v))$.
C. $Mom(x) \equiv \sim\exists(Tr \text{ in } P(-,x))$.

Every individual has momentary parts:

A7. **A.** $(x)(\exists y)(Pyx.Mom(y))$.
C. $\exists(P(-,x).Mom)$.

As in **39a**, so here '$Sli(x,y)$' means "x is a slice of the thing y". This relation holds between x and y provided y is a thing and x is a maximal momentary part of y (i.e. x is a momentary part of y and there is no momentary part of y of which x is a proper part):

D3. $Sli(x,y) \equiv Th(y).Mom(x).Pxy.\sim(\exists z)(Mom(z).Pzy.Pxz.(x \neq y))$.

Theorems. Two different slices of a thing have no parts in common:

T9. **A.** $Sli(x,z).Sli(y,z).(x \neq y) \supset \sim(\exists u)(Pux.Puy)$.
C. $(J \text{ in } Sli(-,z)) \subset \sim(P^{-1}|P)$.

Of two different slices of a thing, one is earlier than the other:

T10. **A.** $Sli(x,z).Sli(y,z).(x \neq y) \supset Tr(x,y) \vee Tr(y,x)$.
C. $Connex(Tr \text{ in } Sli(-,z))$.

A slice x of y which is earlier than all other slices of y we term an *initial slice* of y, and write '$ISli(x,y)$' (D4). A slice of x of y which is later than all other slices of y we term an *end slice* of y, and write '$ESli(x,y)$' (D5).

D4. **A.** $ISli(x,y) \equiv Sli(x,y).(z)[Sli(z,y).(z \neq x) \supset Tr(x,z)]$.
C. $ISli(x,y) \equiv Sli(x,y).(Sli(-,y).\sim\{x\} \subset Tr(x,-))$.

D5. **A.** $ESli(x,y) \equiv Sli(x,y).(z)[Sli(z,y).(z \neq x) \supset Tr(z,x)]$.
C. $ESli(x,y) \equiv Sli(x,y).(Sli(-,y).\sim\{x\} \subset Tr(-,x))$.

Axioms. Every thing has at least one initial slice (A8) and at least one end slice (A9):

A8. **A.** $Th(x) \supset (\exists y)ISli(y,x)$.
C. $Th \subset mem_2(ISli)$.

A9. **A.** $Th(x) \supset (\exists y)ESli(y,x)$.
C. $Th \subset mem_2(ESli)$.

Theorems. Every thing has exactly one initial slice (T11; from A8 and T10) and exactly one end slice (T12; from A9 and T10):

T11. **A.** $Th(x) \supset (\exists y)(z)(ISli(z,x) \equiv (z=y))$.
C. $Th(x) \supset 1(ISli(-,x))$.

T12. **A.** $Th(x) \supset (\exists y)(z)(ESli(z,x) \equiv (z=y))$.
C. $Th(x) \supset 1(ESli(-,x))$.

Every thing has at least one slice (by A8):

T13. **A.** $Th(x) \supset (\exists y)Sli(y,x)$.
C. $Th \subset mem_2(Sli)$.

If y is a momentary part of a thing x, then x has exactly one slice z of which y is a part:

T14. **A.** $Th(x).Pyx.Mom(y) \supset (\exists z)(u)[Sli(u,x).Pyu \equiv (u=z)]$.
C. $Th(x).Pyx.Mom(y) \supset 1(Sli(-,x).P(y,-))$.

Every thing is identical with the sum of its slices:

T15. **A.** $Th(x).(y)(Fy \equiv Sli(y,x)) \supset (z)(Su(z,F) \equiv (z=x))$.
C. $Th(x) \supset (x=Su\text{`}Sli(-,x))$.

52c. The time relation. The following is phrased only in language C, and may be read after **38**.

Theorem. Respecting the slices of a thing, the time relation Tr is a series (from A3, T3, and T10):

T16. **C.** $Ser(Tr\ in\ Sli(-,z))$.

Axioms. Between two different slices of a thing there is always a third slice:

A10. **C.** $(Tr\ in\ Sli(-,z)) \subset (Tr\ in\ Sli(-,z))^2$.

Respecting the slices of a thing, the time relation Tr is a Dedekind relation:

A11. **C.** $Ded(Tr\ in\ Sli(-,x))$.

Theorem. Respecting the slices of a thing, the time relation Tr is a series with Dedekind continuity (from T16, A10 and A11):

T17. **C.** $DedSer(Tr\ in\ Sli(-,x))$.

53. AS INVOLVING BIOLOGICAL CONCEPTS

53a. Division and fusion. Following Woodger [Biology], the AS described in **52** above will now be broadened into a biological AS by the addition of several new primitive signs and axioms. What we give here is only the first part of Woodger's system. Our formulation A in **53a** can be read after **19**, formulation C after **35**.

Additional primitive signs here are: '*Org*', '*Y*', '*Cell*' and '*Orgs*'. Explanations of them run as follows: '$Org(x)$' means "x is an organic unit" (examples of an organic unit are an organism, an organ, a cell); 'Yxy' means "The organic unit x is transformed into the organic unit y" [i.e. x divides into several parts of which one is y (e.g. cell division), or x fuses with one or more other units to produce y (e.g. cell fusion)]; '$Cell(x)$' means "x is a cell"; '$Orgs(x)$' means "x is an organism". A cell is here conceived as a thing, i.e. as temporally extended, in distinction to the slices of cells ($Sli``Cell$); and the same for an organism. The duration of an organic unit —and thus, in particular, of a cell or an organism—is counted from the instant of its production (e.g. by division or fusion) to the instant of its end (e.g. through the instant of its division, or of its fusion with other units of the same kind).

Axioms. Each organic unit is a thing:

A12. A. $Org(x) \supset Th(x)$.
C. $Org \subset Th$.

The members of Y are organic units:

A13. A. $Yxy \supset Org(x) . Org(y)$.
C. $mem(Y) \subset Org$.

Suppose that Yxy, that u is an (the) end slice of x, and that v is an (the) initial slice of y; then u and v are different, and either u is part of v or v is part of u:

A14. A. $Yxy . ESli(u,x) . ISli(v,y) \supset (u \neq v) . (Puv \vee Pvu)$.
C. $(ESli | Y | ISli^{-1}) \subset (P \vee P^{-1}) . J$.

Now we define division ('*Div*') and fusion ('*Fs*'). We say: x is transformed by division into y ('$Dv(x,y)$') provided Yxy and an (the) initial slice of y is part of an (the) end slice of x (D6). Again, we say: x is transformed by fusion into y ('$Fs(x,y)$') provided Yxy and an (the) end slice of x is part of an (the) initial slice of y (D7).

D6. A. $Dv(x,y) \equiv Yxy . (\exists u)(\exists v)[ESli(u,x) . ISli(v,y) . Pvu]$.
C. $Dv = Y . (ESli^{-1} | P^{-1} | ISli)$.

D7. A. $Fs(x,y) \equiv Yxy . (\exists u)(\exists v)[ESli(u,x) . ISli(v,y) . Puv]$.
C. $Fs = \big(Y . (ESli^{-1} | P | ISli)\big)$.

The axioms which follow are formulated more simply with the help of these definitions.

If x is transformed by division into y, then x is the only element which bears the relation Y to y:

A15. **C.** $Dv(x,y) \supset (u)(Yuy \equiv (u=x))$.
C. $Dv(x,y) \supset (x = Y`y)$.

If x is transformed by division into y, then there is a z different from y such that x is transformed by division into z:

A16. **A.** $Dv(x,y) \supset (\exists z)[(z \neq y) . Dv(x,z)]$.
C. $Dv \subset (Dv|J)$.

If x is transformed by fusion into y, then y is the only element to which x bears the relation Y:

A17. **A.** $Fs(x,y) \supset (u)(Yxu \equiv (u=y))$.
C. $Fs(x,y) \supset (y = Y^{-1}`x)$.

If x is transformed by fusion into y, then there is a z different from x which is transformed by fusion into y:

A18. **A.** $Fs(x,y) \supset (\exists z)[(z \neq x) . Fs(z,y)]$.
C. $Fs \subset (J|Fs)$.

Theorems. Relation Y is the union of relations Dv and Fs:

T18. **A.** $Yxy \equiv Dv(x,y) \vee Fs(x,y)$.
C. $Y = Dv \vee Fs$.

Relation Y is irreflexive (T19), intransitive (T20), and asymmetric (T21):

T19. **A.** $\sim Yxx$.
C. $Irr(Y)$.

T20. **A.** $Yxy . Yyz \supset \sim Yxz$.
C. $Intr(Y)$.

T21. **A.** $Yxy \supset \sim Yyx$.
C. $As(Y)$.

Relation Dv is one-many (T22) and asymmetric (T23):

T22. $Un_1(Dv)$.

T23. **A.** $Dv(x,y) \supset \sim Dv(y,x)$.
C. $As(Dv)$.

Relation Fs is many-one (T24) and asymmetric (T25):

T24. $Un_2(Fs)$.

T25. **A.** $Fs(x,y) \supset \sim Fs(y,x)$.
C. $As(Fs)$.

Relations Dv and Fs have no first members in common (i.e. no individual is transformed both by division and by fusion) (T26), and no second members in common (i.e. no individual is produced both by division and by fusion) (T27):

T26. **A.** $\sim(\exists x)(\exists y)(\exists z)[Dv(x,y).Fs(x,z)]$.
C. $\sim\exists(mem_1(Dv).mem_1(Fs))$.

T27. **A.** $\sim(\exists x)(\exists y)(\exists z)[Dv(x,z).Fs(y,z)]$.
C. $\sim\exists(mem_2(Dv).mem_2(Fs))$.

53b. Hierarchies, cells, organisms. (Formulations given here in language A—these occur only in D11 and in the axioms—can be read after **19**; those given in language C, after **36**.) We turn now to the *logical* concept of hierarchy, a concept especially useful in biology. A relation H is called a *hierarchy* ('$Hier(H)$') provided the following three conditions obtain: H is asymmetric and one-many; H has exactly *one* initial member; and every member is only finitely many H-steps removed from this initial member. The concept of hierarchy is related to that of progression (**37a**); the difference is that a progression is also many-one (hence one-one) and has no terminal member, whereas a hierarchy permits bifurcation in the direction away from the initial member and allows the occurrence of terminal members.

D8. **C.** $Hier(H) \equiv As(H).Un_1(H).1(init(H)).(x)(y)[init(H)x.mem_2(H)y \supset H^{>0}(x,y)]$.

If x is a first-place member of Dv, then the relation Dv with respect to the Dv-posterity of x (recall **36c**) is a hierarchy:

T28. **C.** $mem_1(Dv)x \supset Hier(Dv\ in\ Dv^{\geq 0}(x,-))$.

Such a hierarchy is called a "Dv-hierarchy":

D9. **C.** $DvHier(H) \equiv (\exists x)[mem_1(Dv)x.(H = Dv\ in\ Dv^{\geq 0}(x,-))]$.

A subrelation H of Y is called *dendritic*, symbolically '$Dend(H)$', provided H is formed by selecting some Y-member x and by limiting the field of Y to those elements that can be reached from x by a finite chain composed arbitrarily of Y- and Y^{-1}-steps:

D10. **C.** $Dend(H) \equiv (\exists x)[mem(Y)x.(H = Y\ in\ [(Y \vee Y^{-1})^{\geq 0}(x,-)])]$.

If two dendritic relations have a member in common, then they are identical:

T29. **C.** $Dend(H).Dend(K).\exists(mem(H).mem(K)) \supset (H = K)$.

We say x is an organic part of y, and write '$OP(x,y)$', provided: x and y are different organic units; more than one slice of x is a part of y; and if u is a slice of x and v a slice of y such that u is neither earlier nor later than v, then u is a part of v.

D11. $OP(x,y) \equiv Org(x).Org(y).(x \neq y).(\exists w)(\exists z)((w \neq z).Sli(w,x).Sli(z,x).Pwy.Pzy).(u)(v)[Sli(u,x).Sli(v,y).\sim Tr(u,v).\sim Tr(v,u) \supset Puv]$.

If an organic unit is a part of another organic unit, then the first is an organic part of the second:

T30. **C.** $[(P.J)\ in\ Org] \subset OP$.

Below are several axioms involving '*Cell*' (cell) and '*Orgs*' (organism). The first (A19) is to the effect that for every cell y there is a cell x such that Yxy (i.e. y results from x by division or fusion):

A19. **A.** $Cell(y) \supset (\exists x)(Cell(x) . Yxy)$.
C. $Cell \subset mem_2(Y\ in\ Cell)$.

Every organism has a cell as a (proper or improper) part:

A20. **A.** $Orgs(x) \supset (\exists y)(Cell(y) . Pyx)$.
C. $Orgs(x) \supset \exists(Cell . P(-,x))$.

Every cell is an organism or an organic part of an organism:

A21. **A.** $Cell(x) \supset Orgs(x) \vee (\exists y)(Orgs(y) . OP(x,y))$.
C. $Cell \subset (Orgs \vee OP``Orgs)$.

If x is an organism whose initial slice is an initial slice of a cell that has resulted from fusion (i.e. if x begins with a zygote), then x has not resulted from division:

A22. **A.** $Orgs(x) . (\exists y)(\exists z)(\exists u)[ISli(y,x) . Cell(z) . ISli(y,z) . Fs(u,z)] \supset \sim(\exists v)(Dv(v,x)$.
C. $Orgs(x) . (\exists z)[(Cell . mem_2(Fs))z . (ISli`x = ISli`z)] \supset \sim mem_2(Dv)x$.

Organisms are organic units:

A23. **A.** $Orgs(x) \supset Org(x)$.
C. $Orgs \subset Org$.

It now follows (from A21, A23, and D11) that cells are organic units:

T31. **C.** $Cell \subset Org$.

54. AS FOR KINSHIP RELATIONS

54a. Biological concepts of kinship. The AS here presented treats the relations of kinship between humans. The treatment in **54a** considers biological concepts of kinship, that in **54b** deals with legal concepts of the same. Things, humans in particular, are taken as individuals; thus use is made of language form IA explained in **39b**. It is a consequence of this choice that temporal relationships cannot be expressed. (For ASs in which concepts of kinship are further analyzed and time relations are also examined, see **55d**—problems **25, 26, 27.**) The sense intended for the biological concepts introduced below may be more readily grasped if it is understood that we say x is father of y provided x has engendered y; that x is mother of y provided x has borne y; that x is husband of y provided x

has engendered a child by y; etc. [Insofar as **54a** is given in formulation A, it may be read after **17**; in formulation C, after **36**.]

Primitive signs: Signs 'Par' and 'Ml' may be thought to designate respectively the relation Parent and the class of male humans. For definitions of 'Hu' (human), 'Fl' (female), 'Fa' (father), 'Ch' (child), 'Son', '$GrPar$' (grandparent) in language A, see **15c**; for that of 'Bro' (brother), see **17b**. Proceeding similarly, it is an easy matter to define 'Dau' (daughter), '$GrFa$' (grandfather), '$GrMo$' (grandmother), 'Sis' (sister), 'Sib' (sibling); and also grandchild, grandson, grand-daughter, etc. For several of these concepts, and for some additional ones, definitions in language C can be found in **30c**.

We begin with definitions of 'Mo' (mother), 'Anc' (ancestor), 'Des' (descendant), 'Hus' (husband, in the biological sense explained just above) and 'Wif' (wife, in a similar biological sense). [Our definitions of 'Anc' and 'Des' appear only in formulation C; cf. **36b**.]

D1. $Mo(x,y) \equiv Par(x,y) . Fl(x)$.

D2. **C.** $Anc = Par^{>0}$.

D3. **C.** $Des = Ch^{>0}$.

D4. **A.** $Hus(x,y) \equiv (\exists z)(Fa(x,z) . Mo(y,z))$.

C. $Hus = Fa | Mo^{-1}$.

D5. **A.** $Wif(x,y) \equiv Hus(y,x)$.

C. $Wif = Hus^{-1}$.

Several theorems follow at once from these definitions, even before axioms are laid down; such theorems are therefore provable in the basic language (recall **42a**), and hence are L-true.

Every human is male or female; and conversely, every male or female human is a human.

T1. **A.** $Hu(x) \equiv Ml(x) \vee Fl(x)$.

C. $Hu = Ml \vee Fl$.

A parent of someone is either his father or his mother, and conversely:

T2. **A.** $Par(x,y) \equiv Fa(x,y) \vee Mo(x,y)$.

C. $Par = Fa \vee Mo$.

The classes Ml and Fl are mutually exclusive (T3), hence so also are the relations Fa and Mo (T4):

T3. **A.** $\sim(\exists x)(Ml(x) . Fl(x))$.

C. $\sim\exists(Ml . Fl)$.

T4. **A.** $\sim(\exists x)(\exists y)(Fa(x,y) . Mo(x,y))$.

C. $\sim\exists(Fa . Mo)$.

The relation Hus is asymmetric. (The same holds for the relation Wif; consequently, both Hus and Wif are irreflexive.)

T5. **A.** $Hus(x,y) \supset \sim Hus(y,x)$.

C. $As(Hus)$.

Axioms. Relation *Fa* is one-many, i.e. everyone has at most one father (A1). Similarly, *Mo* is one-many (A2). And again, *Anc* is irreflexive, i.e. no one is his own ancestor (A3).

A1. **A.** $Fa(x,z) . Fa(y,z) \supset (x=y)$.
C. $Un_1(Fa)$.

A2. **A.** $Mo(x,z) . Mo(y,z) \supset (x=y)$.
C. $Un_1(Mo)$.

A3. **A.** $\sim Anc(x,x)$.
C. $Irr(Anc)$.

Theorems. From A1 and A2 it follows that everyone has at most two parents (T7), and that if someone has two parents, they are his father and his mother (T8):

T7. **C.** $\sim 3_m(Par(-,x))$.

T8. **C.** $2(Par(-,x)) \supset (\exists u)(\exists v)[(Par(-,x)=\{u,v\}) . (u=Fa\text{‘}x) . (v=Mo\text{‘}x)]$.

From A3 it follows that these relations are irreflexive and asymmetric: Ancestor, Parent, Father, Mother, Descendent, Child, Son, Daughter, and further all powers of these relations (viz. Grandparent, Great-grandparent, Grandfather, etc.):

T9. **C.** $\{Anc,Par,Fa,Mo,Des,Ch,Son,Dau,Par^2,Par^3,...\} \subset (Irr . As)$.

54b. Legal concepts of kinship. Here we extend the system of **54a** by adding to it legal concepts.

Additional primitive signs: '*LPar*' and '*LHus*'. We read '$LPar(x,y)$' as "x is a legal parent of y" (i.e., the parenthood, whether natural or by adoption, is legally recognized); and '$LHus(x,y)$' as "x is a legal husband of y" (i.e. the male x at some time in his life legally married the female y). [With the exception of D41 and D42, **54b** in formulation A can be read after **17**; **54b** in formulation C can be read after **36**.]

We begin with definitions of additional legal concepts: '*LFa*' (legal father), '*LCh*' (legal child), '*LSon*' (legal son), '*LWif*' (legal wife), '*LSp*' (legal spouse), '*EPar*' (x is a legitimate parent of y, i.e. both x and a legal spouse of x are legal parents of y), '*EFa*' (legitimate father), '*ECh*' (legitimate child), '*ESon*' (legitimate son), '*ESib*' (legitimate sibling), '*EBro*' (legitimate brother), '*InPar*' (parent-in-law), '*InFa*' (father-in-law), '*InCh*' (son-in-law or daughter-in-law), '*InSon*' (son-in-law), '*InSib*' (brother-in-law or sister-in-law), '*InBro*' (brother-in-law), '*StPar*' (step-parent), '*StFa*' (stepfather), '*StCh*' (stepchild), '*StSon*' (stepson), '*HSib*' (half sibling, i.e., half brother or half sister), '*HBro*' (half brother), '*StSib*' (step-brother or step-sister), '*StBro* (step-brother), '*UnAn*' (uncle or aunt), '*Un*' (uncle), '*NeNi*' (nephew or niece), '*Ne*' (nephew), '*Co*' (male or female cousin), '*MlCo*' (male cousin), '*EGrPar*' (legitimate grandparent), '*EGrCh*' (legitimate grandchild), '*EGrSon*' (legitimate grandson). [Corresponding relations of female persons ('LMo' (legal mother), etc.) are readily defined in analogy with D6, 8, 12, 14, 16, 18, 20, 22, 24, 26, 28, 30, 32, 34, 36, 38, and 40 by replacing 'Ml' by 'Fl' in the definiens.]

D6. $LFa(x,y) \equiv LPar(x,y)Ml(x)$.

D7. **A.** $LCh(x,y) \equiv LPar(y,x)$.
C. $LCh = LPar^{-1}$.

D8. $LSon(x,y) \equiv LCh(x,y) . Ml(x)$.

D9. $LWif(x,y) \equiv LHus(y,x)$.

D10. **A.** $LSp(x,y) \equiv LHus(x,y) \vee LWif(x,y)$.
C. $LSp = LHus \vee LWif$.

D11. **A.** $EPar(x,y) \equiv LPar(x,y) . (\exists z)(LSp(x,z) . LPar(z,y))$.
C. $EPar = LPar . (LSp | LPar)$.

D12. $EFa(x,y) \equiv EPar(x,y) . Ml(x)$.

D13. $ECh(x,y) \equiv EPar(y,x)$.

D14. $ESon(x,y) \equiv ECh(x,y) . Ml(x)$.

D15. **A.** $ESib(x,y) \equiv (\exists u)(\exists v)(ECh(x,u) . EFa(u,y) . ECh(x,v) . EMo(v,y) . (x \neq y))$.
C. $ESib = (ECh | EFa) . (ECh | EMo) . J$.

D16. $EBro(x,y) \equiv ESib(x,y) . Ml(x)$.

D17. **A.** $InPar(x,y) \equiv (\exists z)(EPar(x,z) . LSp(z,y))$.
C. $InPar = EPar | LSp$.

D18. $InFa(x,y) \equiv InPar(x,y) . Ml(x)$.

D19. $InCh(x,y) \equiv InPar(y,x)$.

D20. $InSon(x,y) \equiv InCh(x,y) . Ml(x)$.

D21. **A.** $InSib(x,y) \equiv (\exists z)[(ESib(x,z) . LSp(z,y)) \vee (LSp(x,z) . ESib(z,y))]$.
C. $InSib = (ESib | LSp) \vee (LSp | ESib)$.

D22. $InBro(x,y) \equiv InSib(x,y) . Ml(x)$.

D23. **A.** $StPar(x,y) \equiv (\exists z)(LSp(x,z) . LPar(z,y) . \sim LPar(x,y))$.
C. $StPar = (LSp | LPar) . \sim LPar$.

D24. $StFa(x,y) \equiv StPar(x,y) . Ml(x)$.

D25. $StCh(x,y) \equiv StPar(y,x)$.

D26. $StSon(x,y) \equiv StCh(x,y) . Ml(x)$.

D27. **A.** $HSib(x,y) \equiv (\exists z)(LCh(x,z) . LPar(z,y)) . (x \neq y) . \sim ESib(x,y)$.
C. $HSib = (LCh | LPar) . J . \sim ESib$.

D28. $HBro(x,y) \equiv HSib(x,y) . Ml(x)$.

D29. **A.** $StSib(x,y) \equiv (\exists z)[LPar(z,x) . StPar(z,y)]$.
C. $StSib = LCh | StPar$.

D30. $StBro(x,y) \equiv StSib(x,y) . Ml(x)$.

D31. **A.** $UnAn(x,y) \equiv (\exists z)[(ESib(x,z) \vee InSib(x,z)) . EPar(z,y)]$.
C. $UnAn = (ESib \vee InSib) | EPar$.

D32. $Un(x,y) \equiv UnAn(x,y) . Ml(x)$.

D33. $NeNi(x,y) \equiv UnAn(y,x)$.

D34. $Ne(x,y) \equiv NeNi(x,y) . Ml(x)$.

D35. **A.** $Co(x,y) \equiv (\exists u)(\exists v)(ECh(x,u) . ESib(u,v) . EPar(v,y))$.
C. $Co = ECh|ESib|EPar$.

D36. $MlCo(x,y) \equiv Co(x,y) . Ml(x)$.

D37. **A.** $EGrPar(x,y) \equiv (\exists z)(EPar(x,z) . EPar(z,y))$.
C. $EGrPar = EPar^2$.

D38. $EGrFa(x,y) \equiv EGrPar(x,y) . Ml(x)$.

D39. $EGrCh(x,y) \equiv EGrPar(y,x)$.

D40. $EGrSon(x,y) \equiv EGrCh(x,y) . Ml(x)$.

The definitions for '*EAnc*' (legitimate ancestor) and '*EDes*' (legitimate descendent) we give only in formulation C; these definitions are analogous to D2 and D3.

D41. **C.** $EAnc = EPar^{>0}$.

D42. **C.** $EDes = EAnc^{-1}$.

As in **54a** so here many theorems follow from the definitions alone, without the intervention of axioms; however, we shall not introduce any of them at this point.

Axioms. At first glance one might think that some of these legal concepts might be regulated by axioms analogous to those laid down for their counterpart biological concepts (A1 through A3 in **54a**). Such is not the case, however. The relations '*LFa*', '*LMo*', '*EFa*', and '*EMo*' are not one-many, for in the course of time these relations can be dissolved and replaced by relations to other persons. Further, the relation '*LAnc*' is not absolutely irreflexive: while it is highly unlikely that at a certain moment a man could be his own legal grandfather, it is not impossible that between two men *a* and *b* of approximately equal age legal paternity by adoption first goes in one direction and then is dissolved and reinstituted in the opposite direction; in this case '*LGrFa(a,a)*' would hold. [This possibility can be excluded only by laying down special legal conditions governing adoptions, e.g. conditions requiring a minimum difference in age.] And again, there are no simple relations between *Fa* and *LFa*, since each of these relations can occur without the other; the same applies to *Mo* and *LMo*, to *Hus* and *LHus*, etc.

Nevertheless, axioms can be extracted from the usual legal conditions which prohibit legal parenthood and legal marriage in certain cases. The axioms A4 through A10 which follow illustrate this possibility.

In a legal marriage, the husband is male (A4) and the wife female (A5):

A4. **A.** $LHus(x,y) \supset Ml(x)$.
C. $Mem_1(LHus) \subset Ml$.

A5. **A.** $LHus(x,y) \supset Fl(y)$.
C. $mem_2(LHus) \subset Fl$.

It is prohibited that x marry y if x is (in the biological sense) father of y (A6), or son of y (A7), or brother of y (A8):

A6. **A.** $Fa(x,y) \supset \sim LHus(x,y)$.
C. $Fa \subset \sim LHus$.

A7 and A8 are formulated similarly.

Legal parenthood is prohibited in the case of identity (A9), the sibling relation (A10), and certain other kinds of kinship:

A9. **A.** $\sim LPar(x,x)$.
C. $Irr(LPar)$.

A10. **A.** $Sib(x,y) \supset \sim LPar(x,y)$.
C. $Sib \subset \sim LPar$.

Many prohibitions against marriage cannot be expressed in the simple system above because they contain temporal specifications. Among these e.g. are the prohibition against bigamy, against marriage between x and y if x is legal father of y—or legal son of y, or legitimate brother or half-brother of y (all such prohibitions involve the concept of simultaneity); also to be mentioned here is the minimum-age requirement for marriage. The same remark applies to similar limitations on legal parenthood (in cases involving adoption). All such conditions require for their formulation a more complicated language form (cf. Problem 27 in **55d**).

Appendix

55. PROBLEMS IN THE APPLICATION OF SYMBOLIC LOGIC

We take "AS" as abbreviation for "axiom system". The *degree of difficulty* of each Problem is specified at the outset by a notation like "[Diff. I]"; I—quite easy; II—easy; III—moderately difficult; IV—quite difficult. In each problem, the aim is to formulate the indicated AS in symbols, e.g. in one of the languages A and C described in this book. The material for the AS is to be found in the publications referred to.

55a. Set theory and arithmetic

Problem 1. [Diff. IV.] AS of *set theory* according to J. von Neumann (see **43**). Instead of '[*x*,*y*]', take either '*R*'*y*' or '*k*(*y*)'. The primitive sign is again '*E*', as in **43a**.

Problem 2. [Diff. III.] Construction of a language form for *rational numbers* by supplementing a previously given coordinate language (cf. the "first way" of **40d**; see also the references to Russell and Waismann):

a. A language form for positive rational numbers as pairs of natural numbers; on the basis of the language form of **40a, b.**

b. A language form for both positive and negative rational numbers as pairs of integers; on the basis of the language form of **40c**.

Problem 3. [Diff. III.] Continuation of *Problem* 2 to the introduction of *real numbers* as classes of rational numbers (cf. **40d**; Russell and Waismann):

a. To positive real numbers, in continuation of Problem 2a.

b. To both positive and negative real numbers, in continuation of Problem 2b.

Problem 4. [Diff. II.] AS of the real numbers, following Hilbert (see **45**).

Problem 5. [Diff. II.] AS of the theory of magnitudes (cf. also **41**):

a. "Relativistic". Russell [Principles] sec. 154, Couturat [Principes] ch. V, sec. A.

b. "Absolutistic". Russell [Principles] sec. 155, Couturat, ibid.

Problem 6. [Diff. III.] AS of extensive magnitudes. Couturat [Principes] ch. V, sec. B. (Based on Burali-Forti.)

55b. Geometry

Problem 7. [Diff. II.] Definitions of additional concepts in *topology* (point set theory) on the basis of the concept of neighborhood, this in connection with **46b** and following Hausdorff [Grundzüge] 221 ff. or Rosser [Logic] ch. IX, sec. 8, or Bohnenblust (see **46**).

Problem 8. [Diff. III.] AS of *topology* on the basis of the concept of convergent sequence of points, after Hausdorff [Grundzüge] 210, 233 ff. Single primitive sign: '*Lim*'. We read '*Lim*(x,f)' as "point x is limit of the convergent sequence f of points". Thus, the convergent sequences constitute $mem_1(Lim)$. A sequence f of points is a function which coordinates points with the natural numbers. Such a sequence is therefore to be designated by a functor, say 'k', and '$k(n)=x$' means "the nth point of the sequence k is x". Here 'n' is a natural number variable, and 'x' is a point variable; a two-sorted language is used, see **21c**.)

Problem 9. [Diff. III.] AS of *metric geometry* (in point set theory), following Hausdorff [Grundzüge] 211 ff., 290 ff. Single primitive sign: '*dis*'. We read '$dis(x,y)=r$' as "the distance between points x and y is r". Here 'x' and 'y' are point variables, and 'r' is a real number variable; and again, as in Problem 8, a two-sorted language is used.

Problem 10. [Diff. III.] Definitions for concepts of *combinatorial topology* (e.g. in connection with L. Vietoris, "Über den höheren Zusammenhang kompakter Räume", *Math. Ann.*, 97 (1927) 454 ff.; cf. also O. Veblen, *Analysis Situs*, Cambridge Colloquium of Amer. Math. Soc., 1916). Single primitive sign: '*Con*'. We read '$Con(x,y)$' as "the points x and y are connected". Relation *Con* is reflexive and symmetric. By '$Si(F)$' we mean "F is a simplex"; it is defined by '$(V_2 \; in \; F) \subset Con$'. The expression '$Si.5$' designates the class of so-called 5-simplices. By '$Side(F,G)$' we mean "F is a side of G"; it is defined by '$Si(F).Si(G).(F \subset G)$'. The class of complexes is defined by '$sub_1(Si).ClsInduct$'. Also to be defined are: edge of a simplex, edge of a complex, cycle, connexity number, etc.

Problem 11. [Diff. III.] AS of *projective geometry*, with lines as relations (this is based on Russell [Principles] ch. XLV). Single primitive sign: '*Lin*'. The sign '*Lin*' is taken to designate the class of lines; and every line is a relation between points. Thus, if $Lin(R)$—i.e. if R is an element of *Lin*—, then R is a line and 'Rxy' says "x and y are two points on line R". Thereupon the class *Po* of points can be defined by '$sm_1(mem``Lin)$'.

Problem 12. [Diff. II.] AS of *projective geometry* without infinitely distant points, i.e. with open lines (such geometry Russell called "descriptive geometry"). (See O. Veblen, "A system of axioms for geometry", *Trans. Amer. Math. Soc.*, 5 (1904) 343–384. See also the presentation in Couturat [Principes] ch. VI, sec. C.) Single primitive sign: '*Bet*'. We read '$Bet(x,y,z)$' as "the point y lies between the points x and z".

Problem 13. [Diff. II.] AS of *projective geometry* without infinitely distant points. (Following Russell [Principles] ch. XLVI, whose basis was the system given by Vailati, "Sui principii fondamentali della geometria della retta, *Riv. Mat.*, 2 (1892) 71–75; see also Couturat [Principes] ch. VI, sec. C.) Single primitive sign: '*Lin*'. If *R* is an element of *Lin*, *R* is a series which orders the points on a line.

Problem 14. [Diff. III.] AS of *projective geometry* with closed lines, via extension of the system of Problem 12. First, define the class *Bun* comprising all bundles of rays; a bundle of rays is a class of lines all through the same point, or all parallel to each other. The elements of *Bun* (i.e. the bundles) are then taken as the elements of a complete projective geometry. The method is described in Russell [Principles] sec. 384 ff.

Problem 15. [Diff. III.] Extend the system of Problem 13 in the same way that Problem 14 extends the system of Problem 12.

Problem 16. [Diff. III.] AS of *metric geometry* on the basis of the concept of *motion*. (See Pieri, "Della geometria elementare come sistema ipotetico deduttivo, Monografia del punto e del moto", *Mem. Accad. Torino*, 1899, and his "Sur la géometrie envisagée comme un système purement logique", *Bibl. Congrès Int. Philos.*, Paris, 1900, vol. III, 367–404; see also the presentation in Couturat [Principes] ch. VI, sec. D.). Single primitive sign: '*Mot*'; if *R* is an element of *Mot* then *R* is a motion, i.e., a one-one relation between points.

Problem 17. [Diff. III.] AS of *metric geometry* on the basis of the concept of *sub-sphere* (following E. V. Huntington, "A set of postulates for abstract geometry", *Math. Ann.*, 73 (1913) 522–559). Single primitive sign: '*S*'; *S* is viewed as the (transitive, irreflexive) relation between two spheres of which the first lies completely within the second. The class *Sph* of spheres is defined by '*mem*(*S*)'. Three forms of this system are possible:

a. *The spheres as point classes.* The class *Po* of points is defined by '$sm_1(Sph)$'. We say: *y* lies between *x* and *z* provided *y* belongs to every sphere containing both *x* and *z*. This definition yields the relation between which is the primitive concept of the system treated in Problem 12. Additional projective concepts can therefore be defined as in Problem 12, and corresponding axioms formulated. The following concepts are also defined: cord, surface, mid-point, diameter (of a sphere), congruence (between segments with an end-point in common, between parallel segments, between segments in general). Thereby, the metric concepts are achieved.

b. *The spheres as individuals*, including point-spheres. Point-spheres are spheres having no sub-spheres. Development of the system goes forward from here in a way analogous to form (a) above. (This system form is the one devised by Huntington.)

c. *The spheres as individuals*, but without point-spheres. Points are defined as certain infinite sequences of spheres each successively lying within its predecessor. (Recall **39b**, the note to language form IB.) Development runs in a way analogous to form (a).

Problem 18. [Diff. III.] AS of *metric geometry* based on the concept of *vector* (the presentation in Couturat [Principes] ch. VI, sec. D is based on the system of Peano, "Analisi della teoria dei vettori", *Atti Accademia Torino*, 1898; see also Russell [Principles] sec. 414). Here two forms of the system:

a. The single primitive sign is a *predicate*: '*Prd*'. We read '*Prd*(*r*,*H*,*K*)' as "(The real number) *r* is the inner product of vectors *H* and *K*". The class *Ve* of vectors is defined by 'mem_2(*Prd*)'. Here a vector is a one-one relation between points; the class *Po* of points is defined by 'sm_1(*mem*"*Ve*)'.

b. The single primitive sign is a *functor*: '*prd*'; '*prd*(k_1,k_2)' is of the same type as the real number expressions, and designates the inner product of k_1 and k_2 when these last are vectors. The class *Ve* of vectors is defined as the class of those functions *k* for which *prd*(*k*,*k*) is a real number. Here vectors are designated by functors; if *k* is a vector, '$k(x)=y$' says that vector *k* runs from point *x* to point *y*.

Problem 19. [Diff. III.] AS of *metric geometry* in the fashion of Hilbert, *Foundations of geometry*. This system employs seven primitive concepts: three classes—of points, of lines, and of planes; and four relations—of incidence ("lies upon"), of betweenness, of segment-congruence, and of angle-congruence. Various versions are possible; see e.g. O. Helmer (loc. cit. in **47**).

Problem 20. [Diff. III.] AS of two-dimensional *Clifford geometry*, following Russell [Principles] sec. 415. (See also W. Killing, *Einführung in die Grundlagen der Geometrie*, vol. 1 (1893) ch. IV.) What is in view here is the geometry of a two-dimensional space analogous to the Clifford surface, i.e. a space having curvature 0 everywhere, but having a finite area. There are two primitive signs: '*Dir*' and '*Sma*'. The sentence '*Dir*(*H*)' is read "*H* is a direction"; these directions are symmetric irreflexive relations between points. The sentence '*Sma*(*H*,*K*)' is read "The distance *H* is smaller than the distance *K*". A distance is a symmetric relation between points. If *R* is a direction, the class comprising a point *x* and the points to which *x* bears the relation *R* is a line through *x*.

55c. Physics

Problem 21. [Diff. IV.] AS of *space-time topology*. Complete the system of **50**, where the single primitive sign is '*S*'.

Problem 22. [Diff. IV.] AS of the *theory of events*, using Whitehead's method of extensive abstraction (presented in *The concept of nature* (1920) ch.IV, and more completely in *An enquiry concerning the principles of natural knowledge* (1919) Part III). The construction has two stages:

a. *Topology*. Here the only primitive sign is '*P*'; the sentence '*Pxy*' reads "The event *x* is a part of the event *y*". Events are thus members of the relation *P*. In the construction, differentiate between abstractive series and abstractive classes, the latter being the fields of the former. The abstractive series represent the point-events (recall the note to language form IB in **39b**). By means of these series, according to Whitehead, all spatial and temporal concepts can be expressed.

b. *Metric.* Here a second primitive sign, '*Cogr*' for cogredience.

55d. Biology

Problem 23. [Diff. III.] Continuation of the AS presented in **52** and **53**, following Woodger [Biology].

Problem 24. [Diff. II.] AS of the *biological concepts of kinship*, without reference to temporal relations (in a fashion similar to **54a**, but with other primitive signs). Use the primitive signs '*Son*' and '*Dau*' for son and daughter.

Problem 25. [Diff. IV.] Definitions of the *biological concepts of kinship* that involve temporal relations, based on the system of **53**.

Problem 26. [Diff. III.] AS of the *biological concepts of kinship* involving temporal relations. Slices of certain things (viz. human organisms, spermatozoa, ova, fertilized ova, embryos) are taken as individuals, in accordance with language form IB presented in **39b**. There are three primitive signs: '*Tr*', the time relation; '*P*', the part relation; and '*Fl*', female—all referred to slices of the kinds named above, and such that both a spermatozoon and an ovum are regarded as genidentical with the embryo and the person which develops from their fusion. With the help of certain facts (viz. that the spermatozoon first is part of the father, later becomes part of the fertilized ovum and so of the mother; and that the unfertilized ovum, the fertilized ovum and the embryo are parts of the mother), the concepts Father and Mother are defined, and thence the remaining biological concepts of kinship (see **54a**).

Problem 27. [Diff. III.] AS of the *legal concepts of kinship* (cf. **54b**), but involving temporal reference. Supplementation of the system of Problem 26 by addition of two other primitive signs '*LMar*' and '*LChi*' for the concepts of legal marriage and of the legal relation of child to parent; here, in contrast to **54b**, these are relations between person-slices, rather than persons. The sentence '*LMar*(x,y)' reads "x is a slice of a male person, y a simultaneous slice of a female person, and x and y are legally married". The sentence '*LChi*(x,y)' reads "x is a legal child of y, x and y being simultaneous slices of two persons". The following concepts can be defined: legally born child, illegitimate child, legitimatized child, adopted child. Thus there can be formulated here definitions and axioms respecting time-dependent relations which are not expressible in the system of **54b**. (Recall the remarks at the end of **54b**).

56. BIBLIOGRAPHY

The abbreviations in square brackets serve as reference within the text of this book. References to [P.M.] (see: Whitehead) and to several of my own books are occasionally given without the author's name.

ACKERMANN, see HILBERT.

BACHMANN, see CARNAP.

BECKER, Oskar: [Logistik], *Einführung in die Logistik*. Meisenheim 1951.

BEHMANN, Heinrich: [Math.], *Mathematik und Logik*. Leipzig and Berlin 1927.

BERNAYS, see HILBERT.

BETH, E. W.: [Logik], *Symbolische Logik und Grundlegung der exakten Wissenschaften* (Bibliographische Einführungen in das Studium der Philosophie, Vol. 3) Bern 1948.

BOCHENSKI, I. M.: [Logica], *Nove lezioni di logica simbolica*, Roma 1938.

—— [Logique], *Précis de logique mathématique*. Bussum, Holland, 1949.

—— [Logik], *Formale Logik*. Freiburg and München 1956.

CARNAP, Rudolf: [Aufbau], *Der logische Aufbau der Welt*. Berlin 1928.

—— [Abriss], *Abriss der Logistik*. Wien 1929.

—— [Neue Logik], "Die alte und die neue Logik", *Erkenntnis* 1, 1930.

—— [Syntax], *Logische Syntax der Sprache*. Wien 1934.

—— [Syntax E], *Logical Syntax of Language*. London and New York 1937. (This book contains also the translations of [Antinomien] and [Gültigkeitskriterium].)

—— [Antinomien], "Die Antinomien und die Unvollständigkeit der Mathematik", *Monatsh. Math. Phys.* 41, 1934.

—— [Gültigkeitskriterium], "Ein Gültigkeitskriterium für die Sätze der klassischen Mathematik", *Monatsh. Math. Phys.* 42, 1935.

—— [Foundations], *Foundations of logic and mathematics*. Encyclopedia of unified science, Vol. I, No. 3, Chicago 1939.

—— [Semantics], *Introduction to semantics*. Cambridge, Mass. 1942.

—— [Formalization], *Formalization of logic*. Cambridge, Mass 1943

—— [Meaning], *Meaning and necessity. A study in semantics and modal logic*. Chicago 1947; 2nd ed., 1956.

—— [Probability], *Logical foundations of probability*. Chicago 1950.

—— [Logik] *Einführung in die symbolische Logik, mit besonderer Berücksichtigung ihrer Anwendungen*. Wien 1954. (The original of the present book.)

—— und BACHMANN, Friedrich: [Extrem.], "Über Extremalaxiome", *Erkenntnis* 6, 1936.

—— und STEGMÜLLER, Wolfgang: *Induktive Logic und Wahrscheinlichkeit*. Wien 1959.

CHURCH, Alonzo: [Bibliogr.], "A bibliography of symbolic logic", *Journal of Symbolic Logic* 1, 1936, and 3, 1938. Also issued separately.

—— [Introduction], *Introduction to mathematical logic*. Vol. I. Princeton, N.J. 1956.

—— [Brief bibliography], "Brief bibliography of formal logic", *Proc. Amer. Acad. of Arts and Sciences* 80, 157–172, 1952.

COOLEY, John: [Logic], *A primer of formal logic*. New York 1942.

COPI, Irving: [Logic], *Symbolic Logic*. New York 1954.

COUTURAT, Louis: [Principes], *Les principes des mathématiques*. Paris 1905. (German translation, 1908.)

CURRY, Haskell B.: [Logique], *Leçons de logique algébrique*. Paris and Louvain 1952.

DOPP, Joseph: [Logique], *Leçons de logique formelle*. (Vol. I: *Logique ancienne*; II and III: *Logique moderne*.) Louvain 1950.

DÜRR, Karl: [Logistik], *Lehrbuch der Logistik*. Basel 1954.

FEYS, Robert: [Logistique], *Principes de logistique*. Louvain 1939.

—— [Logistiek], *Logistiek, geformaliseerde logica*. Vol. I. Antwerpen and Nijmegen 1944.

FITCH, Frederic B.: [Logic], *Symbolic logic*. New York 1952.

FRAENKEL, Abraham A.: [Grundlegung], *Zehn Vorlesungen über die Grundlegung der Mengenlehre*, Leipzig and Berlin 1927.
—— [Einleitung], *Einleitung in die Mengenlehre*. Berlin (1919), 3rd ed. 1928. Reprinted, Dover, New York 1946.
—— *Abstract set theory*. Amsterdam and New York 1953.
FREGE, Gottlob: [Begriffsschrift], *Begriffsschrift. Eine der arithmetischen nachgemachte Formelsprache des reinen Denkens*. Halle 1879.
—— [Grundlagen], *Die Grundlagen der Arithmetik. The foundations of arithmetic*. (Original (1884) plus English translation.) Oxford 1953.
—— [Grundgesetze], *Grundgesetze der Arithmetik*. Jena I, 1893; II, 1903.
GOODMAN, Nelson: [Structure], *The structure of appearance*. Cambridge, Mass. 1951.
HAUSDORFF, Felix: [Grundzüge], *Grundzüge der Mengenlehre*. Leipzig 1914. (Reprinted New York 1949.)
HERMES, Hans, and SCHOLZ, Heinrich: [Logik], "Mathematische Logik" (*Enzyklopädie der mathematischen Wissenschaften*, Band I1, 2nd ed., Heft 1, Teil I.) Leipzig 1952.
HILBERT, David and ACKERMANN, Wilhelm: [Logik], *Grundzüge der theoretischen Logik*. Berlin (1928), 3rd ed. 1949.
—— [Logic], *Principles of mathematical logic*. New York 1950. (Translation of [Logik]).
—— and BERNAYS, Paul: [Grundlagen], *Grundlagen der Mathematik*. Berlin I, 1934, II, 1939. Reprinted Ann Arbor, Mich. 1944.
JØRGENSEN, Jørgen: [Treatise], *A treatise of formal logic. Its evolution and main branches with its relation to mathematics and philosophy*. 3 vols. Copenhagen and London 1931.
KLEENE, Stephen Cole: [Metamathematics], *Introduction to metamathematics*. New York 1952.
KÖNIG, Julius: [Logik], *Neue Grundlagen der Logik, Arithmetik und Mengenlehre*. Leipzig 1914.
LANGER, Susanne K.: [Logic] *Introduction to symbolic logic*, (1937); 2nd ed., New York 1953.
LEBLANC, Hugues: [Logic], *An introduction to deductive logic*. New York and London 1955.
LEWIS, Clarence I.: [Survey], *A survey of symbolic logic*. Berkeley 1918.
—— and LANGFORD, Charles H.: [Logic], *Symbolic logic*. New York and London 1932; revised, New York 1951.
MARC-WOGAU, Konrad: [Logik], *Modern logik. Elementär lärobok*. Stockholm 1950.
MATHEMATICS STAFF OF THE COLLEGE, UNIVERSITY OF CHICAGO: [Fund. Math.], *Fundamental Mathematics*, 3rd ed., Vols. I, II, III. Chicago 1948–49.
MENGER, Karl: [Logic], "The new logic", *Philos. of Science* 4, 1937.
MORRIS, Charles W.: [Foundations], "Foundations of the theory of signs," *Encyclopedia of unified science*, Vol. I, No. 2, Chicago 1938.
—— [Signs], *Signs, language and behavior*. New York 1946.
MOSTOWSKI, Andrzej: [Logika], *Logika matematyczna*. Warszawa and Wroclaw 1948.
PEANO, Giuseppe: [Notations], *Notations de logique mathématique*. Torino 1894.
—— [Formulaire], *Formulaire de mathématiques*. Torino 1895–1908.
PRIOR, Arthur N.: [Logic], *Formal logic*. Oxford 1955.
QUINE, Willard V. O.: [Logistic], *A system of logistic*. Cambridge, Mass., 1934.
—— [Types], "On the theory of types", *Journal of Symb. Logic*. 3, 1938.
—— [Math. Logic], *Mathematical logic*. New York 1940, 3rd ed., Cambridge, Mass., 1951.
—— [Lógica], *O sentido da nova lógica*. São Paulo, Brasil 1944.
—— [Methods], *Methods of logic*. New York 1950.
RAMSEY, Frank P.: [Foundations], *The foundations of mathematics, and other logical essays*. London and New York 1931. Reprinted 1950.

REICHENBACH, Hans: [Logic], *Elements of symbolic logic.* New York 1947.
ROSENBLOOM, Paul C.: [Logic], *The elements of mathematical logic.* New York 1950.
ROSSER, J. Barkley: [Logic], *Logic for mathematicians.* New York 1953.
ROTH, Eugen: [Axiomat.], *Axiomatische Untersuchungen zur projektiven, affinen und metrischen Geometrie.* Leipzig 1937.
RUSSELL, Bertrand: [Principles], *The principles of mathematics* (Cambridge 1903), 2nd ed., text unchanged and new introduction, London 1937, New York 1938.
—— [P.M.], *Principia mathematica*, see WHITEHEAD.
—— [World], *Our knowledge of the external world as a field for scientific method in philosophy.* London and Chicago 1914.
—— [Introduction], *Introduction to mathematical philosophy*, London and New York 1919.
SCHOLZ, Heinrich: [Geschichte], *Geschichte der Logik.* Berlin 1931.
—— [Vorlesungen], *Vorlesungen über Grundzüge der mathematischen Logik.* 2 vols., Münster i. W. (1949), 2nd ed., 1950–51.
SCHRÖDER, Ernst: [Vorlesungen], *Vorlesungen über die Algebra der Logik.* I–III, Leipzig 1890–1905.
TARSKI, Alfred: [Wahrheitsbegriff], "Der Wahrheitsbegriff in den formalisierten Sprachen", *Studia Philosophica* 1, 1936. (English translation in [Metamathematics].)
—— [Logik] *Einführung in die mathematische Logik.* Wien 1937.
—— [Logic], *Introduction to logic and to the methodology of deductive sciences.* New York 1941. (Translation of [Logik], enlarged and revised.)
—— [Metamathematics], *Logic, semantics, metamathematics.* Oxford 1956.
WAISMANN, Friedrich: [Math. Thinking], *Introduction to mathematical thinking: The formation of concepts in modern mathematics.* Unger, New York 1951.
WHITEHEAD, Alfred North and RUSSELL, Bertrand: [P.M.], *Principia mathematica.* Cambridge. (I 1910, II 1912, III 1913); 2nd ed. I 1925 (text unchanged, new introduction and appendices); II, III 1927 (unchanged). Reprinted 1950.
WILDER, Raymond L.: [Foundations], *The foundations of mathematics.* New York 1952.
WITTGENSTEIN, Ludwig: [Tractatus], *Tractatus logico-philosophicus.* (German and English.) With introduction by RUSSELL. London 1922.
WOODGER, Joseph: [Biology], *The axiomatic method in biology.* Cambridge 1937.
—— [Theory construction], *The technique of theory construction.* Encyclopedia of unified science, Vol. II, No. 5, Chicago 1939.

57. GENERAL GUIDE TO THE LITERATURE

On the *history* of symbolic logic: Lewis [Survey], Jörgensen [Treatise], Bochenski [Logik]. History of the development of the formal character of logic, issuing in symbolic logic: Scholz [Geschichte].

Bibliography: Lewis [Survey], to 1917; Fraenkel [Einleitung], to 1928 and with special reference to the logical foundations of mathematics; Church [Bibliogr.], complete bibliography, excellently indexed, running from before Leibniz to 1935 and under continued extension *via Jour. Symb. Logic*; Church [Brief bibliography]; Beth [Logik]; Bochenski [Logik]. For the years 1939–48, see the contributions of E. Beth, R. Feys and F. Gonseth in: *Philosophie, Chronique des années ...*, edited by R. Bayer, vols. XI and XIII (Actual. Scient. Nos. 1089 and 1105) Paris 1950.

Older systems, essentially of historical interest only: A. De Morgan, *Formal logic*, London 1847, reissued 1926. G. Boole, *An investigation of the laws of thought*, London 1854, reissued New York 1951. W. S. Jevons, *Pure Logic*, London 1864, reissued 1890.

Older works, essentially superseded by more recent developments but nevertheless containing some matters still of value: Frege [Begriffsschrift], [Grundlagen], [Grundgesetze]; Schröder [Vorlesungen]; Peano [Notations], [Formulaire]; Charles S. Pierce,

Collected Papers, Cambridge, Mass., 1931 ff., particularly vols. II–IV; Whitehead, *A treatise on universal algebra*, Cambridge, England, 1898.

Modern symbolic logic: The principal work is Whitehead and Russell [P.M.]. Structurally most of the recent systems lean more or less on the system of [P.M.], even though their symbolisms may differ. Most modern systems, however, depart from [P.M.] in two respects: the axiom of reducibility is given up, and the ramified hierarchy of types is replaced by the simplified one (cf. **21c** and [Syntax] §27); further, more stringent demands are made today regarding a full specification of the rules of deduction employed, and regarding a strictly formal presentation.—*Systems with a structure similar to that of* [*P.M.*]: Hilbert [Logik], with a somewhat different symbolism; Behmann [Math.], also with a different symbolism; the present book, whose symbolism in the main is the same. —*Systems with a structure strongly dissimilar to that of* [*P.M.*]: Lewis [Survey] and [Logic], with intensional (i.e. non-extensional) connectives, cf. **29c**; Heyting, "Die formalen Regeln der intuitionistischen Logik", *Sitzungsber. Preuss. Akademie, phys.-math. Klasse*, Berlin 1930 (avoids the law of excluded middle); A. Church [Introduction]. —*Logical systems in which the differentiation of types is entirely avoided or essentially weakened* (cf. **21c**): J. von Neumann, "Zur Hilbertschen Beweistheorie", *Math. Zeitsch.* 26, 1927; Quine, "New foundations for mathematical logic", *Amer. Math. Mon.* 44, 1937, [Types], [Math. Logic]; Ackermann, "Mengentheor. Begründung der Logik", *Math. Ann.* 115, 1937.

Textbooks of symbolic logic: In English: Church [Introduction]; Cooley [Logic]; Copi [Logic]; Fitch [Logic]; Langer [Logic]; Leblanc [Logic]; Lewis and Langford [Logic]; Quine [Methods]; Reichenbach [Logic]; Rosser [Logic]; Tarski [Logic].—*In German*: Becker [Logistik]; Carnap [Logik]; Dürr [Logistik]; Hilbert [Logik].—*In French*: Bochenski [Logique]; Curry [Logique]; Dopp [Logique]; Feys [Logistique].—*In Italian*: Bochenski [Logica].—*In Dutch*: Feys [Logistiek].—*In Portuguese*: Quine [Lógica].—*In Polish*: Mostowski [Logika].—*In Swedish*: Marc-Wogau [Logik].

For advanced study, after working through the present book, the following (among others) should be considered: [P.M.]; Russell [Principles] and [Introduction]; Rosser [Logic]; Quine [Math. Logic] and [Methods]; Kleene [Metamathematics]; Hermes and Scholz [Logik]; Scholz [Vorlesungen]; Carnap [Syntax E] and [Semantics].

Application of symbolic logic, outside the field of deductive logic and mathematics: Woodger [Theory construction] and [Biology], see **52** and **53**; Goodman [Structure]; Clark L. Hull, *Mathematico-deductive theory of rote learning*, 1940; Carnap [Aufbau], [Probability]. *General discussion regarding the possibilities of application*: E. C. Berkeley, "Conditions affecting the application of symbolic logic", *Jour. Symb. Log.* 7, 1942. For additional references, see Quine [Math. Logic] 8.

Index

The numbers refer to pages. The most important references are indicated by boldface type.

Symbols of the symbolic language and of the metalanguage

A CATALOG OF SELECTED
DOVER BOOKS
IN SCIENCE AND MATHEMATICS

THE ELECTROMAGNETIC FIELD, Albert Shadowitz. Comprehensive undergraduate text covers basics of electric and magnetic fields, builds up to electromagnetic theory. Also related topics, including relativity. Over 900 problems. 768pp. 5⅜ × 8¼. 65660-8 Pa. $15.95

FOURIER SERIES, Georgi P. Tolstov. Translated by Richard A. Silverman. A valuable addition to the literature on the subject, moving clearly from subject to subject and theorem to theorem. 107 problems, answers. 336pp. 5⅜ × 8½. 63317-9 Pa. $7.95

THEORY OF ELECTROMAGNETIC WAVE PROPAGATION, Charles Herach Papas. Graduate-level study discusses the Maxwell field equations, radiation from wire antennas, the Doppler effect and more. xiii + 244pp. 5⅜ × 8½. 65678-0 Pa. $6.95

DISTRIBUTION THEORY AND TRANSFORM ANALYSIS: An Introduction to Generalized Functions, with Applications, A.H. Zemanian. Provides basics of distribution theory, describes generalized Fourier and Laplace transformations. Numerous problems. 384pp. 5⅜ × 8½. 65479-6 Pa. $9.95

THE PHYSICS OF WAVES, William C. Elmore and Mark A. Heald. Unique overview of classical wave theory. Acoustics, optics, electromagnetic radiation, more. Ideal as classroom text or for self-study. Problems. 477pp. 5⅜ × 8½. 64926-1 Pa. $10.95

CALCULUS OF VARIATIONS WITH APPLICATIONS, George M. Ewing. Applications-oriented introduction to variational theory develops insight and promotes understanding of specialized books, research papers. Suitable for advanced undergraduate/graduate students as primary, supplementary text. 352pp. 5⅜ × 8½. 64856-7 Pa. $8.50

A TREATISE ON ELECTRICITY AND MAGNETISM, James Clerk Maxwell. Important foundation work of modern physics. Brings to final form Maxwell's theory of electromagnetism and rigorously derives his general equations of field theory. 1,084pp. 5⅜ × 8½. 60636-8, 60637-6 Pa., Two-vol. set $19.90

AN INTRODUCTION TO THE CALCULUS OF VARIATIONS, Charles Fox. Graduate-level text covers variations of an integral, isoperimetrical problems, least action, special relativity, approximations, more. References. 279pp. 5⅜ × 8½. 65499-0 Pa. $7.95

HYDRODYNAMIC AND HYDROMAGNETIC STABILITY, S. Chandrasekhar. Lucid examination of the Rayleigh-Benard problem; clear coverage of the theory of instabilities causing convection. 704pp. 5⅜ × 8¼. 64071-X Pa. $12.95

CALCULUS OF VARIATIONS, Robert Weinstock. Basic introduction covering isoperimetric problems, theory of elasticity, quantum mechanics, electrostatics, etc. Exercises throughout. 326pp. 5⅜ × 8½. 63069-2 Pa. $7.95

DYNAMICS OF FLUIDS IN POROUS MEDIA, Jacob Bear. For advanced students of ground water hydrology, soil mechanics and physics, drainage and irrigation engineering and more. 335 illustrations. Exercises, with answers. 784pp. 6⅛ × 9¼. 65675-6 Pa. $19.95

ROTARY-WING AERODYNAMICS, W.Z. Stepniewski. Clear, concise text covers aerodynamic phenomena of the rotor and offers guidelines for helicopter performance evaluation. Originally prepared for NASA. 537 figures. 640pp. 6⅛ × 9¼.
64647-5 Pa. $14.95

DIFFERENTIAL GEOMETRY, Heinrich W. Guggenheimer. Local differential geometry as an application of advanced calculus and linear algebra. Curvature, transformation groups, surfaces, more. Exercises. 62 figures. 378pp. 5⅜ × 8½.
63433-7 Pa. $7.95

INTRODUCTION TO SPACE DYNAMICS, William Tyrrell Thomson. Comprehensive, classic introduction to space-flight engineering for advanced undergraduate and graduate students. Includes vector algebra, kinematics, transformation of coordinates. Bibliography. Index. 352pp. 5⅜ × 8½. 65113-4 Pa. $8.95

A SURVEY OF MINIMAL SURFACES, Robert Osserman. Up-to-date, in-depth discussion of the field for advanced students. Corrected and enlarged edition covers new developments. Includes numerous problems. 192pp. 5⅜ × 8½.
64998-9 Pa. $8.95

ANALYTICAL MECHANICS OF GEARS, Earle Buckingham. Indispensable reference for modern gear manufacture covers conjugate gear-tooth action, gear-tooth profiles of various gears, many other topics. 263 figures. 102 tables. 546pp. 5⅜ × 8½. 65712-4 Pa. $11.95

SET THEORY AND LOGIC, Robert R. Stoll. Lucid introduction to unified theory of mathematical concepts. Set theory and logic seen as tools for conceptual understanding of real number system. 496pp. 5⅜ × 8¼. 63829-4 Pa. $8.95

A HISTORY OF MECHANICS, René Dugas. Monumental study of mechanical principles from antiquity to quantum mechanics. Contributions of ancient Greeks, Galileo, Leonardo, Kepler, Lagrange, many others. 671pp. 5⅜ × 8½.
65632-2 Pa. $14.95

FAMOUS PROBLEMS OF GEOMETRY AND HOW TO SOLVE THEM, Benjamin Bold. Squaring the circle, trisecting the angle, duplicating the cube: learn their history, why they are impossible to solve, then solve them yourself. 128pp. 5⅜ × 8½. 24297-8 Pa. $3.95

MECHANICAL VIBRATIONS, J.P. Den Hartog. Classic textbook offers lucid explanations and illustrative models, applying theories of vibrations to a variety of practical industrial engineering problems. Numerous figures. 233 problems, solutions. Appendix. Index. Preface. 436pp. 5⅜ × 8½. 64785-4 Pa. $8.95

CURVATURE AND HOMOLOGY, Samuel I. Goldberg. Thorough treatment of specialized branch of differential geometry. Covers Riemannian manifolds, topology of differentiable manifolds, compact Lie groups, other topics. Exercises. 315pp. 5⅜ × 8½. 64314-X Pa. $8.95

HISTORY OF STRENGTH OF MATERIALS, Stephen P. Timoshenko. Excellent historical survey of the strength of materials with many references to the theories of elasticity and structure. 245 figures. 452pp. 5⅜ × 8½. 61187-6 Pa. $10.95